Lecture Notes in Physics

Springer-Verlag Berlin Heidelberg GmbH

The Editorial Policy for Proceedings

The series Lecture Notes in Physics reports new developments in physical research and teaching – quickly, informally, and at a high level. The proceedings to be considered for publication in this series should be limited to only a few areas of research, and these should be closely related to each other. The contributions should be of a high standard and should avoid lengthy redraftings of papers already published or about to be published elsewhere. As a whole, the proceedings should aim for a balanced presentation of the theme of the conference including a description of the techniques used and enough motivation for a broad readership. It should not be assumed that the published proceedings must reflect the conference in its entirety. (A listing or abstracts of papers presented at the meeting but not included in the proceedings could be added as an appendix.)

When applying for publication in the series Lecture Notes in Physics the volume's editor(s) should submit sufficient material to enable the series editors and their referees to make a fairly accurate evaluation (e.g. a complete list of speakers and titles of papers to be presented and abstracts). If, based on this information, the proceedings are (tentatively) accepted, the volume's editor(s), whose name(s) will appear on the title pages, should select the papers suitable for publication and have them refereed (as for a journal) when appropriate. As a rule discussions will not be accepted. The series editors and Springer-Verlag will normally not interfere with the detailed editing except in fairly obvious cases or on technical matters.

Final acceptance is expressed by the series editor in charge, in consultation with Springer-Verlag only after receiving the complete manuscript. It might help to send a copy of the authors' manuscripts in advance to the editor in charge to discuss possible revisions with him. As a general rule, the series editor will confirm his tentative acceptance if the final manuscript corresponds to the original concept discussed, if the quality of the contribution meets the requirements of the series, and if the final size of the manuscript does not greatly exceed the number of pages originally agreed upon. The manuscript should be forwarded to Springer-Verlag shortly after the meeting. In cases of extreme delay (more than six months after the conference) the series editors will check once more the timeliness of the papers. Therefore, the volume's editor(s) should establish strict deadlines, or collect the articles during the conference and have them revised on the spot. If a delay is unavoidable, one should encourage the authors to update their contributions if appropriate. The editors of proceedings are strongly advised to inform contributors about these points at an early stage.

The final manuscript should contain a table of contents and an informative introduction accessible also to readers not particularly familiar with the topic of the conference. The contributions should be in English. The volume's editor(s) should check the contributions for the correct use of language. At Springer-Verlag only the prefaces will be checked by a copy-editor for language and style. Grave linguistic or technical shortcomings may lead to the rejection of contributions by the series editors. A conference report should not exceed a total of 500 pages. Keeping the size within this bound should be achieved by a stricter selection of articles and not by imposing an upper limit to the length of the individual papers. Editors receive jointly 30 complimentary copies of their book. They are entitled to purchase further copies of their book at a reduced rate. As a rule no reprints of individual contributions can be supplied. No royalty is paid on Lecture Notes in Physics volumes. Commitment to publish is made by letter of interest rather than by signing a formal contract. Springer-Verlag secures the copyright for each volume.

The Production Process

The books are hardbound, and the publisher will select quality paper appropriate to the needs of the author(s). Publication time is about ten weeks. More than twenty years of experience guarantee authors the best possible service. To reach the goal of rapid publication at a low price the technique of photographic reproduction from a camera-ready manuscript was chosen. This process shifts the main responsibility for the technical quality considerably from the publisher to the authors. We therefore urge all authors and editors of proceedings to observe very carefully the essentials for the preparation of camera-ready manuscripts, which we will supply on request. This applies especially to the quality of figures and halftones submitted for publication. In addition, it might be useful to look at some of the volumes already published. As a special service, we offer free of charge LaTeX and TeX macro packages to format the text according to Springer-Verlag's quality requirements. We strongly recommend that you make use of this offer, since the result will be a book of considerably improved technical quality. To avoid mistakes and time-consuming correspondence during the production period the conference editors should request special instructions from the publisher well before the beginning of the conference. Manuscripts not meeting the technical standard of the series will have to be returned for improvement.

For further information please contact Springer-Verlag, Physics Editorial Department II, Tiergartenstrasse 17, D-69121 Heidelberg, Germany

Paul Weingartner   Gerhard Schurz (Eds.)

Law and Prediction
in the Light
of Chaos Research

Springer

Editors

Paul Weingartner
Institut für Wissenschaftstheorie
Mönchsberg 2
A-5020 Salzburg, Austria

Gerhard Schurz
Institut für Philosophie
Franziskanergasse 1
A-5020 Salzburg, Austria

Publications of the International Research Center, Salzburg, Austria, No. 67

Cataloging-in-Publication Data applied for.

Die Deutsche Bibliothek - CIP-Einheitsaufnahme

Law and prediction in the light of chaos research / Paul
Weingartner ; Gerhard Schurz (ed.).

(Lecture notes in physics ; Vol. 473) (Publications of the International
Research Center, Salzburg, Austria ; No. 67)
 ISBN 978-3-662-14100-7   ISBN 978-3-540-70693-9 (eBook)
 DOI 10.1007/978-3-540-70693-9
NE: Weingartner, Paul [Hrsg.]; 1. GT; Internationales
 Forschungszentrum für Grundfragen der Wissenschaften <Salzburg>:
 Veröffentlichungen des Internationalen ...

ISSN 0075-8450
ISBN 978-3-662-14100-7

Typesetting: Camera-ready by the authors/editors
Cover design: *design & production* GmbH, Heidelberg
SPIN: 10520094   55/3142-543210 - Printed on acid-free paper

Preface

The contributions in this book are based on the conference *"Law and Prediction in Natural Science in the Light of our New Knowledge from Chaos Research"* held at the *Institut für Wissenschaftstheorie, Internationales Forschungszentrum*, Salzburg, July 1994. The papers given at the conference have been revised and partially extended, taking into account the discussion and scientific communication at the conference. Great emphasis has been given to the discussions among the participants. Preliminary versions of the papers were distributed among the participants one month before the conference. Each participant had about an hour for the presentation of his paper and an hour for discussion afterwards. Since important additions and clarifications have emerged from the discussions, important parts of them have been included in the proceedings.

There were people who were invited to the conference but could not attend because of other commitments (Prigogine, Lighthill, Nicolis and Haken). They agreed to send their papers or comments. Peter Schuster who attended the conference was not able to send his paper.

The editors would like to thank the publisher, especially Prof. Wolf Beiglböck, Gabriele Köbrunner-Krisch for writing the manuscript and transcribing the discussions from the tape and Helmut Prendinger for the conversion into LaTeX. Last but not least the editors would like to express their gratitude in the name of all participants of the conference to those institutions that generously sponsored this research conference: The Austrian *Bundesministerium für Wissenschaft und Forschung*, Vienna and the *Internationales Forschungszentrum*, Salzburg.

Paul Weingartner

Gerhard Schurz

Contents

Part III Chaos, Complexity and Order

List of Participants

Batterman, Robert W.   rbatterm@magnus.acs.ohio-state.edu
  Ohio State University, USA,
Chirikov, Boris   B.V.Chirikov@inp.nsk.su
  Budker Institute of Nuclear Physics, 630090 Novosibirsk, Russia,
Haken, H.   HAKEN@IFTPUS.PHYSIK.UNI-STUTTGART.DE
  Institut für Theoretische Physik und Synergetik, Universität Stuttgart, Pfaf-
  fenwaldring 57/4, D–70569 Stuttgart, Germany,
Lighthill, James   fax: 071-383 5519
  University College London, UK,
Miller, David   PYRAL@snow.csv.warwick.ac.uk
  Department of Philosophy, University of Warwick, Coventry CV4 7AL, UK,
Nicolis, Gregoire   gnicolis@ulb.ac.be
  Center for Nonlinear Phenomena and Complex Systems, Université Libre de
  Bruxelles, Campus Plaine, CP.231, 1050 Bruxelles, Belgium,
Noyes, H. Pierre   Noyes@slacvm.bitnet
  Stanford Linear Accelerator Center, Stanford University, Stanford, Califor-
  nia 94309, USA,
Prigogine, Ilya   annie@physics.utexas.edu
  International Solvay Institutes, CP 231, 1050 Brussels, Belgium; Nobel Lau-
  reate (1977),
Schurz, Gerhard   Gerhard.Schurz@mh.sbg.ac.at
  Institut für Philosophie, Universität Salzburg, Franziskanergasse 1, A-5020
  Salzburg, Austria,
Schuster, Peter   pks@imb-jena.de
  Institut für Molekulare Biotechnologie e.V., Beutenbergstrasse 11, PF 100813,
  D-07708 Jena, Germany,
Stöckler, Manfred   fax: (0421) 218-4265
  Fachbereich 10 (Philosophie), Universität Bremen, Postfach 330440, D-2800
  Bremen 33, Germany,
Suppes, Patrick   suppes@Ockham.Stanford.EDU
  Ventura Hall, Stanford University, Stanford, California 94305-4115, USA,
Weingartner, Paul   Paul.Weingartner@mh.sbg.ac.at
  Institut für Philosophie, Universität Salzburg, Franziskanergasse 1, A-5020
  Salzburg, and Institut für Wissenschaftstheorie Internationales Forschungszen-
  trum Salzburg, Mönchsberg 2a, A-5020 Salzburg, Austria,

Wunderlin, Arne   wum@theo.physik.uni-stuttgart.de
Institut für Theoretische Physik und Synergetik, Universität Stuttgart, Pfaffenwaldring 57/4, D–70569 Stuttgart, Germany

Part I

Laws of Nature in the Light of Chaos

Time, Chaos and the Laws of Nature

I. Prigogine

International Solvay Institutes, Belgium

I

As I was unable to attend the conference "Law and Prediction in (Natural) Science in the Light of our New Knowledge from Chaos Research," I am glad to present here a few remarks. Time, as incorporated in the basic laws of physics, from classical Newtonian dynamics to relativity and quantum physics, does not include any distinction between past and future. These laws are time reversible and deterministic. Yet on all levels, from cosmology to human sciences, future and past play different roles.

For many scientists, we as humans, imperfect observers, would be responsible for the difference between past and future through the approximations (such as "coarse graining") we would introduce in our description of nature. The arrow of time would be part of "phenomenology" and not included in the fundamental description of nature. But over the last decades, physics of nonequilibrium processes has led to spectacular advances.[1] It led to concepts such as *self-organization* and *dissipative structures* which are widely used today. These phenomena illustrate the *constructive* role of the arrow of time. Irreversibility can no longer be identified with a mere appearance that would dissapear if we had "perfect" knowledge; therefore, the claim that the arrow of time is only "phenomenological" becomes absurd. We are the children of time, of evolution and not its progenitors.

There is a second development that is crucial for the problem of time; that is the dynamics of "unstable" systems. This includes deterministic chaos as well as classes of nonintegrable systems in the sense of Poincaré. We shall come back to the definition of such systems in section 3. My point of view has always been that instability forces us to generalize the laws of classical and quantum mechanics.[2]

Thanks to remarkable advances in functional analysis,[3] this program has been largely realized. Once instability is included, the meaning of laws of nature changes radically, as they have to be formulated on the statistical level. They no longer express "certitudes" but "possibilities." In this brief note, I would like to summarize the basic steps in this development.[4]

---

[1] See Nicolis (1989); I may be permitted to mention that my first paper emphasizing self-organization as the result of irreversible processes appeared already in 1945!

[2] See i. e. Prigogine (1980).

[3] Gelfand and Vilenkin (1964); Maurin (1968); Bohm and Gadella (1989).

[4] Prigogine (1993); French translation (1994); Prigogine and Stengers (1993).

II

I shall first consider deterministic chaos. The simplest example is the Bernoulli map. Every second we multiply the number x_n contained in the interval (0-1) by two. After multiplication, we reinject it into the interval. The "equation of motion" is therefore

$$x_{n+1} = 2x_n \qquad \text{mod } 1 \qquad (2.1)$$

It is well-known that this leads to trajectories that diverge exponentially. The exponent is called the Lyapunov exponent. Here it is $lg2$. This is an example of "sensitivity to initial conditions."

Instead of the trajectory description, we can also introduce a probabilistic description in terms of the distribution functions $\rho_n(x)$. Each second $\rho_n(x)$ is modified through the action of an operator U. The map is then described

$$\rho_{n+1}(x) = U\rho_n(x) \qquad (2.2)$$

where U is called the Perron-Frobenius operator.[5] The basic question is then, "Are these two descriptions equivalent?". The answer for chaotic maps is *no*, the statistical description admits *new* solutions which cannot be expressed in terms of trajectories, since quantum mechanics operator calculus has become part of the daily arsenal of physicists. All textbooks in quantum mechanics describe how to obtain eigenfunctions and eigenvalues of operators. We want to apply a similar method U, the Perron-Frobenius operator.

In this simple case, we easily can obtain the explicit form of U. It is

$$U\rho(x) = \frac{1}{2}\left[\rho\left(\frac{x}{2}\right) + \rho\left(\frac{1+x}{2}\right)\right] \qquad (2.3)$$

The solution of the statistical problem requires, as in quantum mechanics, the spectral resolution of U, that is the determination of its eigenfunctions and eigenvalues. But there is a difference. In quantum mechanics, we consider usually operators that act on "nice" normalizable functions. These functions belong to the so-called Hilbert space, which is an extension of the simple vector space. Here, the eigenfunctions do not exist in the usual Hilbert space of normed functions, but only in generalized function spaces.[6] The result is [7]

$$U = \sum B_n(x)\rangle \frac{1}{2^n} \langle \tilde{B}_n(x) \qquad (2.4)$$

The right eigenfunctions $B_n(x)$ are regular functions, the Bernoulli polynomials, but the left eigenfunctions are generalized functions, in this simple case they

[5] Lasota and Mackey (1985).

[6] Gelfand and Vilenkin (1964); Maurin (1968); Bohm and Gadella (1989).

[7] Some recent papers dealing with the spectral decompostion of chaotic maps are: Hasegawa and Saphir (1992), p. 471; Gaspard (1992), p. 303; Antoniou and Tasaki (1993), p. 73; Antoniou and Tasaki (1992); Hasegawa and Driebe (1994).

are derivatives of the delta function. This is the function $\delta(x)$ which is infinite for $x = 0$ and vanishing elsewhere. Such spectral decompositions have now been obtained for a large class of chaotic systems, including invertible systems (such as the so-called baker transformation).[8] In general, eigenfunctions of maps are "fractals" that are outside the Hilbert space. The main conclusion is that expressions such as (2.4) for the evolution operator (called spectral decomposition) can only be applied to distribution functions ρ, which are "test functions" (such as continuous, differentiable functions) and not to individual trajectories, which are themselves singular functions represented by $\delta(x - x_n)$ where $x = x_n$, which are *singular* functions (therefore the scalar product $\langle \tilde{B}_n(x) | \delta(x - x_n) \rangle$ diverges. The basic conclusion is that deterministic chaos leads to a *statistical formulation*. The equivalence between the individual description (2.1) and the statistical description (2.2) is broken.

We have indeed new solutions for ensembles that contain the approach to equilibrium (at each iteration, the Bernouilli polynomial $B_n(x)$ is multiplied by $\frac{1}{2^n}$), so in the limit of infinite number of iterations only $B_0(x)$, which is equal to one, survive; this is the equilibrium distribution. The Bernouilli map is not invertible (if instead of (2.1) we would consider $x_{n+1} = \frac{1}{2}x_n$, we would simply reach the origin, but there are also time-invertible maps (such as the Baker transformation), which can be analyzed in the same way. In simple cases, we have both a spectral representation in the Hilbert space, which is time reversible and corresponds to a dynamical group, and a spectral representation in generalized function spaces, which includes time symmetry breaking. The characteristic rates at which the distribution function reaches equilibrium (as well as other characteristics of irreversible processes) now appear in the spectrum. The basic laws of chaos can be expressed in terms of statistical laws once we go beyond the Hilbert space. We see already in these examples that the breaking of time symmetry and therefore the extension of dynamics to include irreversibility is associated to a precise problem of modern mathematics (called functional analysis).

III

We come now to Hamiltonian systems, be it in classical or quantum mechanics. Here, also there are two descriptions, the "individual description" in terms of trajectories in classical mechanics, in terms of wave functions (Schrödinger equation) in quantum mechanics and the "statistical description" in terms of probability distributions ρ or density matrices. The statistical description is in terms of the Liouville operator L

$$i\frac{\partial \rho}{\partial t} = L\rho \qquad (3.1)$$

---

[8] Prigogine (1993), french transl. (1994); Prigogine and Stengers (1993).

here $L\rho$ is the Poisson bracket in classical mechanics, the commutator with the Hamiltonian in quantum mechanics. It has always been assumed that the two descriptions were equivalent. Probability distributions were introduced for "practical reasons" or as approximations. Indeed, at first it seems that nothing is gained. The Liouville operator L belongs to the class of so-called "hermitian operators", whose eigenvalues are real in the Hilbert space, but this is no longer so when we can extend L outside the Hilbert space. There are well defined conditions to be able to do so. There is first the question of integrability.[9] The problem of integrability is at the center of the fundamental work of Henri Poincaré. In short, whenever the Hamiltonian can be expressed in terms only of the momenta (often called "action variables"), the system is integrable; moreover, Poincaré identified the reason for nonintegrability: The existence of resonances between the various degrees of freedom. For example, consider a system characterized by two degrees of freedom. The corresponding frequencies are ω_1, ω_2. Whenever $n_1\omega_1 + n_2\omega_2 = 0$ with $n_1 n_2$ non vanishing integers, we have resonance. These resonances lead to the problem of small denominators as they show up in perturbation calculations through dangerous denominators $\frac{1}{n_1\omega_1 + n_2\omega_2}$.

The importance of Poincaré resonances is well recognized today (it led to the so-called KAM theory). They give rise to random trajectories. In this sense we may say that Poincaré resonances are also associated to chaos. We shall be mainly interested in so-called "Large Poincaré Systems" (LPS), in which the frequency ω_k depends continuously on the wave vector k.

A simple example is provided by the interaction between an oscillator ω_1 coupled to a field. Resonances appear when the field frequencies ω_k are equal to the oscillator frequency ω_1. Resonances in LPS are responsible for fundamental phenomena such as emission or absortion of light, decay of unstable particles...they play a fundamental role both in classical and quantum physics.

Let us now come back to our basic problem, the equivalence between the "individual" and the "statistical" description. Whatever the solution of the problem of integrability, (3.1) leads only to irreversible processes as long as we remain in the Hilbert space. As mentioned, the operator (better the superoperator) L is hermitian and its eigenvalues are real in the Hilbert space. We need, as in deterministic chaos (see 2.4), an extension Hilbert space.

Here we come to an essential point. When we consider this room, molecules in the atmosphere continously collide one with the other. We have "persistent" interactions. This is in contrast with transitory interactions as considered, i.e. in ordinary scattering experiments (described by the "S-matrix" theory) in which we have free asymptotic "in" and "out" states. Now to describe persistent interactions we have to introduce *singular* distribution functions. For example, if we consider a distribution function $\rho(q, p)$ where q is the coordinate and p the momentum and impose boundary conditions such as

$$\rho(q, p) \rightarrow \rho_0(p) \quad \text{for } q \rightarrow \pm\infty \tag{3.2}$$

---

[9] Prigogine (1962).

this requires, as shown in elementary mathematics, singular Fourier transforms with a delta singularity in the Fourier index k

$$\rho_k(p) = \rho_0(p)\delta(k) + \rho_k(p) \tag{3.3}$$

Already in equilibrium statistical mechanics where the distribution function is a function of the Hamiltonian H, ρ is a singular function (note that the Hamiltonian contains the kinetic energy which is also "delocalized").

Our basic mathematical problem is the derivation of the spectral representation of the Liouville operator L for LPS and singular delocalized distribution functions. This problem has been recently solved.[10] For example we may look for eigenfunctions $F(q, p; \alpha)$ of L such that their Fourier components have the same singularity as in (3.3)

$$F_k(p; \alpha) = F_k^0(p, \alpha)\delta(k) + F_k'(p; \alpha) \tag{3.4}$$

where $F_k'(p; \alpha)$ is a regular function of k. This problem has a meaning both in classical and quantum mechanics. In quantum mechanics we start with the momentum representation $\langle p'|\rho|p''\rangle$ of the density matrix and introduce new variables $p' - p'' = k$ and $\frac{p'+p''}{2} = P$ (note the relation with the so-called Wigner representation).

Complete sets of eigenfunctions have been obtained. There appear diffusive "non-Newtonian" and "non-Schrödinger" types of contributions. In the extension of L beyond the Hilbert space L is no more a commutator, as it is for integrable quantum systems, the eigenvalues are complex and the eigenfunctions are not implantable by wave functions (or trajectories).

The origin of the new contributions is the coupling of dynamical events through Poincaré resonances. This coupling leads to diffusive effects of the Fokker-Planck type in classical mechanics, of the Pauli type in quantum mechanics. We call more generally these operators associated to diffusion "collision operators". The collision operators play a dominant role in the dynamics as described in generalized function spaces (often called "rigged" Hilbert spaces). They are responsible for the time symmetry breaking. It can be shown that the eigenvalues of the Liouville operator are equal to the eigenvalues of the collision operators and the eigenfunctions are obtained by operators acting on the eigenfunctions of the collsion operators. For example, for persistent one-dimensional quantum scattering, we obtain eigenfunctions whose momentum representation are

$$\frac{1}{\sqrt{2}}[\delta(p - \alpha) + \delta(p + \alpha)] \tag{3.5}$$

and

$$\frac{1}{\sqrt{2}}[\delta(p - \alpha) - \delta(p + \alpha)] \tag{3.6}$$

---

[10] Petrosky and Prigogine (1995).

These eigenfunctions are superpositions of distributions and cannot be written as equilibrium, its eigenvalue is zero, while the eigenvalue corresponding to (3.6) is the cross section (which is not a difference between two numbers). The usual structure of quantum mechanics is destroyed. (Note that in the Hilbert space eigenfunctions of the Liouvillian are products of wave functions, and the eigenvalues are different.)

The physics of nonequilibrium processes lies in the new solutions of the Liouville equation, which are irreducible to wave functions or trajectories. These deviations from classical or quantum mechanics vanish for localized distribution functions, which correspond to situations described in terms of the Hilbert space.

This conclusion is in agreement with our intuition and with physical experience; indeed dissipation appears in "*large*" Hamiltonian systems.[11] There is an analogy with phase transitions. Here also the "whole" is more than its "parts." Irreversibility is an emerging property, as are also phase transitions, it can only be defined on the level of "populations". Once we have dynamics with broken time symmetry, we can easily construct dynamic expressions for entropy. The message of entropy is not "ignorance", on the contrary, it is the expression that we live in a world of broken time symmetry. Our approach leads to a number of predictions. Delocalized distribution functions can be prepared in computer simulations and our theoretical predictions have been tested successfully.[12]

It is precisely in macroscopic physics, which deals with delocalized situations where irreversiblity is so obvious, that the traditional description in terms of trajectories or wave functions fails. Note also that our approach leads to a realistic formulation of quantum theory, as there is no longer any need for extradynamical features to take into account the collapse of the wave functions.[13] No special role has to be attributed to the observer. There have been many proposals in the past to extend quantum mechanics (many world theories, alternative histories) but they remained on a purely verbal level. In contrast, our approach is a "down to earth" one based on dynamical considerations. In the past, physical ideas have always been associated with new mathematical developments. This is true for quantum mechanics and for relativity. This is also true here; the problem of irreversibility can be solved only by an approximate extension of functional analysis, which makes apparent properties of matter hidden until today. Let us also notice that this approach is of great interest in other fields, such as consensed matter and cosmology. But it is time to conclude. Dynamical instability coupled to persistent interactions changes the very meaning of laws of nature. The future is no more given, following the expression of the French poet Paul Valéry, future is construction, a construction in which we participate.

---

[11] In classical dynamics even "small" Hamiltonian systems (corresponding to maps or Poincaré systems) may already lead to irreversible processes. A well documented example is the kicked rotator. But we wanted to concentrate here on situations which are of importance both in classical and qunatum mechanics.

[12] Petrosky and Prigogine (1993a); Petrosky and Prigogine (1993b; Petrosky and Prigogine (1994); Prigogine and Petrosky (to appear).

[13] Rae (1989).

Acknowledgments

This is the outcome of the work of the Austin-Brussels group. It is impossible to thank individually all my young colleagues and co-workers, but I want to make an exception in favor of Dr. T. Petrosky and Dr. I. Antoniou, whose contributions to this subject have been outstanding. I acknowledge the support of the U. S. Department of Energy, the Robert A. Welch Foundation and the European Community.

References

Antoniou, I., Tasaki, S. (1992): Physica **190A** 313.

Antoniou, I., Tasaki, S. (1993): J. of Physics A, 73

Bohm, A., Gadella, M. (1989): Direac Kets, Gamow Vectors and Gelfand Triplets. Lecture Notes in Physics, vol. 348 (Springer, Berlin)

Gaspard, P. (1992): J. of Physics **A25**, 303

Gelfand, I.M., Vilenkin, N. Ya. (1964): *Generalized Functions*, vol. 4, (Academic Press, New York)

Hasegawa, H. H., Saphir, W. (1992): Physics Letters **A161**, 471

Hasegawa, H. H., Driebe, D. (1994): Phys. Rev.

Lasota, A., Mackey, M. (1985): *Probabilistic Properties of Deterministic Systems* (Cambridge University Press, Cambridge)

Maurin, K. (1968): *Generalized Eigenfunctions, Expansions and Unitary Representations of Topological Groups* (Polish Scientific Publishers, Warzawa)

Nicolis, G. (1989): Physics of far from equilibrium systems and self-organization. In P. Davies, *The New Physics* (Cambridge University Press)

Petrosky, T., Prigogine, I. (1993a): Proc. National Acad. Sci. USA **90**, 9393

Petrosky, T., Prigogine, I. (1993b): Phys. Lett. **A182**, 5

Petrosky, T., Prigogine, I. (1994): Chaos, Solitons and Fractals **4**, 311

Petrosky, T., Prigogine, I. (1995): Chaos, Solitons and Fractals

Prigogine, I. (1962): *Nonequilibrium Statistical Mechanics* (Wiley-Interscience, New York)

Prigogine, I. (1980): *From Being to Becoming* (Freeman, New York)

Prigogine, I. (1993): *Le leggi del caos* (Laterza, Rome); French translation *Les Lois du Chaos* (Flammarion, Paris)

Prigogine, I., Stengers, I. (1993): *Das Paradox der Zeit* (Piper, München)

Prigogine, I., Petrosky, T. (to appear): Proceedings of the Clausthal Symposium. Nonlinear, Dissipative, Irreversible Quantum Systems

Rae, A. (1989): *Quantum, Illusion or Reality?* (Cambridge University Press)

Natural Laws and Human Prediction

B. Chirikov

Budker Institute of Nuclear Physics, Russia

Abstract. Interrelations between dynamical and statistical laws in physics, on the one hand, and between classical and quantum mechanics, on the other hand, are discussed within the philosophy of separating the natural from the human, as a very specific part of Nature, and with emphasis on the new phenomenon of dynamical chaos.

The principal results of the studies of chaos in classical mechanics are presented in some detail, including the strong local instability and robustness of motion, continuity of both phase space and the motion spectrum, and the time reversibility but nonrecurrency of statistical evolution, within the general picture of chaos as a specific case of dynamical behavior.

Analysis of the apparently very deep and challenging contradictions of this picture with the quantum principles is given. The quantum view of dynamical chaos, as an attempt to resolve these contradictions guided by the correspondence principle and based upon the characteristic time scales of quantum evolution, is explained. The picture of quantum chaos as a new generic dynamical phenomenon is outlined together with a few other examples of such chaos: linear (classical) waves, the (many-dimensional) harmonic oscillator, the (completely integrable) Toda lattice, and the digital computer.

I conclude with discussion of the two fundamental physical problems: quantum measurement (ψ-collapse), and the causality principle, which both appear to be related to the phenomenon of dynamical chaos.

1 Philosophical Introduction: Separation of the Natural from the Human

The main purpose of this paper is the analysis of conceptual implications from the studies of a new phenomenon (or rather a whole new field of phenomena) known as *dynamical chaos* both in classical and especially in quantum mechanics. The concept of dynamical chaos resolves (or, at least, helps to do so) the two fundamental problems in physics and, hence, in all the natural sciences:

- are the dynamical and statistical laws of a different nature or does one of them, and which one, follow from the other;
- are classical and quantum mechanics of a different nature or is the latter the most universal and general theory currently available to describe the whole empirical evidence including the classical mechanics as the limiting case.

The essence of my debut philosophy is the separation of the human from the natural following Einstein's approach to the science – *building up a model of the real world*. Clearly, the human is also a part of the world, and moreover the most important part for us as human beings but not as physicists. The whole phenomenon of life is extremely specific, and one should not transfer its peculiarities into other fields of natural sciences as was erroneously (in my opinion) done in almost all major philosophical systems. One exception is positivism, which seems to me rather dull; it looks only at Nature but does not even want to see its internal mechanics. Striking examples of the former are Hegel's 'Philosophy of Nature' (Naturphilosophie) and its 'development', Engels' 'Dialectic of Nature'.

Another notorious confusion of such a 'human-oriented' physics was Wigner's claim that quantum mechanics is incompatible with the existence of self-reproducing systems (Wigner (1961)). The resolution of this 'paradox' is just in that Wigner assumed the Hamiltonian of such a system to be arbitrary, whereas it is actually highly specific (Schuster (1994)).

A more hidden human-oriented philosophy in physics, rather popular nowadays, is the information-based representation of natural laws, particularly when information is substituted for entropy (with opposite sign). In the most general way such a philosophy was recently presented by Kadomtsev (1994). That approach is possible and might be done in a self-consistent way, but one should be very careful to avoid many confusions. In my opinion, the information is an adequate conception for only the special systems that actually use and process the information like various automata, both natural (living systems) and man-made ones. In this case the information becomes a physical notion rather than a human view of natural phenomena. The same is also true in the theory of measurement, which is again a very specific physical process, the basic one in our studies of Nature but still not a typical one for Nature itself. This is crucially important in quantum mechanics as will be discussed in some detail below (Sections 2.4 and 3.1).

One of the major implications from studies of dynamical chaos is the conception of statistical laws as an intrinsic part of dynamics without any additional statistical hypotheses [for the current state of the theory see, e.g., Lichtenberg and Lieberman (1992) and recent collection of papers by Casati and Chirikov (1995) as well as the introduction to this collection by Casati and Chirikov (1995a)]. This basic idea can be traced back to Poincaré (1908) and Hadamard (1898), and even to Maxwell (1873); the principal condition for dynamical chaos being strong local instability of motion (Section 2.4). In this picture the statistical laws are considered as *secondary* with respect to more fundamental and general *primary* dynamical laws.

Yet, this is not the whole story. Surprisingly, the opposite is also true! Namely, under certain conditions the dynamical laws were found to be completely contained in the statistical ones. Nowadays this is called 'synergetics' (Haken (1987), Wunderlin (these proceedings)) but the principal idea goes back to Jeans (1929) who discovered the instability of gravitating gas (a typical example of a statistical system), which is the basic mechanism for the formation of galaxies and stars in modern cosmology, and eventually the Solar system, a classical example of a dynamical system. In this case the resulting dynamical laws proved to be secondary with respect to the primary statistical laws which include the former.

Thus, the whole picture can be represented as a chain of dynamical–statistical inclusions:

$$...?...\boxed{D \supset S} \supset D \supset S...?... \qquad (1.1)$$

Both ends of this chain, if any, remain unclear. So far the most fundamental (elementary) laws of physics seem to be dynamical (see, however, the discussion of quantum measurement in Sections 3 and 4). This is why I begin chain (1.1) with some primary dymamical laws.

The strict inclusion on each step of the chain has a very important consequence allowing for the so-called numerical experiments, or computer simulation, of a broad range of natural processes. As a matter of fact the former (not laboratory experiments) are now the main source of new information in the studies of the secondary laws for both dynamical chaos and synergetics. This might be called *the third way of cognition*, in addition to laboratory experiments and theoretical analysis.

In what follows I restrict myself to the discussion of just a single ring of the chain as marked in (1.1). Here I will consider the dynamical chaos separately in classical and quantum mechanics. In the former case the chaos explains the origin and mechanism of random processes in Nature (within the classical approximation). Moreover, that deterministic randomness may occur (and is typical as a matter of fact) even for a minimal number of degress of freedom $N > 1$ (for Hamiltonian systems), thus enormously expanding the domain for the application of the powerful methods of statistical analysis.

In quantum mechanics the whole situation is much more tricky and still remains rather controversial. Here we encounter an intricate tangle of various apparent contradictions between the correspondence principle, classical chaotic behavior, and the very foundations of quantum physics. This will be the main topic of my discussions below (Section 3).

One way to untangle this tangle is the new general conception, *pseudochaos*, of which quantum chaos is the most important example. Another interesting example is the digital computer, also very important in view of the broad application of numerical experiments in the studies of dynamical systems. On the other hand, pseudochaos in computers will hopefully help us to understand quantum pseudochaos and to accept it as a sort of *chaos* rather than a sort of regular motion, as many researchers, even in this field, still do believe.

The new and surprising phenomenon of dynamical chaos, especially in quantum mechanics, holds out new hopes for eventually solving some old, long-standing, fundamental problems in physics. In Section 4, I will briefly discuss two of them:

– the causality principle (time ordering of cause and effect), and
– ψ-collapse in the quantum measurement.

The conception of dynamical chaos I am going to present here, which is not common as yet, was the result of the long-term Siberian–Italian (SI) collaboration including Giulio Casati and Italo Guarneri (Como), and Felix Izrailev and

Dima Shepelyansky (Novosibirsk) with whom I share the responsibility for our joint scientific results and the conceptual interpretation.

2 Scientific Results and Conceptual Implications: the Classical Limit

Classical dynamical chaos, as a part of classical mechanics, was historically the first to have been studied simply because in the time of Boltzmann, Maxwell, Poincaré and other founders, statistical mechanics and quantum mechanics did not exist. No doubt, the general mathematical theory of dynamical systems, including the ergodic theory as its modern part describing various statistical properties of the motion, has arisen from (and is still conceptually based on) classical mechanics (Kornfeld et al. (1982), Katok and Hasselblatt (1994)). Yet, upon construction, it is not necessarily restricted to the latter and can be applied to a much broader class of dynamical phenomena, for example, in quantum mechanics (Section 3).

2.1 What is a Dynamical System?

In classical mechanics, 'dynamical system' means an object whose motion in some *dynamical space* is *completely* determined by a given interaction and the *initial conditions*. Hence, the synonym *deterministic system*. The motion of such a system can be described in two seemingly different ways which, however, prove to be essentially equivalent.

The first one is through the *motion equations* of the form

$$\frac{d\mathbf{x}}{dt} = \mathbf{v}(\mathbf{x}, t), \tag{2.1}$$

which always have a unique solution

$$\mathbf{x} = \mathbf{x}(t, \mathbf{x_0}) \tag{2.2}$$

Here $\mathbf{x}$ is a finite-dimensional vector in the dynamical space and $\mathbf{x_0}$ is the initial condition $[\mathbf{x_0} = \mathbf{x}(0)]$. A possible explicit time-dependence in the right-hand side of (2.1) is assumed to be a regular, e.g., periodic, one or, at least, one with a discrete spectrum.

The most important feature of dynamical (deterministic) systems is the *absence of any random parameters or any noise* in the motion equations. Particularly for this reason I will consider a special class of dynamical systems, the so-called *Hamiltonian (nondissipative) systems*, which are most fundamental in physics.

Dissipative systems, being very important in many applications, are neither fundamental (because the dissipation is introduced via a crude approximation of the very complicated interaction with some 'heat bath') nor purely dynamical in view of principally inevitable random noise in the heat bath (fluctuation–dissipation theorem). In a more accurate and natural way the dissipative systems can be described in the frames of the secondary dynamics ($S \supset D$ inclusion in (1.1)) when both dissipation and fluctuations are present from the beginning in the primary statistical laws.

A purely dynamical system is necessarily the *closed* one, which is the main object in fundamental physics. Thus, any coupling to the environment is completely neglected. I will come back to this important question below (Section 2.4).

In Hamiltonian mechanics the dynamical space, called *phase space*, is an even-dimensional one composed of N pairs of canonically conjugated 'coordinates' and 'momenta', each pair corresponding to one freedom of motion.

In the problem of dynamical chaos the initial conditions play a special role: they completely determine a particular trajectory, for a given interaction, or a particular realization of a dynamical process which may happen to be a very specific, nontypical, one. To get rid of such singularities another description is useful, namely the Liouville partial differential equation for the *phase space density*, or distribution function $f(\mathbf{x}, t)$:

$$\frac{\partial f}{\partial t} = \hat{L} f \tag{2.3}$$

with the solution

$$f = f(\mathbf{x}, t; f_0(\mathbf{x})). \tag{2.4}$$

Here $\hat{L}$ is a *linear* differential operator, and $f_0(\mathbf{x}) = f(\mathbf{x}, 0)$ is the initial density. For any smooth f_0 this description provides the generic behavior of a dynamical system via a continuum of trajectories. In the special case $f_0 = \delta(\mathbf{x} - \mathbf{x_0})$ the density describes a single trajectory like the motion equations (2.1).

In any case the phase space itself is assumed to be *continuous*, which is the most important feature of the classical picture of motion and the main obstacle in the understanding of quantum chaos (Section 3).

2.2   What is Dynamical Chaos?

Dynamical chaos can be characterized in terms of both the individual trajectories and the trajectory ensembles, or phase density. Almost all trajectories of a chaotic system are in a sense most complicated (they are *unpredictable* from observation of any preceding motion to use this familiar human term). Exceptional, e.g., periodic trajectories form a set of zero invariant measure, yet it might be everywhere dense.

An appropriate notion in the theory of chaos is the *symbolic trajectory* first introduced by Hadamard (1898). The theory of symbolic dynamics was developed further by Morse (1966), Bowen (1973), and Alekseev and Yakobson (1981). The symbolic trajectory is a projection of the true (exact) trajectory on to a discrete partition of the phase space at discrete instants of time t_n, e.g., such that $t_{n+1} - t_n = T$ fixed. In other words, to obtain a symbolic trajectory we first turn from the motion differential equations (2.1) to the difference equations over a certain time interval T:

$$\mathbf{x}(t_{n+1}) \equiv \mathbf{x}_{n+1} = M(\mathbf{x}_n, t_n). \qquad (2.5)$$

This is usually called *mapping* or *map*: $\mathbf{x}_n \to \mathbf{x}_{n+1}$. Then, while running a (theoretically) *exact* trajectory we record each $\mathbf{x}_n$ to a *finite* accuracy: $\mathbf{x}_n \approx m_n$. For a finite partition each m_n can be chosen to be integer. Hence, the whole infinite symbolic trajectory

$$\sigma \equiv ...m_{-n}...m_{-1}m_0m_1...m_n... = S(\mathbf{x}_0; T), \qquad (2.6)$$

can be represented by a *single* number σ, which is generally irrational and which is some function of the *exact* initial conditions. The symbolic trajectory may be also called a *coarse-grained trajectory*. I remind you that the latter is a *projection* of (not substitution for) the exact trajectory to represent in compact form the global dynamical behavior without unimportant microdetails.

A remarkable property of chaotic dynamics is that the set of its symbolic trajectories is *complete*; that is, it actually contains all possible sequences (2.6). Apparently, this is related to continuity of function $S(\mathbf{x}_0)$ (2.6). On the contrary, for a regular motion this function is everywhere discontinuous.

In a similar way the *coarse-grained phase density* $\overline{f}(m_n, t)$ is introduced, in addition to the exact, or *fine-grained density*, which is also a projection of the latter on to some partition of the phase space.

The coarse-grained density represents the global dynamical behavior, particularly the most important process of *statistical relaxation*, for chaotic motion, to some *steady state* $f_s(m_n)$ (statistical equilibrium) independent of the initial $f_0(\mathbf{x})$ if the steady state is *stable*. Otherwise, synergetics comes into play giving rise to a secondary dynamics. As the relaxation is an aperiodic process the spectrum of chaotic motion is *continuous*, which is another obstacle for the theory of quantum chaos (Section 3).

Relaxation is one of the characteristic properties of statistical behavior. Another is *fluctuation*. Chaotic motion is a generator of noise which is purely *intrinsic* by definition of the dynamical system. Such noise is a particular manifestation of the complicated dynamics as represented by the symbolic trajectories or by the difference

$$f(\mathbf{x}, t) - \overline{f}(m_n, t) \equiv \tilde{f}(\mathbf{x}, t). \qquad (2.7)$$

The relaxation $\overline{f} \to f_s$, apparently asymmetric with respect to time reversal $t \to -t$, gave rise to a long-standing misconception of the notorious *time arrow*. Even now some very complicated mathematical constructions are still being erected (see, e.g., Misra et al. (1979), Goldstein et al. (1981)) in attempts to extract somehow statistical irreversibility from the reversible mechanics. In the theory of dynamical chaos there is no such problem. The answer turns out to be conceptual rather than physical: one should separate two similar but different notions, *reversibility* and *recurrency*. The exact density $f(\mathbf{x}, t)$ is always *time-reversible* but *nonrecurrent* for chaotic motion; that is, it will never come back to the initial $f_0(\mathbf{x})$ in *both directions of time* $t \to \pm\infty$. In other words, the relaxation, also present in f, is time–symmetric. The projection of f, coarse-grained $\overline{f}$, which is both nonrecurrent and irreversible, emphasizes nonrecurrency of the exact solution. The apparent violation of the statistical relaxation upon time reversal, as described by the exact $f(\mathbf{x}, t)$, represents in fact the growth of a big fluctuation which will eventually be followed by the same relaxation in the opposite direction of time. This apparently surprising symmetry of the statistical behavior was discovered long ago by Kolmogorov (1937). One can say that instead of an imagionary time arrow there exists a *process arrow* pointing always to the steady state. The following simple example would help, perhaps, to overcome this conceptual difficulty. Consider the hyperbolic one-dimensional (1D) motion:

$$x(t) = a \cdot \exp{(\Lambda t)} + b \cdot \exp{(-\Lambda t)}, \qquad (2.8)$$

which is obviously time-reversible yet remains *unstable* in both directions of time ($t \to \pm\infty$). Besides its immediate appeal, this example is closely related to the mechanism of chaos which is the motion instability.

2.3   A Few Physical Examples of Low-Dimensional Chaos

In this paper I restrict myself to finite-dimensional systems where the peculiarities of dynamical chaos are most clear (see Section 3.2 for some brief remarks on infinite systems). Consider now a few examples of chaos in minimal dimensionality.

Billiards (2 degrees of freedom). The ball motion here is chaotic for almost any shape of the boundary except special cases like circle, ellipse, rectangle and some other (see, e.g., Lichtenberg and Lieberman (1992), Kornfeld et al. (1982), Katok and Hasselblatt (1994)). However, the ergodicity (on the energy surface) is only known for singular boundaries. If the latter is smooth enough the structure of motion becomes a very complicated admixture of chaotic and regular domains of various sizes (the so-called divided phase space). Another version of billiards is the wave cavity in the geometric optics approximation. This provides a helpful bridge between classical and quantum chaos.

Perturbed Kepler motion is a particular case of the famous 3-body problem. Now we understand why it has not been solved since Newton: chaos is generally present in such a system. One particular example is the motion of comet Halley perturbed by Jupiter which was found to be chaotic with an estimated life time in the Solar system of the order of 10 Myrs (Chirikov and Vecheslavov (1989); 2 degrees of freedom in the model used, divided phase space).

Another example is a new, diffusive, mechanism of ionization of the Rydberg (highly excited) hydrogen atom in the external monochromatic electric field. It was discovered in laboratory experiments (Bayfield and Koch (1974)) and was explained by dynamical chaos in a classical approximation (Delone et al. (1983)).

In this system a given field plays the role of the third body. The simplest model of the diffusive photoelectric effect has 1.5 degrees of freedom (1D Kepler motion and the external periodic perturbation), and is also characterized by a divided phase space.

Budker's problem: charged particle confinement in an adiabatic magnetic trap (Chirikov (1987)). A simple model of two freedoms (axisymmetric magnetic field) is described by the Hamiltonian:

$$H = \frac{p^2}{2} + \frac{(1 + x^2)\, y^2}{2}.$$
(2.9)

Here magnetic field $B = \sqrt{1 + x^2}$; $p^2 = \dot{x}^2 + \dot{y}^2$; x describes the motion along magnetic line, and y does so accross the line (a projection of Larmor's rotation). At small pitch angles $\beta \approx |\dot{y}/\dot{x}|$ the motion is chaotic with the chaos border being at roughly

$$p \sim \frac{1}{|\ln \beta|}$$
(2.10)

and being very complicated, so-called critical, structure (Section 2.5).

Matinyan's problem: internal dynamics of the Yang–Mills (gauge) fields in classical approximation (Matinyan (1979), Matinyan (1981)). Surprisingly, this completely different physical system can be also represented by Hamiltonian (2.9) with a symmetrized 'potential energy':

$$U = \frac{(1 + x^2)\, y^2 + (1 + y^2)\, x^2}{2}.$$
(2.11)

Dynamics is always chaotic with a divided phase space similar to model (2.9) (Chirikov and Shepelyansky (1982)). Model (2.11) describes the so-called massive gauge field; that is, one with the quanta of nonzero mass. The massless field corresponds to the 'potential energy'

$$U = \frac{x^2 y^2}{2}$$
(2.12)

and looks ergodic in numerical experiments.

2.4   Instability and Chaos

Local instability of motion responsible for a very complicated dynamical behavior is described by the *linearized equations*:

$$\frac{d\mathbf{u}}{dt} = \mathbf{u} \cdot \frac{\partial \mathbf{v}(\mathbf{x}^0(t),\, t)}{\partial \mathbf{x}}.$$
(2.13)

Here $\mathbf{x}^0(t)$ is a reference trajectory satisfying (2.1), and $\mathbf{u} = \mathbf{x}(t) - \mathbf{x}^0(t)$ is the deviation of a close trajectory $\mathbf{x}(t)$. On average, the solution of (2.13) has the form

$$|\mathbf{u}| \sim \exp{(\Lambda t)},$$
(2.14)

where Λ is *Lyapunov's exponent*. The motion is (exponentially) unstable if $\Lambda > 0$. In the Hamiltonian system of N degrees of freedom there are $2N$ Lyapunov's exponents satisfying the condition $\sum \Lambda = 0$. The partial sum of all positive exponents $\Lambda_+ > 0$,

$$h = \sum \Lambda_+ \tag{2.15}$$

is called the (dynamical) *metric entropy*. Notice that it has the dimensions of frequency and characterises the instability rate.

The motion instability is only a necessary but not sufficient condition for chaos. Another important condition is *boundedness* of the motion, or its oscillatory (in a broad sense) character. The chaos is produced by the combination of these two conditions (also called stretching and folding). Let us again consider an elementary example of a 1D map

$$x_{n+1} = 2x_n \mod 1, \tag{2.16}$$

where operation mod 1 restricts (folds) x to the interval (0,1). This is not a Hamiltonian system but it can be interpreted as a 'half' of that; namely, as the dynamics of the oscillation phase. This motion is unstable with $\Lambda = \ln 2$ because the linearized equation is the same except for the fractional part (mod 1). The explicit solution for both reads

$$u_n = 2^n u_0,$$

$$x_n = 2^n x_0 \mod 1. \tag{2.17}$$

The first (linearized) motion is unbounded, like Hamiltonian hyperbolic motion, (2.8) and is perfectly regular. The second one is not only unstable but also chaotic just because of the additional operation mod 1, which makes the motion bounded, and which mixes up the points within a finite interval.

We may look at this example from a different viewpoint. Let us express the initial x_0 in the binary code as the sequence of two symbols, 0 and 1, and let us make the partition of the unit x interval also in two equal halves marked by the same symbols. Then, the symbolic trajectory will simply repeat x_0; that is, (2.6) takes the form

$$\sigma = x_0. \tag{2.18}$$

It implies that, as time goes on, the global motion will eventually depend on ever-diminishing details of the initial conditions. In other words, when we formally fix the *exact* x_0 we 'supply' the system with infinite complexity, which arises due to the strong motion instability. Still another interpretation is that the exact x_0 is the source of *intrinsic noise* amplified by the instability. For this noise to be *stationary* the string of x_0 digits has to be infinite, which is only possible in *continuous* phase space.

A nontrivial part of this picture of chaos is that the instability must be *exponential* because a power-law instability is insufficient for chaos. For example, the linear instability ($|u| \sim t$) is a generic property of perfectly regular motion of the completely integrable system whose motion equations are *nonlinear* and, hence, whose oscillation frequencies depend on the initial conditions (Born (1958), Casati et al. (1980)). The character of motion for a faster instability ($|u| \sim t^{\alpha}$, $\alpha > 1$) is unknown.

On the other hand, the exponential instability ($h > 0$) is not invariant with respect to the change of time variable (Casati and Chirikov (1995a), Batterman (these proceedings)); in this respect the only invariant statistical property is ergodicity, Kornfeld et al. (1982), Katok and Hasselblatt (1994)). A possible resolution of this difficulty is that the proper characteristic of motion instability, important for dynamical chaos, should be taken with respect to the oscillation phases whose dynamics determines the nature of motion. It implies that the proper time variable must change proportionally with the phases so that the oscillations become stationary (Casati and Chirikov (1995a)). A simple example is harmonic oscillation with frequency ω recorded at the instances of time $t_n = 2^n t_0$. Then, oscillation phase $x = \omega t / 2\pi$ obeys map (2.16), which is chaotic. Clearly, the origin of chaos here is not in the dynamical system but in the recording procedure (random t_0). Now, if ω is a parameter (linear oscillator), then the oscillation is exponentially unstable (in new time n) but only with respect to the change of parameter ω, not of the initial x_0 ($x \to x + x_0$). In a slightly 'camouflaged' way, essentially the same effect was considered by Blümel (1994) with far-reaching conclusions on quantum chaos (Section 3.2).

Rigorous results concerning the relation between instability and chaos are concentrated in the Alekseev–Brudno theorem (see Alekseev and Yakobson (1981), Batterman (these proceedings), White (1993)), which states that the complexity per unit time of almost any symbolic trajectory is asymptotically equal to the metric entropy:

$$\frac{C(t)}{|t|} \to h, \qquad |t| \to \infty. \tag{2.19}$$

Here $C(t)$ is the so-called algorithmic complexity, or in more familiar terms, the information associated with a trajectory segment of length $|t|$.

The transition time from dynamical to statistical behavior according to (2.19) depends on the partition of the phase space, namely, on the size of a cell μ, which is inversely proportional to the biggest integer $M \geq m_n$ in symbolic trajectory (2.6). The transition is controlled by the *randomness parameter* (Chirikov (1985)):

$$r = \frac{h|t|}{\ln M} \sim \frac{|t|}{t_r}, \tag{2.20}$$

where t_r is the *dynamical time scale*. As both $|t|$, $M \to \infty$ we have a somewhat confusing situation, typical in the theory of dynamical chaos, in which two limits do not commute:

$$M \to \infty, |t| \to \infty \neq |t| \to \infty, M \to \infty. \tag{2.21}$$

For the left order ($M \to \infty$ first) parameter $r \to 0$, and we have *temporary determinism* ($|t| \lesssim t_r$), while for the right order $r \to \infty$, and we arrive at *asymptotic randomness* ($|t| \gtrsim t_r$).

Instead of the above double limit we may consider the *conditional limit*

$$|t|, \ M \rightarrow \infty, \qquad r = \text{const}, \qquad (2.22)$$

which is also a useful method in the theory of chaotic processes. Particularly for $r \lesssim 1$, strong dynamical correlations persist in a symbolic trajectory, which allows for the prediction of trajectory from a finite-accuracy observation. This is no longer the case for $r \gtrsim 1$ when only a statistical description is possible. Nevertheless, the motion equations can still be used to completely derive all the statistical properties without any *ad hoc* hypotheses. Here the exact trajectory *does exist* as well but becomes the Kantian *thing-in-itself*, which can be neither predicted nor reproduced in any other way.

The mathematical origin of this peculiar property goes back to the famous Gödel theorem (Gödel (1931)), which states (in a modern formulation) that *most* theorems in a given mathematical system are unprovable, and which forms the basis of contemporary mathematical logic (see Chaitin (1987) for a detailed explanation and interesting applications of this relatively less-known mathematical achievement). A particular corollary, directly related to symbolic trajectories (2.6), is that almost all real numbers are uncomputable by any finite algorithm. Besides rational numbers some irrationals like π or e are also known to be computable. Hence, their total complexity, e.g., $C(\pi)$, is finite, and the complexity per digit is zero (cf. (2.19)).

The main object of my discussion here, as well as of the whole physics, is a closed system that requires neglection of the external perturbations. However, in case of strong motion instability this is no longer possible, at least dynamically. What is the impact of a weak perturbation on the statistical properties of a chaotic system? The rigorous answer was given by the robustness theorem due to Anosov (1962): not only do statistical properties remain unchanged but, moreover, the trajectories get only slightly deformed providing (and due to) the same strong motion instability. The explanation of this striking peculiarity is that the trajectories are simply transposed and, moreover, the less the stronger is instability.

In conclusion let me make a very general remark, far beyond the particular problem of chaotic dynamics. According to the Alekseev–Brudno theorem (2.19) the source of stationary (new) information is always chaotic. Assuming farther that any creative activity, science including, is such a source we come to an interesting conclusion that any such activity has to be (partly!) chaotic. This is the creative side of chaos.

2.5 Statistical Complexity

The theory of dynamical chaos does not need any statistical hypotheses, nor does it allow for arbitrary ones. Everything is to be deduced from the dynamical equations. Sometimes the statistical properties turn out to be quite simple and familiar (Lichtenberg and Lieberman (1992), Chirikov (1979)). This is usually the case if the chaotic motion is also ergodic (on the energy surface), like in some billiards and other simple models (Section 2.3). However, quite often, and even typically for a few-freedom chaos, the phase space is divided, and the chaotic component of the motion has a very complicated structure.

One beautiful example is the so-called Arnold diffusion driven by a weak ($\epsilon \to 0$) perturbation of a completely integrable system with $N > 2$ degrees of freedom (Lichtenberg and Lieberman (1992), Chirikov (1979)). The phase space of such a system is pierced by the everywhere-dense set of nonlinear resonances

$$\sum_n m_n \cdot \omega_n^0(I) \approx 0, \qquad (2.23)$$

where m_n are integers, and ω_n^0 are the unperturbed frequences depending on dynamical variables (usually actions I). Each resonance is surrounded by a separatrix, the singular highly unstable trajectory with zero motion frequency. As a result, no matter how weak the perturbation ($\epsilon \to 0$) is, a narrow chaotic layer always arises around the separatrix. The whole set of chaotic layers is everywhere dense as is the set of resonances. For $N > 2$ the layers form a united connected chaotic component of the motion supporting the diffusion over the whole energy surface. Both the total measure of the chaotic component and the rate of Arnold diffusion are exponentially small ($\sim \exp\left(-C/\sqrt{\epsilon}\right)$) and can be neglected in most cases; hence the term *KAM integrability* (Chirikov and Vecheslavov (1990)) for such a structure (after Kolmogorov, Arnold and Moser who rigorously analysed some features of this structure). This quasi-integrability has the nature and quality of adiabatic invariance. However, on a very big time scale this weak but universal instability may essentially affect the motion.

One notable example is celestial mechanics, particularly the stability of the Solar system (Wisdom (1987) Laskar (1989), Laskar (1990), Laskar (1994)). Surprisingly, this 'cradle' of classical determinism and the exemplar case of dynamical behavior proves to be unstable and chaotic. The instability time of the Solar system was found to be rather long ($\Lambda^{-1} \sim 10$ Myrs), and its life time is still many orders of magnitude larger. It has not been estimated as yet, and might well exceed the cosmological time ~ 10 Byrs.

Another interesting example of complicated statistics is the so–called *critical structure* near the chaos border which is a necessary element of divided phase space (Chirikov (1991)). The critical structure is a hierarchy of chaotic and regular domains on ever decreasing spatial and frequency scales. It can be universally described in terms of the *renormalization group*, which proved to be so efficient in other branches of theoretical physics. In turn, the renormalization group may be considered as an abstract dynamical system that describes the variation of the whole motion structure, for the original dynamical system, in dependence of its spatial and temporal scale. Logarithm of the latter plays a role of 'time' (renormtime) in that renormdynamics. At the chaos border the latter is determined by the motion frequencies. The simplest renormdynamics is a periodic variation of the structure or, for a renorm-map, the invariance of the structure with respect to the scale (MacKay (1983)). Surprisingly, this scale invariance includes the chaotic trajectories as well. The opposite limit—renormchaos—is also possible, and was found in several models (see Chirikov (1991)).

Even though the critical structure occupies a very narrow strip along the chaos border it may qualitatively change the statistical properties of the whole chaotic component. This is because a chaotic trajectory unavoidably enters from time to time the critical region and 'sticks' there for a time that is longer the closer it comes to the chaos border. The sticking results in a slow power-law correlation decay for large time, in a singular motion spectrum for low frequency, and even in the superdiffusion when the phase-density dispersion $\sigma^2 \sim t^\alpha$ ($\alpha > 1$) grows faster than time (Chirikov (1987), Chirikov (1991)).

3   Scientific Results and Conceptual Implications: Quantum Chaos

The mathematical theory of dynamical chaos—ergodic theory—is self-consistent. However, this is not the case for the physical theory unless we accept the philosophy of the two separate mechanics: classical and quantum. Even though such a view cannot be excluded at the moment it has a profound difficulty concerning the border between the two. Nor is it necessary according to recent intensive studies of quantum dynamics. Then, we have to understand the mechanics of dynamical chaos from a quantum point of view. Our guiding star will be the *correspondence principle* which requires the complete quantum theory of any classical phenomenon, in the quasiclassical limit, assuming that the whole classical mechanics is but a special part (the limiting case) of the currently most general and fundamental physical theory: quantum mechanics. Now it would be more correct to speek about quantum field theory but here I restrict myself to finite-dimensional systems only (see Sections 3.2 and 3.4).

3.1   The Correspondence Principle

In attempts to build up the quantum theory of dynamical chaos we immediately encounter a number of apparently very deep contradictions between the well-established properties of classical dynamical chaos and the most fundamental principles of quantum mechanics.

To begin with, quantum mechanics is commonly understood as a *fundamentally statistical theory*, which seems to imply *always some quantum chaos*, independent of the behavior in the classical limit. This is certainly true but in some restricted sense only. A novel developement here is the *isolation* of this fundamental quantum randomness as solely the characteristic of the very specific quantum process, measurement, and even as the particular part of that—the so-called ψ-collapse which, indeed, has so far no dynamical description (see Section 4 for further discussion of this problem).

No doubt, quantum measurement is absolutely necessary for the study of the microworld by us, the macroscopic human beings. Yet, the measurement is, in a sense, foreign to the proper microworld that might (and should) be described separately from the former. Explicitly (Casati and Chirikov (1995a)) or, more often, implicitly such a philosophy has become common in studies of chaos but not yet beyond this field of research (see, e.g., Shimony (1994)).

This approach allows us to single out the dynamical part of quantum mechanics as represented by a *specific dynamical variable* $\psi(t)$ in *Hilbert space*, satisfying some *deterministic equation of motion*, e.g., the Schrödinger equation. The more difficult and vague statistical part is left for a better time. Thus, we temporarily bypass (not resolve!) the first serious difficulty in the theory of quantum chaos (see also Section 4). The separation of the first part of quantum dynamics, which is very natural from a mathematical viewpoint, was first introduced and emphasized by Schrödinger, who, however, certainly underestimated the importance of the second part in physics.

However, another principal difficulty arises. As is well known, the energy (and frequency) spectrum of any quantum motion *bounded in phase space* is always *discrete*. And this is not the property of a particular equation but rather a consequence of the fundamental quantum principle—the *discreteness of phase space* itself, or in a more formal language, the noncommutative geometry of quantum phase space. Indeed, according to another fundamental quantum principle—the uncertainty principle—a single quantum state cannot occupy the phase space volume $V_1 \lesssim \hbar^N \equiv 1$ [in what follows I set $\hbar = 1$, particularly, not to confuse it with metric entropy h (2.15)]. Hence, the motion bounded in a domain of volume V is represented by $V/V_1 \sim V$ eigenstates, a property even stronger than the general discrete spectrum (almost periodic motion).

According to the existing ergodic theory such a motion is considered to be *regular*, which is something opposite to the known chaotic motion with a continuous spectrum and exponential instability (Section 2.2), again independent of the classical behavior. This seems to *never imply any chaos* or, to be more precise, any *classical-like chaos* as defined in the ergodic theory. Meanwhile, the correspondence principle requires *conditional chaos* related to the nature of motion in the classical limit.

3.2   Pseudochaos

Now the principal question to be answered reads: where is the expected quantum chaos in the ergodic theory? Our answer to this question (Chirikov et al. (1981), Chirikov et al. (1988); not commonly accepted as yet) was concluded from a simple observation (principally well known but never comprehended enough) that the sharp border between the discrete and continuous spectrum is physically meaningful in the limit $|t| \to \infty$ only, the condition actually assumed in the ergodic theory. Hence, to understand quantum chaos the existing ergodic theory needs modification by the introduction of a new 'dimension', the time. In other words, a new and central problem in the ergodic theory is the *finite-time statistical properties* of a dynamical system, both quantum as well as classical (Section 3.4).

Within a finite time the discrete spectrum is dynamically equivalent to the continuous one, thus providing much stronger statistical properties of the motion than was (and still is) expected in the ergodic theory for the case of a discrete spectrum. In short, motion with a discrete spectrum may exhibit *all the statistical properties* of classical chaos but only on some *finite time scales* (Section 3.3). Thus, the conception of a time scale becomes fundamental in our theory of quantum chaos (Chirikov et al. (1981), Chirikov et al. (1988)). This is certainly a *new dynamical phenomenon*, related but not identical at all to classical dynamical chaos. We call it *pseudochaos*; the term *pseudo* is used to emphasize the difference from the asymptotic (in time) chaos in the ergodic theory. Yet, from the physical point of view, we accept here that the latter, strictly speaking, does not exist in Nature. So, in the common philosophy of the universal quantum mechanics *pseudochaos is the only true dynamical chaos* (cf. the term 'pseudoeuclidian geometry' in special relativity). Asymptotic chaos is but a limiting pattern which is, nevertheless, important both in theory, to compare with the real chaos, and in applications, as a very good approximation in a macroscopic domain, as is the whole classical mechanics. Ford describes the former *mathematical chaos* as contrasted to the *real physical chaos* in quantum mechanics (Ford (1994)). Another curious but impressive term is *artificial reality* (Kaneko and Tsuda (1994)), which is, of course, a self-contradictory notion reflecting, particularly, confusion in the interpretation of surprising phenomena such as chaos.

The statistical properties of the discrete-spectrum motion are not completely new subjects of research, such research goes back to the time of intensive studies in the mathematical foundations of statistical mechanics *before* dynamical chaos was discovered or, better to say, understood (see, e.g., Kac (1959)). We call this early stage of the theory *traditional statistical mechanics* (TSM). It is equally applicable to both classical as well as quantum systems. For the problem under consideration here, one of the most important rigorous results with far-reaching consequences was the *statistical independence* of oscillations with incommensurable (linearly independent) frequencies ω_n, such that the only solution of the resonance equation,

$$\sum_{n}^{N} m_n \cdot \omega_n = 0, \tag{3.1}$$

in integers is $m_n \equiv 0$ for all n. This is a generic property of the real numbers; that is, the resonant frequencies (3.1) form a set of zero Lebesgue measure. If we define now $y_n = \cos(\omega_n t)$, the statistical independence of y_n means that trajectory $y_n(t)$ is ergodic in N-cube $|y_n| \leq 1$. This is a consequence of ergodicity of the phase trajectory $\phi_n(t) = \omega_n t \bmod 2\pi$ in N-cube $|\phi_n| \leq \pi$.

Statistical independence is a basic property of a set to which the probability theory is to be applied. Particularly, the sum of statistically independent quantities,

$$x(t) = \sum_{n}^{N} A_n \cdot \cos(\omega_n t + \phi_n), \tag{3.2}$$

which is motion with a discrete spectrum, is the main object of this theory. However, the familiar statistical properties such as Gaussian fluctuations, postulated (directly or indirectly) in TSM, are reached in the limit $N \to \infty$ only, which is called the *thermodynamical limit*. In TSM this limit corresponds to infinite-dimensional models (Kornfeld et al. (1982), Katok and Hasselblatt (1994)), which provide a very good approximation for macroscopic systems, both classical and quantal.

However, what is really necessary for good statistical properties of sum (3.2) is a large number of frequencies $N_\omega \to \infty$, which makes the discrete spectrum continuous (in the limit). In TSM the latter condition is satisfied by setting $N_\omega = N$. The same holds true for quantum fields which are infinite-dimensional. In quantum mechanics another mechanism, independent of N, works in the quasiclassical region $q \gg 1$ where $q = I/\hbar \equiv I$ is some big quantum parameter, e.g., quantum number, and I stands for a characteristic action of the system. Indeed, if the quantum motion (3.2) [with $\psi(t)$ instead of $x(t)$] is determined by many ($\sim q$) eigenstates we can set $N_\omega = q$ independent of N. The actual number of terms in expansion (3.2) depends, of course, on a particular state $\psi(t)$ under consideration. For example, if it is just an eigenstate the sum reduces to a single term. This corresponds to the special peculiar trajectories of classical chaotic motion whose total measure is *zero*. Similarly, in quantum mechanics $N_\omega \sim q$ for *most states* if the system is *classically chaotic*. This important condition was found to be certainly *sufficient* for good quantum statistical properties (see Chirikov et al. (1981), Chirikov et al. (1988) and Section 3.3 below). Whether it is also the necessary condition remains as yet unclear.

Thus, with respect to the mechanism of the quantum chaos we essentially *come back* to TSM with an exchange of the number of freedoms N for the quantum parameter q. However, in quantum mechanics we are not interested, unlike in TSM, in the limit $q \to \infty$, which is simply the classical mechanics. Here, the central problem is the statistical properties for *large but finite q*. This problem does not exist in TSM describing macroscopic systems. Thus, with an old mechanism the new phenomena were understood in quantum mechanics.

3.3   Characteristic Time Scales in Quantum Chaos

The existing ergodic theory is asymptotic in time, and hence contains no time scales at all. There are two reasons for this. One is technical: it is much simplier to derive the asymptotic relations than to obtain rigorous finite-time estimates. Another reason is more profound. All statements in the ergodic theory hold true up to measure zero, that is, excluding some peculiar nongeneric sets of zero measure. Even this minimal imperfection of the theory did not seem completely satisfactory but has been 'swallowed' eventually and is now commonly tolerated even among mathematicians, to say nothing about physicists. In a finite-time theory all these exceptions acquire a *small but finite* measure which would be already 'unbearable' (for mathematicians). Yet, there is a standard mathematical trick, to be discussed below, for avoiding both these difficulties.

The most important time scale t_R in quantum chaos is given by the general estimate

$$\ln t_R \sim \ln q, \qquad t_R \sim q^\alpha \sim \rho_0 \lesssim \rho_H, \qquad (3.3)$$

where $\alpha \sim 1$ is a system-dependent parameter. This is called the *relaxation time scale* refering to one of the principal properties of chaos: *statistical relaxation* to some steady state (statistical equilibrium). The physical meaning of this scale is principally simple and is directly related to the fundamental uncertainty principle ($\Delta t \cdot \Delta E \sim 1$) as implemented in the second equation in (3.3), where ρ_H is the *full* average energy level density (also called the Heisenberg time). For $t \lesssim t_R$ the discrete spectrum is not resolved, and the statistical relaxation follows the classical (limiting) behavior. This is just the 'gap' in the ergodic theory (supplemented with the additional, time, dimension) where pseudochaos, particularly quantum chaos, dwells. A more accurate estimate relates t_R to a *part* ρ_0 of the level density. This is the density of the so-called *operative eigenstates*; that is, only those that are actually present in a particular quantum state ψ and actually control its dynamics.

The formal trick mentioned above is to consider not the finite-time relations we really need but rather the special *conditional limit* (cf. (2.22)):

$$t, q \to \infty \qquad \tau = \frac{t}{t_R(q)} = const \qquad (3.4)$$

Quantity τ is a new rescaled time which is, of course, nonphysical but very helpful technically. The *double* limit (3.4) (unlike the single one $q \to \infty$) is *not* the classical mechanics which holds true, in this representation, for $\tau \lesssim 1$ and with respect to the statistical relaxation only. For $\tau \gtrsim 1$ the behavior becomes essentially quantum (even in the limit $q \to \infty$!) and is called nowadays *mesoscopic phenomena*. Particularly, the quantum steady state is quite different from the classical statistical equilibrium in that the former may be *localized* (under certain conditions) that is *nonergodic* in spite of classical ergodicity.

Another important difference is in *fluctuations*, which are also a characteristic property of chaotic behavior. In comparison with classical mechanics quantum $\psi(t)$ plays, in this respect, an intermediate role between the classical trajectory (exact or symbolic) with big relative fluctuations ~ 1 and the coarse-grained classical phase space density with no fluctuations at all. Unlike both the fluctuations of $\psi(t)$ are $\sim N_\omega^{-1/2}$, which are another manifistation of statistical independence, or *decoherence*, of even pure quantum state (3.2) in case of quantum chaos. In other words, chaotic $\psi(t)$ represents statistically a *finite ensemble* of $\sim N_\omega$ systems even though formally $\psi(t)$ describes a single system. Quantum fluctuations clearly demonstrate also the difference between physical time t and auxillary variable τ: in the double limit ($t, q \to \infty$) the fluctuations vanish and one needs a new trick to recover them.

The relaxation time scale should be distinguished from the *Poincaré recurrence time* $t_P \gg t_R$, which is typically much longer, and which sharply increases with a decrease in the recurrence domain. Time scale t_P characterizes big fluctuations (for both the classical trajectory, but not the phase space density, and quantum ψ) of which recurrences is a particular case. Unlike this, t_R describes the average relaxation process.

Stronger statistical properties than relaxation and fluctuations are related in the ergodic theory to the exponential instability of motion. Their importance for statistical mechanics is not completely clear. Nevertheless, in accordance with the correspondence principle, those stronger properties are also present in quantum chaos as well, but on a *much shorter* time scale,

$$t_r \sim \frac{\ln q}{h}, \tag{3.5}$$

where h is classical metric entropy (2.15). This time scale was discovered and partly explained by Berman and Zaslavsky (1978) (see also Chirikov et al. (1981), Chirikov et al. (1988), Casati and Chirikov (1995a)). Being very short, t_r grows indefinitely as $q \to \infty$.

The simplest example of quantum dynamics on this scale is the stretching/squeezing of an initially narrow wave packet, with the conservation of the phase space volume like in classical mechanics, followed by the packet inflation (increasing phase space volume), and eventually by the complete destruction of the packet, its splitting into many irregular subpackets (Casati and Chirikov (1995a)).

In a quasiclassical region ($q \gg 1$), $t_r \ll t_R$ (3.3). This leads to an interesting conclusion that the quantum diffusion and relaxation are *dynamically stable* contrary to the classical behavior. It suggests, in turn, that the motion instability is not important *during* statistical relaxation. However, the *foregoing* correlation decay on a short time scale t_r is *crucial* for the statistical properties of quantum dynamics.

3.4   Examples of Pseudochaos in Classical Mechanics

Pseudochaos is a new generic dynamical phenomenon missed in the ergodic theory. No doubt, the most important particular case of pseudochaos is quantum chaos. Nevertheless, pseudochaos occurs in classical mechanics as well. Here are a few examples of classical pseudochaos, which may help us to understand the physical nature of quantum chaos, my primary goal in this paper. Besides, this unveils new features of classical dynamics as well.

Linear waves is the example of pseudochaos (see, e.g., Chirikov (1992)) that is closest to quantum mechanics. I remind you that here only a part of quantum dynamics is discussed, the one described, e.g., by the Schrödinger equation, which is a linear wave equation. For this reason quantum chaos is sometimes called wave chaos (Šeba (1990)). Classical electromagnetic waves are used in laboratory experiments as a physical model for quantum chaos (Stöckmann and Stein (1990), Weidenmüller et al. (1992)). The 'classical' limit corresponds here to the geometrical 'optics', and the 'quantum' parameter $q = L/\lambda$ is the ratio of a characteristic size L of the system to the wave length λ.

The linear oscillator (many-dimensional) is a particular case of waves (without dispersion). A broad class of quantum systems can be reduced to this model (Eckhardt (1988)). Statistical properties of linear oscillators, particularly in the thermodynamic limit ($N \to \infty$), were studied by Bogolyubov (1945) in the framework of TSM. On the other hand, the theory of quantum chaos suggests richer behavior for a large but finite N, particularly, the characteristic time scales for the harmonic oscillator motion (Chirikov (1986)) and the number of degrees of freedom N playing the role of the 'quantum' parameter.

Completely integrable nonlinear systems also reveal pseudochaotic behavior. An example of statistical relaxation in the Toda lattice had been presented in Ford et al. (1973) much before the problem of quantum chaos arose. Moreover, the strongest statistical properties in the limit $N \to \infty$, including one equivalent to the exponential instability (the so-called K-property) were rigorously proved just for the (infinite) completely integrable systems (see Kornfeld et al. (1982), Katok and Hasselblatt (1994)).

The digital computer is a very specific classical dynamical system whose dynamics is extremely important in view of the ever increasing application in numerical experiments covering now all branches of science and beyond. The computer is an 'overquantized' system in that *any* quantity here is *discrete*, whereas in quantum mechanics only the product of two conjugated variables is. The 'quantum' parameter here is $q = M$, which is the largest computer integer, and the short time scale (3.5) is $t_r \sim \ln M$, which is the number of digits in the computer word (Chirikov et al. (1981), Chirikov et al. (1988)). Owing to the discreteness, any dynamical trajectory in the computer eventually becomes periodic, an effect well known in the theory and practice of the so-called pseudo-random number generators. One should take all necessary precautions to exclude this computer artifact in numerical experiments. On the mathematical part, the periodic approximations in dynamical systems are also studied in ergodic theory, apparently without any relation to pseudochaos in quantum mechanics or computers.

Computer pseudochaos is the best answer to those who refuse accept the quantum chaos as, at least, a kind of chaos, and who still insist that only the classical-like (asymptotic) chaos deserves this name, the same chaos that was (and is) studied to a large extent just on computers; that is, the chaos inferred from a pseudochaos!

4   Conclusion: Old Challenges and New Hopes

The discovery and understanding of the new surprising phenomenon—dynamical chaos—opened up new horizons in solving many other problems including some long-standing ones. Unlike in previous sections, here I can give only a preliminary consideration of possible new approaches to such problems, together with some plausible conjectures (see also Casati and Chirikov (1995a)).

Let us begin with the problem directly related to quantum dynamics, namely the quantum measurement or, to be more correct, the specific stage of the latter:

ψ-*collapse*. This is just the part of quantum dynamics I bypassed above in the report on scientific results. This part still remains very vague to the extent that there is no common agreement even on the question of whether it is a real physical problem or an ill-posed one so that the Copenhagen interpretation of (or convention in) quantum mechanics gives satisfactory answers to all the *admissible* questions. In any event there exists as yet no dynamical description of the quantum measurement including ψ-collapse. The quantum measurement, as far as the result is concerned, is fundamentally a random process. However, there are good reasons to hope that this randomness can be interpreted as a particular manifestation of dynamical chaos (Cvitanović et al. (1992)).

The Copenhagen convention was (and still remains) very important as a phenomenological link between very specific quantum theory and laboratory experiments. Without this link studies of the microworld would be simply impossible. The Copenhagen philosophy perfectly matches the standard experimental setup of two measurements: the first one fixes the initial quantum state, and the second records the changes in the system. However, it is less clear how to deal with *natural processes* without any man-made measurements that is without the notorious *observer*. Since the beginning of quantum mechanics such a question has been considered ill-posed (meaning nasty). However, now there is a revival of interest in a deeper insight into this problem (see, e.g., Cvitanović et al. (1992)). Particularly, Gell-Mann and Hartle put a similar question, true, in the context of a very specific and global problem—the quantum birth of the Universe (Gell-Mann and Hartle (1989)). In my understanding, such a question arises as well in much simpler problems concerning any natural quantum processes. What is more important, the answer from Gell-Mann and Hartle (1989) does not seem satisfactory. Essentially, it is the substitution of the automaton (information gathering and utilizing system) for the standard human observer. Neither seems to be a generic construction in the microworld.

The theory of quantum chaos allows us to solve, at least (the simpler) half of the ψ-collapse problem. Indeed, the measurement device is by purpose a macroscopic system for which the classical description is a very good approximation. In such a system strong chaos with exponential instability is quite possible. The chaos in the classical measurment device is not only possible but unavoidable since the measurement system has to be, by purpose again, a highly unstable system where a microscopic intervention produces the macroscopic effect. The importance of chaos for the quantum measurement is that it destroys the coherence of the initial pure quantum state to be measured converting it into the incoherent mixture. In the present theories of quantum measurement this is described as the effect of external noise (see, e.g., Wheeler and Zureck (1983)). True, the noise is sufficient to destroy the quantum coherence, yet it is not necessary at all. Chaos theory allows us to get rid of the unsatisfactory effect of the external noise and to develop a purely dynamical theory for the loss of quantum coherence. Unfortunately, this is not yet the whole story. If we are satisfied with the *statistical* desciption of quantum dynamics (measurement including) then the decoherence is all we need. However, the *individual* behavior includes the second (main) part of ψ-collapse: namely, the *concentration* of ψ in a single state of the original superposition

$$\psi = \sum_n c_n \psi_n \rightarrow \psi_k, \qquad \sum_n |c_n|^2 = 1. \qquad (4.1)$$

This is the proper ψ-collapse to be understood.

Also, it is another challenge to the correspondence principle. For quantum mechanics to be universal it must explain as well the very specific classical phenomenon of the *event* that does happen and remains for ever in the classical records, and is completely foreign to the proper quantum mechanics. It is just the effect of ψ-collapse.

All these problems could be resolved by a hypotetical phenomenon of *self-collapse*; that is, the collapse without any 'observer', human or automatic. Unfortunately, it seems that any physical explanation of ψ-collapse requires some changes in the existing quantum mechanics, and this is the main difficulty both technical and philosophical.

Now we come to the even more difficult problem of the *causality principle*: the universal time ordering of the events. This principle has been well confirmed by numerous experiments in all branches of physics. It is frequently used in the construction of various theories but, to my knowledge, no general relation of causality to the rest of physics was ever studied.

This principle looks like a statistical law (another time arrow), hence a new hope to understand the mechanism of causality via dynamical chaos. Yet, it directly enters the dynamics as the additional constraint on the interaction and/or the solutions of dynamical equations. A well-known and quite general example is in keeping the retarded solutions of a wave equation, only discarding advanced ones as 'nonphysical'. However, this is generally impossible for a bounded dynamics because of the boundary conditions. Still, causality holds true as well.

In some simple classical *dissipative* models, such as a driven damping oscillator, the dissipation was shown to imply causality (Youla et al. (1959), Dolph (1963), Zemanian (1965), Güttinger (1966), Nussenzveig (1972)). However, such results were formulated as the restriction on a class of systems showing causality rather than the foundations of the causality principle. Nevertheless, it was already some indication of a possible physical connection between dynamical causality and statistical behavior. To my knowledge, this connection was never studied further. To the contrary, the developement of the theory went the opposite way: taking for granted the causality to deduce all possible consequences, particularly various dispersion relations (Nussenzveig (1972)).

Causality relates two qualitatively different kinds of events: *causes* and *effects*. The former may be simply the initial conditions of motion, the point missed in the above-mentioned examples of the causality-dissipation relation. The initial conditions not only formally fix a particular trajectory but also are *arbitrary*, which is, perhaps, the key point in the causality problem. Also, this may shed some light on another puzzling peculiarity of *all* known dynamical laws: they discribe the motion up to arbitrary initial conditions only (cf. Weingartner (these proceedings)). It looks like the dynamical laws already include the causality implicitly even though they do not this explicitly. In any event, something arbitrary suggests chaos is around.

Again, we arrive at a tangle of interrelated problems. A plausible conjecture for how to resolve them might be as follows. An arbitrary cause indicates some statistical behavior, while the cause–effect relation points out a dynamical law. Then, we may conjecture that when the cause acts the transition from statistical to dynamical behavior occurs, which statistically separates the cause from the 'past' and dynamically fixes the effect in the 'future'. In this imaginory picture the 'past' and 'future' are related not to time but rather to cause and effect, respectively. Thus, the causality might be not time ordering (time arrow) but *cause–effect ordering*, or the *causality arrow*. The latter is very similar to the process arrow discussed in Section 2.2. Now, the central point is that the cause is arbitrary while the effect is not, whatever the time ordering.

This is, of course, but a raw guess to be developed, carefully analysed, and eventually confirmed or disproved experimentally.

Also, this picture seems to be closer to the statistical (secondary) dynamics [synergetics, or $S \supset D$ inclusion in (1.1)] rather than to dynamical chaos. Does it mean that the primary physical laws are statistical or, instead, that the chain of inclusions (1.1) is actually a closed ring with a 'feedback' coupling the secondary statistics to the primary dynamics?

We don't know.

In all this long lecture I have never given the definition of dynamical chaos, either classical or quantal, restricting myself to informal explanations (see Casati and Chirikov (1995a) for some current definitions of chaos). In a mathematical theory the definition of the main object of the theory precedes the results; in physics, expecially in new fields, it is quite often vice versa. First, one studies a new phenomenon such as dynamical chaos and only at a later stage, after understanding it sufficiently, we try to classify it, to find its proper place in the existing theories and eventually to choose the most reasonable definition. This time has not yet come.

References

Alekseev V. M., Yakobson, M. V. (1981): Phys. Reports **75**, 287

Anosov, D. D. (1962): Dokl. AN SSSR **145**, 707

Batterman, R. (these proceedings): Chaos: Algorithmic Complexity vs. Dynamical Instability

Bayfield, J., Koch, P. (1974): Phys. Rev. Lett. **33**, 258

Berman G. P., Zaslavsky, G. M. (1978): Physica A **91**, 450

Blümel, R. (1994): Phys. Rev. Lett. **73**, 428

Bogolyubov, N. N. (1945): *On Some Statistical Methods in Mathematical Physics* (Kiev) p. 115; Selected Papers, Naukova Dumka, Kiev, 1970, Vol. 2, p. 77 (in Russian)

Born, M. (1958): Z.Phys. **153**, 372

Bowen, R. (1973): Am. J. Math. **95**, 429

Casati, G., Chirikov, B. V., Ford, J. (1980): Phys. Lett. A **77**, 91

Casati, G., Chirikov, B. V., Eds. (1995): *Quantum Chaos: Between Order and Disorder* (Cambridge Univ. Press)

Casati, G., Chirikov, B. V. (1995a): *The Legacy of Chaos in Quantum Mechanics*, in Casati and Chirikov (1995).

Chaitin, G. (1987): Information, Randomness and Incompleteness, World Scientific

Chirikov, B. V. (1979): Phys. Reports **52**, 263

Chirikov, B. V. (1985): in *Proc. 2d Intern. Seminar on Group Theory Methods in Physics*, Harwood, Vol. 1, p. 553

Chirikov, B. V. (1986): Foundations of Physics **16**, 39

Chirikov, B. V. (1987): Proc. Roy. Soc. Lon. A **413**, 145

Chirikov, B. V., Izrailev, F. M., Shepelyansky, D. L. (1988): Physica D **33**, 77

Chirikov, B. V. (1991): Chaos. Solitons and Fractals **1**, 79

Chirikov, B. V. (1992): Linear Chaos. In: Nonlinearity with Disorder, Springer Proc. in Physics **67**, Springer, p. 3.

Chirikov, B. V., Izrailev, F. M., Shepelyansky, D. L. (1981): Sov.Sci.Rev. C **2**, 209

Chirikov, B. V., Shepelyansky, D. L. (1982): Yadernaya Fiz. **36**, 1563

Chirikov, B. V., Vecheslavov, V. V. (1989): Astron.Astrophys. **221**, 146

Chirikov, B. V., Vecheslavov, V. V. (1990): in *Analysis etc.*, Eds. P. Rabinowitz and E. Zehnder, Academic Press, p. 219

Cvitanović, P., Percival, I., Wirzba, A., Eds. (1992): *Quantum Chaos - Quantum Measurement* (Kluwer)

Delone, N. B., Krainov, V. P., Shepelyansky, D. L. (1983): Usp.Fiz.Nauk **140**, 335

Dolph, L. (1963): Ann. Acad. Sci. Fenn, Ser. AI **336/9**

Eckhardt, B. (1988): Phys. Reports **163**, 205

Ford, J. (1994): private communication

Ford J. et al. (1973): Prog. Theor. Phys. **50**, 1547

Gell-Mann, M., Hartle, J. (1989): Quantum Mechanics in the Light of Quantum Cosmology. Proc. 3rd Int. Symposium on the Foundations of Quantum Mechanics in the Light of New Technology, Tokyo

Gödel, K. (1931): Monatshefte für Mathematik und Physik **38**, 173

Goldstein, S., Misra, B., Courbage, M. (1981): J. Stat. Phys. **25**, 111

Güttinger, W. (1966): Fortschr. Phys. **14**, 483, 567

Hadamard, J. (1898): J. Math. Pures et Appl. **4**, 27

Haken, H. (1987): *Advanced Synergetics* (Springer)

Jeans, J. (1929): Phil. Trans. Roy. Soc. A **199**, 1

Katok, A., Hasselblatt, B. (1994): *Introduction to the Modern Theory of Dynamical Systems* (Cambridge Univ. Press)

Kac, M. (1959): Statistical Independence in Probability, Analysis and Number Theory, Math. Ass. of America

Kadomtsev, B. B. (1994): Usp. Fiz. Nauk **164**, 449

Kaneko, K., Tsuda, I. (1994): Physica D **75**, 1

Kolmogorov, A. N. (1937): Math. Ann. **113**, 766

Kornfeld, I., Fomin, S., Sinai, Ya. (1982): *Ergodic Theory* (Springer)

Laskar, J. (1989): Nature **338**, 237

Laskar, J. (1990): Icarus **88**, 266

Laskar, J. (1994): Astron. Astrophys. **287**, L9

Lichtenberg, A., Lieberman, M. (1992): *Regular and Chaotic Dynamics* (Springer)

MacKay, R. (1983): Physica D **7**, 283

Matinyan S. et al. (1979): Zh. Eksp. Teor. Fiz. (Pisma) **29**, 641

Matinyan S. et al. (1981): Zh. Eksp. Teor. Fiz. (Pisma) **34**, 613

Maxwell, C. (1873) *Matter and Motion* (London)

Misra, B., Prigogine, I., Courbage, M. (1979): Physica A **98**, 1

Morse, M. (1966): Symbolic Dynamics. Lecture Notes, Inst. for Advanced Study (Princeton)

Nussenzveig, H. (1972): *Causality and Dispersion Relations* (Academic Press)

Poincaré, H. (1908): *Science et Methode* (Flammarion)

Schuster, P. (1994): private communication

Šeba, P. (1990): Phys. Rev. Lett. **64**, 1855

Sheynin, O. B. (1985): Archive for History of Exact Sciences **33**, 351

Shimony, A. (1994): The Relation Between Physics and Philosophy, in: Proc. 3d Int. Workshop on Squeezed States and Uncertainty Relations (Baltimore, 1993), NASA, p.617

Stöckmann, H., Stein, J. (1990): Phys. Rev. Lett. **64**, 2215

Weidenmüller H. et al., Phys. Rev. Lett. **69**, 1296

Weingartner, P. (these proceedings): *Under What Transformations are Laws Invariant?*

Wheeler, J., Zurek, W., Eds. (1983): *Quantum Theory and Measurement* (Princeton Univ. Press)

White, H. (1993): Ergodic Theory and Dynamical Systems **13**, 807

Wigner, E. (1961): In *The Logic of Personal Knowledge*, London, Ch. 19

Wisdom, J. (1987): Icarus **72**, 241

Wunderlin, A. (these proceedings): On the Foundations of Synergetics

Youla, D., Castriota, L., Carlin, H. (1959): IRE Trans. **CT-6**, 102

Zemanian, A. (1965): *Distribution Theory and Transform Analysis* (McGraw-Hill)

Comment on Boris Chirikov's Paper "Natural Laws and Human Prediction"

J. Lighthill

University College London, UK

I enjoyed all parts of this paper, but the part on which I should especially like to comment is the part dealing with the history of investigations of chaos in systems subject to the Hamiltonian equations of classical mechanics. I believe that this history has been described by Professor Chirikov, perhaps through modesty, in a way which does not fully bring out the importance of contributions to the study of chaos in such systems which were made by Professor Chirikov himself. Those classical authors which are cited in the paper, including especially Poincaré, had admittedly achieved an understanding of the possibilities of chaotic behaviour that may arise in Hamiltonian systems. On the other hand, their attempts at rigorous mathematical proof of the properties of such systems came up against some very severs difficulties. Necessarily, such proofs were attempted by means of perturbation theory, for sufficiently small departures from a regular (periodic-orbits) solution. Nevertheless, many formidable obstacles (including the famous "small divisors" problem, for example) opposed the development of their arguments into a "watertight" mathematical proof. Against this background, one of the vitally important contributions of the famous "KAM" papers of Kolmogorov (1954), Arnold (1963) and Moser (1962) referred to in section 2.5 of Professor Chirikov's paper was their success in overcoming all the obstacles, and in achieving a first rigorous demonstration, for sufficiently small values of a perturbation amplitude, of the properties of such classical systems.

Even in those regions of parameter space (involving e.g. near-coincidence of resonance frequencies) where the difficulties were most formidable, the KAM methods produced completely reliable results. As far as chaos was concerned, these results demonstrated beyond any doubt that it could arise in such a system. Nevertheless, they showed that regular behaviour of the system was enormously more common. Indeed, it was only in regions of parameter space whose total measure was of smaller order than any algebraic power of a perturbation amplitude that this regular behaviour was replaced by chaotic behaviour. The presence of those "microscopic" gaps in parameter space where chaotic behaviour could be shown to come about was of course of the greatest physical as well as mathematical interest. On the other hand, a group of "die-hard" mathematicians who had long argued that behaviour was an essentially unproven hypothesis could still claim that the demonstration of its absence except in a region of parameter space of such exceedingly small measure had at least identified it as just "a rarity". It has been against that background that the 1979 paper of Professor Chirikov (see Chirikov (1979)) has required to be seen as of the utmost importance. By using computational methods of extreme precision to derive accurate numerical solutions for a Hamiltonian system, he was able first of all to verify for small amplitudes the transition from regular to chaotic behaviour in those extremely narrow regions of parameter space that are predicted by the KAM theory. There his methods were deriving identical results to those based upon a perturbation-theory approach. Then, he investigated what happened to those extremely narrow regions when the computations were carried out with progressively increasing perturbation amplitudes. It was above all this investigation which convinced the exponents of classical mechanics that chaos is not "a mere curiosity" - and, above all, not just "a rarity". On the contrary, as the perturbation amplitude increased, there appeared a steep widening of the regions of parameter space within which computed solutions exhibited the behaviour characteristic of chaotic systems. With a further increase of amplitude, chaotic behaviour from being exceedingly rare had become extremely normal. For many systems, furthermore, the computations indicated a transition to globally chaotic behaviour, sometimes called global stochasticity. Some other work at that time, being carried out independently in the USA by J.M. Greene (see Greene (1979)), was leading to rather similar conclusions, which have of course been strongly reinforced in many subsequent investigations. Nevertheless, it is no exaggeration for the friends of Professor Chirikov to claim, and moreover to wish to emphasize on an occasion like this, that it was his work above all which led to a full recognition of how, for conservative dynamical systems in classical mechanics, chaotic behaviour is the rule rather than the exception. In relation to the subject of this Symposium (the relation between knowledge of laws governing natural phenomena and the possibilities of prediction of those phenomena) this conclusion has, needless to say, proved to be of fundamental importance.

References

Arnold, V. I. (1963): Usp. Mat. Nauk **18**(13)
Chirikov, B. V. (1979): Phys. Reports **52**(263)
Greene, J. M. (1979): J. Math. Phys. **20**(1183)
Kolmogorov, A. N. (1954): Dokl. Akad. Nauk **98**(527)
Moser, J. (1962): Nach. Akad. Wiss. Gttingen, Math. Phys. Kl. 2, 1

Natural Laws and the Physics of Complex Systems
Comments on the Report by B.Chirikov

G. Nicolis

Université Libre de Bruxelles, Belgium

Abstract. In his report Professor Chirikov raises a number of important scientific and epistemological issues stemming from recent developments of chaos theory. In this short comment I would like to take up three items which I regard of special interest from the perspective of the implications of chaos research in different branches of science :

- The status of biological processes with respect to the laws of physics ;
- The status of the statistical description in general;
- The statistical properties of complex systems giving rise to bifurcations and chaos.

1 Biological processes and the laws of physics

This point, although present in the very title of the report, is subsequently only briefly treated in Section 1. Chirikov insists on the high specificity of the phenomenon of life. This is certainly true, but in this context it is worth drawing attention on the phenomenon of *self-organization* (Nicolis (1977)), whereby individual subunits achieve, through their cooperative interactions, states characterized by new, *emergent properties* transending the properties of their constitutive parts. Self-organization processes are ubiquitous in physics and chemistry, where large classes of systems obeying to nonlinear evolution laws and subjected to a constraint give rise spontaneously and under well-defined laboratory conditions to complex behavior in the form of abrupt transitions, a multiplicity of states, periodic or aperiodic oscillations, regular space patterning or spatio-temporal chaos. Many of these phenomena are observed in *in vitro* experiments on biochemical reactions. They also present appealing analogies with well-known manifestations of life such as biological rhythms, regulation at enzymatic level or at the level of the immune response, morphogenesis during embryonic development, or propagation of information through the nerve impulse (see, for instance Peliti (1991)). This poses the question of genericity and universality of life on a new basis and raises a number of concrete and challenging questions for future investigations.

2 Statistical description

In my view the inclusion of statistical laws into the dynamical ones or vice versa (Chirikov's eq. (1.1)) is not the real issue. At the deterministic level of description the evolution laws (Chirikov's eq. (2.1))

$$\frac{d\mathbf{x}}{dt} = \mathbf{v}(\mathbf{x}, t) \tag{1}$$

can be embedded in phase space. Phase space density f obeys then to the Liouville equation

$$\frac{\partial f}{\partial t} = -div\ \mathbf{v}f = \hat{L}f \tag{2}$$

whose characteristics are, in turn, nothing but eqs. (1). This close correspondence of the two descriptions also holds for discrete time dynamical systems for which eq. (2) must be replaced by the Frobenius-Perron equation.

The real separation between deterministic and statistical views starts when the eigenvalue problem of $\hat{L}$ (or of the Frobenius-Perron operator $\hat{P}$) is addressed, since in this case one must specify the space of functions in which this problem is to be embedded. Depending on the smoothness of the admissible functions one may derive, then, from the statistical description properties that were not built in an obvious manner into the deterministic description. A concrete example will be mentioned in Sec. 3 of this comment.

In many instances there exists an additional motivation for undertaking a statistical description. Macroscopic systems are usually coupled to a complex environment inflicting on them a variety of perturbations, which in many instances can be assimilated to an (external) *noise* process. In addition they themselves generate spontaneously variablity resembling in many respects to a noise process - the thermodynamic *fluctuations*. To account for these phenomena eqs. (1) must be augmented and one is led to a Langevin-type dynamics (Nicolis (1977), Gardiner (1983))

$$\frac{d\mathbf{x}}{dt} = \mathbf{v}(\mathbf{x}, t) + \mathbf{F}(\mathbf{x}, t) \tag{3}$$

where $\mathbf{F}$ is the *random force*. Eqs (3) have been analyzed in detail in the literature in the double limit of weak, white noise. As it turns out, the deterministic description is recovered as the most probable path of the full stochastic process. In this special sense the stochastic description would therefore appear to contain the deterministic one as mentioned briefly by Chirikov in his Sec. 1, although I do not see what "synergetics" has to do with this particular point. Still, some additional comments are in order :

- Basically fluctuations are nothing but deterministic chaos in the high-dimensional ($N \sim 10^{23}$) phase space of a macroscopic system. Eqs. (3) are therefore a shortcut to a full-fledged Liouville equation approach.
- In deriving the properties of the random force $\mathbf{F}$ use has been made of the deterministic properties, notably through the fluctuation-dissipation theorem (Callen and Welton (1951)).

3 Statistical properties of complex systems

In the last years the full solution of the eigenvalue problem of the Liouville equation associated to systems giving rise to bifurcations and chaos has been achieved. The most transparent case is that of *dissipative* systems, for which even 1-dimensional dynamics generates complexity resembling the one one is accustomed to find in many-body systems of interest in statistical mechanics. A simple example is provided by the discrete time chaotic map

$$x_{n+1} = r x_n \quad (mod\ 1), \quad r > 1 \tag{4}$$

for which a full spectral decomposition of the Frobenius-Perron operator has been obtained. As it turns out (Gaspard (1992), Antoniou and Tasaki (1993)):

- for $r = 2$, the eigenvalues are

$$S_k = 2^{-k} \quad k = 0, 1, 2, \dots \tag{5}$$

  The spectrum is thus discrete, even though chaos is an aperiodic process. This is at variance with the statement made by Chirikov in his Sec. 2.2.
- The right eigenfunctions are the Bernoulli polynomials while the left ones are δ-functions and derivatives thereof (notice that $\hat{P}$ is not self-adjoint here).

Similar analysis has been carried out for continuous time dynamical systems undergoing pitchfork bifurcation. It has been shown that at the bifurcation point the spectrum of the Liouville operator becomes continuous, but remains discrete and confined to the negative real axis before and after bifurcation. Furthermore the symmetry-breaking character of the pitchfork bifurcation shows up through the appearance of degeneracies in the spectrum in the critical and post-critical cases (Gaspard et al. (1995)).

Common to both studies mentioned above is the observation that the probabilistic description is stable in the sense that the probability density is driven irreversibly to its invariant form. This is to be contrasted with the instability of motion inherent in the deterministic prediction.

In conlusion the connection between statistical and deterministic description is quite intricate indeed. For certain (perhaps even for most) types of our predictions statistical description is operationally more meaningful, since it reflects the finite precision of measurement process and bypasses the fundamental limitations associated with the instability of motion.

References

Nicolis, G., Prigogine, I. (1977): *Self-organization in nonequilibrium systems* (Wiley, New York)

Peliti, L., ed. (1991): *Biologically inspired physics* (Plenum, New York)

Gardiner, C. (1983): *Handbook of stochastic methods* (Springer, Berlin)

Callen, H., Welton, T. (1951): Phys.Rev. **83**, 34

Gaspard, P. (1992): J. Phys. **A 25**, L 483
Antoniou, I., Tasaki, S. (1993): J.Phys. **A 26**, 73
Gaspard, P., Nicolis, G., Provata, A., Tasaki, S. (1995): Phys. Rev. **E 51**, 74

Discussion of Boris Chirikov's Paper

Batterman, Chirkov, Noyes, Schurz, Suppes, Weingartner

Noyes: I would really like to hear what you have to say about the wave function collapse. This is a problem I have thought quite a bit about. I'd like to know what kind of a position you take on it. You did not have much time to mention it in your talk so I would like you to say something about it now.

Chirikov: I actually have no beforehand solution of this problem, I simply see that there is such a problem and it would be interesting to solve it somehow. But this is a very subtle problem, a very old one, from the beginning of quantum mechanics. And so the main question you need to solve for yourself or try to convince other people is whether this ψ-collapse is a real physical problem. In other words: is it a physical problem or a philosophical problem? You know that there is no such problem in the common (Copenhagen) interpretation of quantum mechanics. Rather, it is a convention necessary to do real research in quantum physics. You need to understand how to relate this ψ to a result of experiment, and how to interpret the experiment and derive a particular fundamental law of physics. In my opinion, I don't know an answer of course and cannot make any strong argument in support, but nevertheless my opinion is that it might be a physical problem. You should distinguish two types of physical processes. One is what you have in your laboratory. You have a particular device and you fix the initial conditions which is not your immediate physical problem. The problem of initial conditions, also very interesting is another part of physics because you choose certain initial conditions by the quantum measurement. Anyway, you make some complete quantum measurement which fixes the ψ-function exactly. Then, to study something you make another measurement, and from the statistical results you derive some fundamental law. This is O.K.: all you need to know is that the modulus of the ψ-function squared is the probability of particular results. This was very important at the beginning of quantum mechanics to have a clear idea how to interpret experimental results, and how to recalculate from them a fundamental law of quantum interaction. But you may consider a different type of processes, I would say. Something just happens around, and nobody is interested to measure what has happened. But something very important does happen, for example, the first living molecule does appear. How would you describe this? It is not clear. You need this ψ-collapse without special measurement in the usual sense. Some physicists, including myself, think that it must be something which might be called the self-collapse that is a collapse without special measurement, something that from time to time produces the event. In the standard quantum mechanics there is no such conception. You have probabilities of everything but nothing happens. But many things do happen, and the problem is how to understand it from quantum mechanics? It seems to me that we need the mechanism of such a self-collapse, some dynamical theory of the ψ-collapse. Again there are two different situations. In many cases you are satisfied with the statistical description of events. So, all you need is decoherence of the ψ-function. You begin with a pure state, not a mixture. One statistical effect of the ψ-collapse is in that you obtain an incoherent mixture of probabilities instead of the coherent superposition of probability amplitudes. This problem is solved in the theory of dynamical chaos. The principal result of this theory - statistical relaxation - means a general decoherence. Even the simple diffusion implies already decorrelation and decoherence, otherwise you would not have this characteristic linear dependence on time of second moment of distribution function. So, principally the problem of quantum decoherence is solved. Of course, it might be very difficult technically but principally it is solved. What remains unsolved is if you are not satisfied with the statistical output of the process but are interested in individual events, for example, the living molecule with the particular chirality, by the way, for some reason. Then the decoherence is not sufficient. You need to describe or to find a mechanism what is called the probability redistribution. It means that not only the different initial states in superposition become uncorrelated, but all the probability goes to a particular state in a particular event. This is, of course, a much more difficult problem to be solved. There are some attempts to find how it may happen but, of course, they do not rely upon the decoherence within the existing quantum mechanics. Particularly, the quantum chaos is a part of quantum mechanics, it is nothing beside the Schrödinger equation, a special solution of this equation. This redistribution of probabilities, if it is a real process which is the question at the moment, if there is no other explanation of the whole problem, requires unfortunately (or fortunately, I don't know) some changes in quantum mechanics because the Schrdinger equation does not describe this. And it this the main difficulty. So what is your opinion?

Noyes: For me, this problem can only be discussed in the framework of a physical cosmology which sets the boundary conditions in such a way that they are not arbitrary. My own cosmological model does just this in a way that is briefly discussed in Chapter 5 of my contribution to this conference. However, for the problem of the origin of biomolecular chirality which you mentioned, my cosmology coincides with the conventional view that earth-type planets are formed from the debris of supernovae, and hence are formed in a specific and necessarily chiral environment. This idea is due to Ed Rubenstein and is discussed briefly in "Supernovae and Life" by E. Rubenstein, W.A. Bonner, H.P. Noyes and G.S. Brown in *Nature* **306**, 118 (1983). When a star goes supernova it leaves behind a neutron star which traps the magnetic field and can be detected observationally as a pulsar. Out to the radius where rigid body motion would exceed the velocity of light, the ionized plasma is locked into this rotating magnetic field and emits synchrotron radiation. This chiral radiation has opposite chirality above and below the plane of rotation. Bill Bonner has shown that 200 nanometer chiral radiation, which is plentiful in the pulsar spectrum, decomposes a racemic mixture of leucine molecules leaving behind a 4% enatiomeric excess of left- or right- handed molecules depending on the chirality of the radiation. Thus any organic molecules in the interstellar medium from which the planets are formed necessarily are handed from the start. We now know that interstellar organic molecules cling to interstellar dust grains in forms that support this mechanism for the production of pre-planetary and planetary biomolecular chirality. Thus this particular problem is solved by embedding it in a well understood cosmological evolutionary scenario. This moves the statistical problem from the quantum mechanical to the macroscopic level, which is part of the problem my contribution to this conference addresses. No "observer" is needed.

Chirikov: I agree, perhaps, it is not the best example. Nevertheless, we need not only statistical results, but also individual ones to understand the events around. As to the cosmology, perhaps you know that Gell-Mann has studied recently this problem in cosmology, I give you a reference. The problem is the quantum birth of the universe. And the question for him is: who was the observer at that time? So he tried to develop a kind of automatic observer, a very particular type of information system. But, in my opinion, you don't need to study such a great problem to understand this. The electron diffraction on two slits is quite sufficient to understand all these difficulties, and to find a solution.

Weingartner: I have two other questions: The one concerns that example of you with the comet Halley. And you said that this just is a chaotic behavior. Now my question is whether this behaviour is chaotic only in the sense that it is a simple bifurcation that it is oscillating this way or does the comet break out from the plane of the ellipse?

Chirikov: All this is known more or less in some detail because the comet Halley is not only a famous event but the only comet for which the most information was updated during many years. So, it is most simple to calculate its trajectory and everything not only presently but over 2000 years or so. And then it was found from numerical simulation, very simple by the way, that the orbit is chaotic. It is within a chaotic component of motion in the sense that everything, period, excentricity, inclination, fluctuate chaotically. So, for example, the period and semi-major axis are diffusing and, moreover, in both directions of time. One interesting thing is that we are interested not so much in forward diffusion because this comet has no future. It will simply disappear, melt and evaporate. What is much more interesting - the diffusion backward in time because, then, you can estimate how long it is within the solar system. The answer - 10 million years - is not clear how to interprete. This is very small compared to the cosmological scale, and so you need to understand the origin of, at least, this particular comet: where it came from, how it appeared within the solar system. Maybe not all of you know that recently it was found that not only particular parts of the solar system like comets and asteroids are chaotic but the whole solar system is also chaotic. Planetary motion is chaotic. But we are not in immediate danger because even the instability scale involving Lyapunov's exponent is about 5 million years, so we have enough time!

Weingartner: When you mentioned this exponential dependency, Lyapunov exponent then, you mentioned also that this is a sign not only for the diverging adjacent points for instance, but also for a loss of information. I always thought that this could be interpreted with Shannon's definition of information but you mentioned that it is not like this. Do you have a reason why you cannot interpret it that way?

Chirikov: You must understand me: I am not an expert in this mathematical theory. But what I actually mean, Shannon's conception is the statistical information for a given ensemble. It is calculated via the distribution function, for example, of many trajectories. Suppose, that within this ensemble there is a symbolic trajectory: 0,0,0,... The statistical definition would not distinguish it from any other trajectory but it is obvious that the former has not the average information, its information is zero. Kolmogorov and other researchers developed a new conception - the information on an individual trajectory. And this information can be interpreted in two opposite ways. One is how much information you need to predict. You need some information from somewhere if you want to predict. And since this is an information flow, information per unit time, you cannot do prediction with any finite algorithm, for example. You need a permanent flow of information from the observation or whatever. Another way is how much information you can obtain if you follow the trajectory, record it. Particularly, you can obtain the information about the initial conditions, actual initial conditions, because in this picture, in this theory you assume the trajectory itself is exact, it is not a beam of trajectories, not a distribution function. You have a single trajectory.

Batterman: I have a question about what you were saying about the correspondence principle. One of your transparencies, you mentioned something called "conditional chaos". I was just wondering whether by that you mean what is sometimes called "quantum pseudo-chaos" or whether you mean something else.

Chirikov: Conditional chaos is the chaos which arises under certain special conditions. On both sides of this transparency there are the unconditional statements: never chaos, and always chaos. But the correspondence principle requires quantum chaos in some sense, some quantum chaos, if and only if there is chaos in the classical limit. Another way to describe this situation is the conditional double limit which is, perhaps, not a common term. You have several limits and take them simultaneously but under a certain special relation between the variables.

Batterman: That raises an interesting questions about the view that quantum mechanics is the true and fundamental theory, and that classical mechanics is completely replaced or superseded. Because, it seems to me that one needs to make sense of the notion that quantum chaos is conditional upon what happens in the classical phase space. At least in the quasi-classical limit, it looks like reference to classical structures are necessary to explain or account for certain apparently quantum mechanical phenomena (such as the statistics of spectra). If this is so, then in what sense is quantum mechanics basic? How has it superseded classical mechanics? Why shouldn't one say that quantum mechanics is, in part, dependent upon classical mechanics?

Chirikov: This is one of the most deep questions. The other way around, I never considered the possibility that quantum mechanics would be not the basic theory. Even though there are such theories which try to derive the quantum properties from classical ones.

Batterman: Right. Using Maslov's techniques and so on.

Chirikov: No, no. This is technical. When I say that there is a single mechanics I mean that generally everything must be described or can be described by the quantum equations while the classical mechanics is an approximation, the limiting case. When you said the way around, you meant that everything can be described by Newtonian classical equations? There are such theories: Bohm's theory, and some versions of that. I don't believe at all in this possibility, for me it is not a question. But there is another question: do you have two separate mechanics? I don't know. Quantum mechanics for micro-world is not necessarily strictly related to the classical mechanics of the macro-world. This essentially is the difficulty of the statement that the correspondence principle fails. Some people say this. But they should understand the implications of the statement on two different mechanics. You can take such point of view but then there is a difficulty (I never thought about it very much because my preference is different): where is the borderline and how to divide the world between these two mechanics. If they are separate we need to divide. Instead, you may try to take the point of view that only quantum mechanics is fundamental. That does not mean you cannot use the classical mechanics, it is a perfect approximation in the macro-world but not in fundamental problems. Everything should have quantum explanation principally. If you cannot find one or the quantum theory gives you a different result as, for example, it may seem in the dynamical chaos, a very unusual and relatively new phenomenon, then there is a problem. Now, my statement, not opinion but statement, is: so far we have no contradiction with the idea that there is only quantum mechanics. So far I don't know any contradiction. We may find one later on but so far no contradiction exists, and in this sense I said that the correspondence principle was confirmed.

Schurz: Did I understand you right that already in classical mechanics you have two possibilities of representation, via trajectories and via phase density. And you said the representation via a phase density is always linear.

Chririkov: The equation is linear.

Schurz: Yes. This happens also in classical mechanics. So, the question of getting chaos from non-chaos could be studied already in classical mechanics because if you use a linear phase density equation in classical mechanics it should not be chaotic. If you describe it by trajectories you have a system which behaves chaotic. But if you describe it via phase density the equation is linear and it does not exhibit chaotic behavior.

Chirikov: It does!

Schurz: But it is linear. You said a necessary condition for chaos is non-linearity.

Chirikov: No, for trajectory equation.

Schurz: How does such a linear phase density equation produce chaos in classical mechanics?

Chirikov: The question is how to imagine this because, first of all, this is a rigorous result, both types of description are completely equivalent. The situation is the following. Sometimes people say: quantum equations are linear, hence, the quantum chaos is not classical. This is a completely wrong statement because we can refer to the Liouville equation. So, the question is not whether the equation is linear or nonlinear but what kind? There are different kinds of linear equations with qualitatively different properties. Now about the relation of nonlinearity to chaos. There are different definitions of chaos we discussed yesterday in relation to the motion stability. The most common definition is related to the exponential instability. Now, the exponential instability is a property of the linearized equations of motion, by the way. Generally, you don't need nonlinear equations. Nevertheless, we do need some nonlinearity even in terms of the linear equations for Lyapunov exponent or in terms of linear Liouville equation. What is the role of nonlinearity then? To make the motion bounded. Because if it is unbounded then unstable motion is not necessarily chaotic. If you have simple exponential instability, you would never call it chaotic. Why? It is just an explosion of trajectories. So, from this point of view nonlinearity simply restricts the motion to a finite phase volume, makes it bounded in phase space. Then, in combination with instability, you obtain the mixing of trajectories. This is a graphical view of the chaos mechanism. Now, another interesting question: what means exponential instability in terms of wave or Liouville equation? It depends in which space you consider the instability. If you consider the space of density itself, slightly change the distribution function or ψ-function, then the difference is described by the same equation and you have no instability at all. Nevertheless, in Schrödinger equation, in classically chaotic case, there is a relatively short time interval when the quantum motion is exponentially unstable if you do not use the Hilbert space but the phase space of classical mechanics. Consider, for example, two narrow wave packets or two distribution functions very close to each other. Because of the instability each of them is spreading, and also they diverge from each other very quickly. But why is it different? Because in terms of linear equation it is not a small change of the wave function: you have one wave packet and then another one. Let me show briefly a picture how it looks in a quantum system, on a simple model. This is phase space and this is the initial wave packet, the so-called coherent state, which is the most narrow wave packet. You see the wave packet is spreading very quickly as under the classical exponential instability. And then, after three steps of the map, everything is destroyed, and there is no more relation to the classical picture. Instead, you see wild fluctuations of the wave function but nevertheless during a much longer time the diffusion and relaxation still remain classical even though the ψ-function itself has almost no resemblance to the classical distribution.

Noyes: Just a quick comment on this question of correspondence principle. This is a question which is being investigated empirically by Tony Leggett at the University of Illinois. He asks whether a system containing 10^{15} atoms still behaves as a quantum system, or if for such a large number quantum coherence has to disappear. The system he uses is two superconducting flux loops ("SQUID's") can be shown to contain either zero or one flux quantum. By coupling them through a Josephson junction and insuring that the system contains a single flux quantum in one loop and none in the other, one can then ask whether the "tunneling" of this quantum state from one macroscopic loop to the other fits the quantum mechanical prediction or not. If it does not, he will have discovered a complexity parameter which could be interpreted as showing where the transition from quantum to classical occurs. Then there would have to be a new theory which might show that there actually is a classical regime. Despite the success of the usual assumption that quantum mechanics rather than classical physics *has* to be the fundamental theory, it's an open question from the point of view of experiment.

Chirikov: I agree. Such a question you would never answer completely. We have a common solution and you must be ready to change this solution and not necessarily to follow your preferences. I would like to mention that the phenomenon you spoke about is a particular case in the very intensively studied field. Now it is called mesoscopics. Maybe you heard the word: mesoscopic is something intermediate. But what people have in mind is that you may be very far in the quasi-classical region, with quantum numbers arbitrarily large, but nevertheless, under some additional conditions, the behavior may be essentially quantum. This is called mesoscopic phenomena. Of course, the extreme case of this is well known since long ago: superfluidity and superconductivity. Mesoscopic phenomena are called also the intermediate asymptotics. It means that the quantum numbers may be arbitrarily large but still it is not the final answer for the correspondence principle, you must go further and you will reach the quantum chaos, it is a theorem. But the example you mentioned is more complicated and more interesting.

Suppes: I'd like to use the privilege of the chair to ask one quick question. In your example a minute ago of the two wave functions that separate exponentially, if you take the expectations of the wave functions then you get a path trajectory. Are those two paths classically scar-paths - in the language of quantum optics, are those expected paths chaotic in the classical sense?

Chirikov: You mean some average in the spirit of the Ehrenfest theorem or something like this? No. Unfortunately, I had no time. But I mentioned that there are characteristic time scales of quantum motion. The most important one I mentioned is the scale on which classical diffusion and relaxation proceed. But there is another one, very short, proportional only to logarithm of quantum number or of Planck's constant on which initially narrow wave packet remains relatively narrow and simply follows the classical trajectory. So, if this trajectory is random, then the motion of the packet on this time scale is equally random. But it terminates because of the spreading of the wave packet and of its eventual destruction I showed. Then, you can no longer follow the wave packet as you have instead a very complicated structure of ψ-function but, nevertheless, the classical diffusion still persists.

Under What Transformations Are Laws Invariant?*

P. Weingartner

Universität Salzburg, Austria

0 Introduction

In order to be able to describe and explain movement we need to distinguish something which changes relative to something which does not change. This important distinction is pointed out by Aristotle[1] also as a criticism of Parmenides' theory of the universe which assumes only one being and nothing else.[2] That what changes, moves was thought to be contingent (not necessary) in respect to the not changing (or even not changeable) necessary principle or law. In general this idea belongs to the Greek Ideal of Science which was more or less manifest in several greek thinkers from Thales on but was elaborated in detail by Plato and Aristotle:

To describe and explain the visible (observable), concrete, particular, changing, material world by non-visible (non-observable) abstract, universal, non changing and immaterial principles.

The problematic character of the distinction between abstract unchangeable laws and particular and contingent changing events and conditions is stated in modern terms very well by Wigner:

"The world is very complicated and it is clearly impossible for the human mind to understand it completely. Man has therefore devised an artifice which permits the complicated nature of the world to be blamed on something which is called accidental and thus permits him to abstract a domain in which simple laws can be found. The complications are called initial conditions; the domains of regularities, laws of nature. ... The artificial nature of the division of information into "initial conditions" and "laws of nature" is perhaps most evident in the realm of cosmology. Equations of motion which purport to be able to predict the future of a universe from an arbitrary present state clearly cannot have an empirical basis. It is, in fact, impossible to adduce reasons against the assumption that the laws of nature would be different even in small domains if the universe had a radically different structure. One cannot help agreeing to a certain degree with E.A. Milne, who reminds us (Kinematic Relativity, Oxford Univ. Press, 1948, page 4) that, according to Mach, the laws of nature are a consequence of

---

* The author is indebted to Boris Chirikov for a number of valuable remarks concerning an earlier version of the paper.
[1] Aristotle (Phys), 190a17f.
[2] Aristotle (Met), 986b15f and Aristotle (Phys), 186a24ff..

the contents of the universe. The remarkable fact is that this point of view could be so successfully disregarded and that the distinction between initial conditions and laws of nature has proved so fruitful."[3]

This problematic character has appeared more clearly within the last 30 years when new phenomena like chaos, self similarity and structure and order were discovered.

The paper will concentrate on the relation between laws and initial (and boundary) conditions in the light of new discoveries when non-linearity is present, i.e. phenomena of order and structure, selfsimilarity and chaos. This will be done by discussing important properties of laws. In this connection I want to underline that when speaking of properties of laws or of conditions which are satisfied by laws what one is searching for in a sense is to find "laws" about laws. Thus the symmetry principles are "metalaws" in this sense: "It is good to emphasize at this point the fact that the laws of nature that is, the correlations between events, are the entities to which the symmetry laws apply, not the events themselves."[4] "Nevertheless, there is a structure in the laws of nature which we call the laws of invariance. This structure is so far-reaching in some cases that laws of nature were guessed on the basis of the postulate that they fit into the invariance structure."[5]

1   Question one:

Are all laws deterministic in the sense that - given an initial state - any state in the future can be predicted and any state in the past can be retrodicted?

1.1   The Idea of Laplace

A positive answer to the above question was the idea of Laplace.[6] It can be illustrated by the following example:

Assume a film is made of the world, i.e. of the events happening in the whole universe. After the film is developed we cut it into pieces corresponding to single film-pictures. Now we put the single pictures successively in time (in the order of time) into a long card index box like the cards of a library catalogue. Then one special state of the universe at a certain time t corresponds to one such card (film picture) of the catalogue. One can follow one trajectory across the (perpendicular to the) catalogue-cards.

Interpreted with the help of this illustration Laplace's determinism means that it suffices to know the law(s) of nature and one single catalogue card (film picture) corresponding to one state (of the universe) at a certain time t in order

---

[3] Wigner (1967), p. 3.

[4] Wigner (1967), p. 19. Cf. also p. 16f.

[5] Ibid. p. 29.

[6] Cf. Laplace (1814), Ch. 2. For a discussion of determinism and indeterminism cf. Van Fraassen (1991), chapters 2 and 3.

to construct all other cards of the catalogue, i.e. to predict and to retrodict all the other states of the universe.

That Laplace's idea is not satisfied in certain areas was discovered a long time ago. Thermodynamics is one example, friction and diffusion are others. Such discoveries led to another global question: Are all laws statistical?

The mechanistic world view underlying Laplace's determinism was based on the belief that all physical systems are - if analyzed in its inmost structure - ultimately mechanical systems. Since a clock was understood as a paradigm example of a mechanical system the main thesis of the mechanistic world view could be expressed by saying that all complex systems (things) of the world - even most complicated ones like gases, swarms of moscitos or clouds - are ultimately (i.e. if we would have enough knowledge of the detailed interaction of the particles) clocks. Or to put it in Popper's words: "All clouds are clocks."[7]

After the discovery of statistical laws in thermodynamics and later in other areas there was a general doubt with respect to the mechanistic and deterministic interpretation of the world. One of the first philosophers to notice that a certain imperfection in all clocks allows to enter chance and randomness and that even the most perfect clock is, taken in respect to its molecular structure, somewhat cloudy, was Charles Sanders Peirce[8]:

"But it may be asked whether if there were an element of real chance in the universe it must not occasionally be productive of signal effects such as could not pass unobserved. In answer to this question, without stopping to point out that there is an abundance of great events which one might to be tempted to suppose were of that nature, it will be simplest to remark that physicists hold that the particles of gases are moving about irregularly, substantially as if by real chance, and that by the principles of probabilities there must occasionally happen to be concentrations of heat in the gases contrary to the second law of thermodynamics, and these concentrations, occurring in explosive mixtures, must sometimes have tremendous effects."

The question was now: Could it not be the case that all laws are statistical and the deterministic outlook is only on the surface of macroscopic phenomena? That is all complex systems (things) of the world are in fact - in its inmost structure, i.e. on the atomic level - like gases or swarms of moskitos or clouds. This led to another extreme picture "All clocks are clouds."

But neither of these extreme pictures proved satisfactory as an explanation of everything. The heroic ideal to explain everything by one (or one kind of) principle had to be replaced by the aim to find relatively few (kinds of) principles (laws) for relatively many facts. "We do not have one structure from which all is deduced, we have several pieces that do not quite fit exactly yet."[9]

By the mid of the twentieth century many physicists accepted a view which

---

[7] cf. Popper (1965), p. 210.

[8] Peirce (1935, 1960), 6.47. Cf. Popper (1965), p. 213. Popper in this essay called my attention to the passage of Peirce.

[9] Feynman (1967), p. 30.

can be roughly stated as follows:

(1) In respect to some areas (mainly macroscopic) deterministic laws with good predictability for single events give an adequate description and explanation.

(2) In respect to other areas (thermodynamics, friction, diffusion and microscopic ones) statistical laws with no good predictability for the single event but with predictability for the whole aggregate give an adequate description and explanation.[10]

Understood in this way it was compatible that for example the pendulum interpreted as a macroscopic dynamical system obeys Newton's laws and allows strict prediction whereas interpreted as a microscopic system, i.e. in respect to its atomic structure behaves in some of its features like a cloud and can then be adequately described by statistical laws without strict predictions for single particles.

1.2   The New Discovery

The above mentioned compromise was that a physical system obeys a certain type of law in respect to a certain area of application (for instance as a macroscopic dynamical system) but obeys a different type of law in respect to another area (for instance if its atomic structure is analyzed).

The new discovery now was that even *within* one such area the behaviour of the system can change radically such that a "clock" can become a stormy "cloud". Thus a dynamical system obeying Newton's laws with strict predictability can become chaotic in its behaviour and practically unpredictable just by changing slightly some initial conditions. Experiments which prove such a behaviour of dynamical systems have been made since the seventies. A special kind of very simple arrangements are experiments with the socalled forced pendulum or with the kicked rotator. One type of such an experiment is described below:[11]

The socalled spherical pendulum consisting of a small weight attached to the lower end of a string (of length l) has a period $T_0 = 2\pi\sqrt{l/g}$ of sinusodial oscillations (provided the oscillations are small). The spherical pendulum (under normal conditions with small amplitudes) shows a regular behaviour with at least

---

[10] The notion of predictability is of course to some extend independent of the notion of law (whether deterministic or stochastic). That even deterministic (dynamical) laws do not automatically imply predictability was shown again by such phenomena as chaotic motion. On the other hand the noition of predictability is not a completely epistemic notion. If it is characterized in an adequate way, it has to be dependent on important properties of laws as necessary preconditions.

[11] For details see Lighthill (1986). For the kicked rotator see Chirikov (1979). The first forerunner of such an experiment was the Galton Board (cf. Galton (1889)). Theoretically chaotic behaviour was investigated already by Hadamard and Poincaré and more specifically in the area of metereology by Lorenz (1963).

three important characteristics:[12]

1.2.1 It is periodic, i.e. the state of the system repeats itself after a finite period of time and continues to do so in the absence of external disturbing forces.

1.2.2 The state of the system at any given time t_i is a definite function of its state at an earlier time t_{i-1}. A unique earlier state (corresponding to a unique solution of the equation) leads under the time evolution to a unique final state (again corresponding to a unique solution of the equation).

This property is a more precise description of Laplace's idea illustrated by the film-pictures as definite states (or unique solutions). It is usually taken as the defining condition for determinism.

1.2.3 The spherical pendulum has a certain type of stability. Assume we make very small changes in the initial states, say within a neighbourhood distance of ε. Then the distance of the state $h(\varepsilon)$ is proportionally small (no more than a linearly increasing function of time). This kind of stability with respect to small perturbations is called "perturbative stability" which holds in many linear systems. The very important false belief of most scientists until 1970 was that this holds also for the general case.

The important new discovery is now that this simple physical system becomes chaotic if the top end is forced to move back and forth (maximal displacement Δ) with a slightly different period T greater than T_0, provided that Δ is about $1/64$ of l and not more than about a tenth of the energy of motion is dissipated by damping (air resistance etc.). Miles (1984) showed experimentally that the system is chaotic for values of $T = 1,00234T_0$. It has to be emphasized however that this does not just mean that the system becomes unstable in the sense of simple bifurcation. Unstability in the sense of simple bifurcation has been known for a long time. In this case the pendulum weight makes a back and forth oszillation in the same plane and by forcing the upper end this movement begins to be unstable. Such a simple bifurcation where the plane is not changed occurs when $T = 0,989T_0$ and slightly above. But for $T = 1,00234T_0$ the pendulum is breaking out of the plane, the number of further bifurcations are arbitrarily increasing, the dependence on initial conditions is completely random such that there is no predictability (or only for very short times).

In view of the new discoveries of this sort Lighthill made the following acknowledgement which shows a change of view under physicists:

"Here I have to pause, and to speak once again on behalf of the broad global fraternity of practitioners of mechanics. We are all deeply conscious today that the enthusiasm of our forebears for the marvellous achievements of Newtonian mechanics led them to make generalizations in this area of predictability which, indeed, we may have generally tended to believe before 1960, but which we now recognize were false. We collectively wish to apologize for having misled the general educated public by spreading ideas about the determinism of sys-

---

[12] These three characteristics are pointed out very clearly in Holt and Holt (1993), p. 716f.

tems satisfying Newton's laws of motion that, after 1960, were to be proved incorrect."[13]

Before finishing this section I want to mention that despite this new situation in physics unpredictable chaotic phenomena in practical and active life have been known by experienced people ever since (see chapter 2.1).

1.3   Chaotic Behaviour

Subsequently I shall give seven characteristics of chaotic behaviour which are necessary conditions for most kinds of chaotic motion (exceptions will be mentioned):

1.3.1 The Local Instability Is Exponential

Small changes in the initial conditions lead to exponentially increasing bifurcations. Not just simple bifurcations which have been known so far as unstable behaviour or as small perturbations i.e. it is not a case of "perturbative stability" (cf. 1.2.3). This property of being sensitively dependent on initial conditions is measured by the (positive) Lyapunov exponent (see chapter 2). This condition is taken by some as the defining property of chaotic behaviour: "Chaos is thus the prevalence of sensitive dependence on initial conditions, whatever the initial condition is."[14]

But there is sensitive dependence on initial conditions where we could not speak of chaotic behaviour as the example of Maxwell shows (see chapter 2). Thus it is better to add more conditions for further specification.

1.3.2 No Predictability

The randomness in the output is so strong that no prediction is possible for the system (except for a very short time). Prediction (and also explanation of a certain state which corresponds to a unique solution of the equation) is based on some kind of regularity described in 1.2.1 - 1.2.3. But if all the three conditions or kinds of regularity fail - as is the case with chaotic behaviour - then there is no predictability (except for a very short time). Also there is no explanation in the sense of being symmetrical to prediction.

It should be noted that the underlying laws for the dynamical system in question (in the above example the pendulum) are deterministic laws in the sense of condition 1.2.2. That means that for a certain state at time t_i (represented by a unique solution of the differential equation) the law (differential equation) gives ("predicts") a unique solution representing a certain state at t_{i+1}. With small amplitudes the pendulum obeys all three conditions 1.2.1 - 1.2.3 perfectly well. But even the forced pendulum obeys these conditions for most of the initial conditions. Should we now say that for very special asymmetrical initial condition the (still) symmetric laws lead to completely asymmetrical outputs? (Cf. chapter 4). Or should we say that the behaviour of the forced pendulum shows

---

[13] Lighthill (1986), p. 38.

[14] Ruelle (1990), p. 242.

that we do not know the deeper underlying laws which might not have those symmetries which we assume for the dynamical laws. To recall the introduction: Is it just a question of random complications of initial conditions or a question of the appropriate laws (despite Wigner's critical attitude concerning this very distinction)? (Cf. Chapter 7). Chirikov (these proceedings), ch. 1 seems to interpret it as a question of laws in the sense that the respective statistical laws as secondary laws form an intrinsic part of the primary dynamical laws.

1.3.3 Superposition Does Not Hold

Chaotic behaviour in the sense of classical dynamical chaos requires physical systems whose equations are non-linear, i.e. the superposition principle does not hold (cf. chapter 3). This is connected with 1.3.1 since the exponential instability (positive Lyapunov exponent) is non-linear.[15]

It is worth mentioning though that not every chaotic behaviour is non-linear. An example is linear wave chaos in quantum mechanics.[16] Thus non-linearity is a necessary condition for classical dynamical chaos but for chaotic behaviour in general it is neither necessary nor sufficient.

1.3.4 Non-periodicity

Chaotic behaviour is non-periodic. And this holds without any external disturbance (cf. 1.2.1). A consequence of that is a further characteristic of chaotic motion: The Poincaré map shows space-filling points. This is a method introduced by Poincaré about 100 years ago which considers the points in which the trajectory cuts a certain plane. If the motion is chaotic there will be no immediate recurrence that is the plane will always be cut at new points and as time goes on will be filled with points. But if the phase space is small there will be recurrance of the trajectory after some finite period of time. To give an illustration: skiing in fresh powder snow is a great pleasure. But if the slope is small and one is skiing down frequently the slope will be filled with traces and after some time no new space is left and thus one has to use ones own traces again (recurrence). If the system is Hamiltonian and area preserving (finite region) then the Poincaré-recurrance theorem holds. It says that the trajectory returns to a given neighbourhood of a point an infinite number of times. If it is ergodic then the system explores the entire region of phase space and eventually covers it uniformly (this implies also recurrence). In stronger kinds of chaotic motion the trajectory might not cover the whole phase space and neither stay in a local area. The description is then more complicated.

---

[15] A non-linear function is a function which contains a variable raised to a power other than one or zero; or a product of two (or more) variables; or a variable as the argument of a transcentendal function (sin, cos). An equation containing one or more of such non-linear terms is a non-linear equation. It has to be mentioned however that the question of linear or non-linear equations depends in a sense also on the kind of description which is choosen as the most suitable in the particular case. Thus to describe the phase space density a Liouville equation can be used which is linear (cf. Chirikov (these proceedings), ch. 2.1.

[16] Cf. Chirikov (1992).

In general it has to be observed that Poincaré's recurrence theorem is sensitive with respect to certain conditions. Already Boltzmann[17] understood very well that it could not be applied to a gas on the assumption that the number of its molecules is infinite and time becomes very long. On the other hand if the number of molecules is very large but finite and time is infinite then recurrence takes place. Zermelo[18] thought that Poincaré's recurrence theorem contradicts Boltzmann's statistical mechanics as an interpretation of thermodynamics, but in fact he had misunderstandings concerning important conditions. As Boltzmann points out the essential fact is that the length of time after which the old state of a gas should recur according to the recurrence theorem is not observable (measurable) - because it is much too long, i.e. the recurrence is extremely unprobable (though not impossible). To illustrate: Increase the number of skiers and enlarge the slope: it will be more and more unprobable that the initial state of all skiers being in a certain position recurs. Or use the example which is given in chapter 5.4: One passenger might come back to his deadend airport by chance after a long time. But if millions of flight passengers fly around without any flight information the probability of recurrence - say all are again in the original starting position - will be very low. Imagine now a one-litre of gas with about $2,7 \cdot 10^{22}$ molecules! These considerations show that it depends very much on the complexity of the chaotic system what kind of recurrence (theoretical, after an infinite length of time or observable) we have.

For other important parameters recurrency does not hold in chaotic motion. For example phase density is nonrecurrent, it will never come back to its initial state, independently of the direction of time. Thus we have non-recurrency and time-reversibility (the latter also for the relaxation property). In consequence it is important to notice that non-recurrence is not sufficient to derive time irreversibility. Non-recurrence and time-irreversibility are not equivalent notions.[19]

The non-periodicity can also be measured by the invariant density which measures how the iterations become distributed over the unit interval and by the correlation function $f(m)$ which measures the correlation between iterations which are m steps apart.

1.3.5 Bounded Motion

Chaotic motion is bounded. That means that the number of degrees of freedom is limited in different ways depending on the kind of chaotic motion. These limitations make the motion - roughly speaking - "oscillatory" in time between stable points (fixed points) whose number multiplies in dependence of certain parameters. One speaks of "folding" with respect to an interval, whereas the exponential separtion of adjacent conjugate points (Lyapunov exponent > 0) is called "stretching". Another possibility is that a trajectory becomes attracted to a bounded area of phase space, i.e. to a socalled "strange attractor",[20] within

---

[17] Cf. Boltzmann (1897a), Boltzmann (1897b).

[18] Zermelo (1896a), Zermelo (1896b).

[19] Cf. Chirikov (these proceedings), ch. 2.2.

[20] Cf. 2.2.2 and Ruelle (1980).

which there is exponential separation of adjacent conjugate points.

Theoretically the boundaries are usually described as Neumann or Dirichlet conditions. In the practical experimental application they may mean for example the size of the system, the number of rolls in a fluid layer of a Bénard-experiment, the diffusion coefficients in a chemical experiment etc.

1.3.6 Change of the Variables of the System

A further necessary condition for chaotic behaviour is the permanent change of the important variables of the system. For example in the case of the forced pendulum the change of the amplitude, in the case of a fluid under heat the change of the conductivity of heat in the layers of the fluid. Although these magnitudes remain within a minimum and maximum value they do not recur in the course of time (cf. 1.3.4.).

1.3.7 Non-integrability

Let $x(t, x_0)$ be a function describing the motion of a dynamical system, where x_0 are the initial conditions (the position of the system at $t = 0$). Let the function $x(t, x_0)$ have a pole at $t_p = t_1 \pm i\Lambda$ in the complex t plane, where Λ is the Lyapunov-exponent. Then the system is integrable - according to a criterion of Kowalevskaya[21] - if every t_p depends on x_0. Now chaotic behaviour (motion) is non-integrable and therefore its poles do not depend on x_0 and specifically Λ does not depend on x_0 (cf. chapter 2).

It should be mentioned however that there are weak kinds of chaos or "quasi-chaos" where integrability holds like in Quantum Chaos. A somewhat stronger case is partial integrability (KAM-integrability) when an integrable system is exposed to weak perturbation but is resistent. In this sense integrability (non-integrability) can be used to distinguish levels of disorder in an arrangement beginning with full integrability via KAM-integrability to chaos.[22]

1.3.8 Continuous Spectrum

The Fourier spectrum of the chaotic motion which is aperiodic (cf. 1.3.4) is continuous and its phase space is continuous, whereas regular motion (i.e. motion which obeys at least one of the 3 conditions 1.2.1 - 1.2.3) has a discrete spectrum. If we count the number of degrees of freedom (for example these may correspond to the number of rolls of a fluid layer in the Bénard experiment) by the number of Fourier components then already in an unstable motion at least one such component is continuous.

It should be added that Quantum chaos violates the condition of the continuous spectrum of the motion and that of continuous phase space (cf. 1.3.3). In connection with the uncertainty principle (which allows only a finite size of the elementary cells of phase-space) the frequency spectrum of quantum motion is discrete for the motion bounded in phase space. In order to do justice to both - to this difference in respect to classical dynamical chaos and to the randomness

---

[21] The russian mathematician Sofia Kowalevskaya formulated the criterion in 1890. Cf. Chirikov (1991a), p. 450.

[22] Cf. Chirikov (1991b) and chapter 2.4 of this essay.

of the quantum mechanical measurement process - Casati and Chirikov suggest to divide the whole physical problem of Quantum Dynamics in two qualitatively different parts:

"(1) The proper quantum dynamics as described by a specific dynamical variable, the wavefunction $\psi(t)$ (for example by the Schrödinger equation)

(2) The quantum measurement including the registration of the result and hence the collapse of the ψ function."[23]

According to this distinction the violation of certain important conditions of classical chaos holds for the first part. The result of a measurement process in quantum mechanics however is a random process and may be interpreted as a process of dynamical chaos.

2   Question two

Do all laws obey the principle "Similar causes lead to similar effects"?

2.1   Aristotle and Maxwell

A first warning with respect to such a principle in the area of epistemology or methodology we find already in Aristotle:

"... the least initial deviation from the truth is multiplied later a thousandfold."[24]

A specific warning with a counterexample is due to Maxwell:

"There is another maxime which must not be confounded with that quoted at the beginning of this article [25], which asserts 'That like causes produce like effects'. This is only true when small variations in the initial circumstances produce only small variations in the final state of the system. In a great many physical phenomena this condition is satisfied; but there are other cases in which a small initial variation may produce a very great change in the final state of the system, as when the displacement of the "points" causes a railway train to run into another instead of keeping its proper course."[26]

Experienced highlanders in mountainous countries like Tyrol know very well that extremely small events can lead to a bursting of an avalanche which might destroy huge forests and even a city.

It should be noted that the unproportional effect need not to be chaotic. In the example of Maxwell, the running of the train in a different direction is certainly not but certain phenomena of the crash might be. Avalanches on the other hand have always been very unpredictable events at least and seem to be quite good examples for chaotic behaviour.

---

[23] Casati and Chirikov (1994), p. 11.

[24] Aristotle (Heav), 271b8.

[25] The one to which Maxwell refers to is "The same causes will always produce the same effects" which he discusses earlier.

[26] Maxwell (MaM), p. 13.

I want to finish this chapter with a short poem by Emily Dickinson which describes geniously both the big effect caused by small deviations and the time-irreversibility of such processes:

> The brain within its groove
> Runs evenly and true;
> But let a splinter swerve,
> 'Twere easier for you
> To put the water back
> When floods have slit the hills,
> And scooped a turnpike for themselves,
> And blotted out the mills.[27]

2.2 Sensitive Dependence on Initial Conditions

This is certainly one of the most important characteristics of chaotic behaviour. And it is taken by some as its defining property (cf. 1.3.1). But there are phenomena which have that property without being chaotic as one can see from Maxwell's example in 2.1. Therefore this condition (sensitive dependence on initial conditions) cannot be a sufficient condition of chaotic behaviour (motion). But it is certainly an important necessary condition. This important property is measured by the so-called Lyapunov-exponent Λ. In fact the Lyapunov exponent measures two things which are described in 2.2.1 and 2.2.4.

2.2.1 It measures the (exponential) separation of adjacent conjugate points (conjugate in respect to the starting point x_0):

$$x_0 \qquad x_0+\varepsilon \qquad \text{N iterations} \qquad f^N(x_0) \qquad f^N(x_0+\varepsilon)$$

$$\vdash\!\!-\!\!-\!\!-\!\!-\!\!\dashv \qquad \Rightarrow \ldots \Rightarrow \qquad \vdash\!\!-\!\!-\!\!-\!\!-\!\!-\!\!-\!\!\dashv$$

$$\varepsilon \qquad\qquad\qquad\qquad\qquad\qquad \varepsilon \cdot e^{N\Lambda(x_0)}$$

This description of stretching of the distance between closely (ε) adjacent points corresponds to a one dimensional Poincaré map. In real motion the stretching occurs in three dimensions.[28]

An enlightening example was calculated by Berry.[29] Assume that an electron somewhere in the universe (say 10^{10} light years away) looses its attraction (gravitational force). That means a slight deviation in the initial conditions. Can we calculate after how many (elastic) bumps one airmolecule on the earth (understood as an elastic ball) fails another one as an effect of that initial change (assuming that it would have hit the other one if the electron wouldn't have lost its attraction)? Berry calculates that this is the case after about 56 to 60 bumps. If we take a man's attraction on the billiard balls at the distance of one meter

---

[27] I am grateful to John Bacon who called my attention to this poem.

[28] The transparent exposition in 2.2.1 and also in 2.2.4 is due to Schuster (1989). The respective concept of information has been worked out especially by Shannon. Cf. Shannon and Weaver (1949).

[29] Cf. Berry (1978).

from a billiard table and billiard balls (as the elastic balls) then it needs only about 9 bumps.

2.2.2 The Strange Attractor

A good example for the exponential separation is a socalled *strange attractor* which can be characterized in the following way.[30] A bounded set A (in m-dimensional space) is a strange attractor for the map F if there is a set U with the following four conditions [where F maps x with coordinates $(x_1...x_m)$ ... $F_m(x_1...x_m)$]:

(a) Neighbourhood conditon: For each point X of A there is a little ball centered at X such that $X \in A \subseteq U$ and every such ball is also contained in U.

(b) Attracting condition: For every initial point X_0 in U, the point X_t with coordinates $x_1(t)...x_m(t)$ remains in U for positive t. Even for large t it stays as close as one wants to A.

(c) Strangeness condition: X_0 is in U. If X_0 is in U there is sensitive dependence on the initial condition.

(d) Indecomposability condition: One can always choose a point X_0 in A such that arbitrarily close to each other point Y in A there is a point X_t for some positive t. This implies that A cannot be split into two different attractors.

2.2.3 Increasing Error

The Hénon Attractor[31] is a particular example of a strange attractor which describes the increasing error or better the sensitive dependence on initial small errors. Thus the Hénon Attractor can be viewed as one possible interpretation of Aristotles' observation (cf. 2.1).

If x_t and x_t' correspond to initial data x_0 and x_0' close to each other, the distance $d(x_t, x_t')$ increases exponentially with t.[32]

$d(x_t, x_t') \sim d(x_0, x_0') \cdot a^t$ (where $a \approx 1,52$).

Since $a > 1$, a^t increases exponentially with t, i.e. the error $d(x_t, x_t')$ increases exponentially with time. This means that small initial errors (small errors in the beginning) which are never completely avoidable in the case of experimental data increase exponentially with time.

2.2.4 The Lyapunov exponent measures also the average loss of information (I_0) about the position of a point in an inverval $[0, 1]$ after one iteration. Assume $[0, 1]$ separated into n equal intervals such that x_0 occurs in each of them with probability $\frac{1}{n}$. The answer to the question which interval contains x_0 is then:

$$I_0 = -\sum_{i=1}^{n} \frac{1}{n} ld \frac{1}{n} = ldn \qquad \text{(where } ld \text{ is the logarithm to the base 2)}$$

---

[30] This characterization is due to Ruelle (1980). That such attractors exist (this was not clear when Ruelle wrote his paper, cf. p. 131) has been proved for the Hénon attractor by Benedicks and Carleson (1991).

[31] Cf. Hénon (1976).

[32] As long as the distance is small. If the distance reaches the order of the attractor it cannot increase anymore.

With decreasing n the information I_0 is decreasing too and $I_0 = 0$ for $n = 1$.

Concerning the information represented by a trajectory Alekseev-Brudno's theorem is worth mentioning: The information given with a trajectory of a certain length of time (its algorithmic complexity per time unit) is asymptotically equal to the metric entropy.[33]

2.3   Kolmogorov-Entropy

The loss of information about the state of a dynamical system in the course of time is connected with the socalled Kolmogorov-entropy ('K' for short). The wellknown entropy in thermodynamics is a measure for the disorder of a system and by the Second Law of Thermodynamics this disorder increases (even if locally the entropy can decrease for example due to living systems). The increase of disorder is also connected by a loss of information about the state of the system (for instance about the positions of molecules of a gas if it mixes with another one, cf. 2.2.4). K is a measure of the degree to which a dynamical system is chaotic. For one-dimensional maps K measures the same as the positive Lyapunov-exponent. For higher dimensional systems K is equal to the average sum of all the positive Lyapunov exponents.[34] It was already mentioned that the Lyapunov-exponent measures also the loss of information about the system (cf. 2.2.4). Moreover K can be defined by Shannon's measure of information in such a way that K is proportional to the degree of loss of information of the state of the dynamical system in the course of time.[35] Thus it is plain that K is also a measure of the (average) rate for the loss of information of a dynamical system with the evolution of time. In consequence of that K is also a measure of predictability: it is inversely proportional to the length of time over which the state of a chaotic dynamical system can be predicted.

Points 2.2 (2.2.1 - 2.2.4) and 2.3 show that the metaprinciple about laws "similar causes lead to similar effects" is violated in two ways when chaotic behaviour occurs:

(1) On the level of physical reality. This is shown in 2.2.1, 2.2.2 and 2.3 when K measures how chaotic a motion is.

(2) On the epistemic level. This is shown in 2.2.3, 2.2.4 and 2.3 when K measures the loss of information.[36]

---

[33] Cf. Brudno (1983) and Chirikov (these proceedings), ch. 2.4.

[34] This was shown by Pesin (1977).

[35] Cf. Farmer (1982).

[36] It should be emphasized however that "epistemic level" does not mean here a kind of subjective measurement. As is clear from 2.2.3, 2.2.4 and 2.3 the measurement is based on objective criteria and the probability used is also not a "subjective interpretation" of probability.

2.4   The Prize-Question of King Oscar II

In 1885 King Oscar II of Sweden announced the following prize question:[37]

"For an arbitrary system of mass points which attract each other according to Newton's laws, assuming that no two points ever collide, give the coordinates of the individual points for all time as the sum of a uniformly convergent series whose terms are made up of known functions."

The prize was given to Poincaré for his great work "Les Méthodes Nouvelles de la Méchanique Celeste". However he did not really solve the problem but gave reasons that such series do not exist i.e. that contrary to the expectation these series of perturbation theory in fact diverge.

The prize-question was partially answered by Kolmogorov in 1954 and solved by his pupil Arnold in 1963. A special case of it was answered by Moser. Hence the name KAM-theorem. It gives an answer to the question whether an integrable system (with an arbitrary number of degrees of freedom) survives weak perturbation. The theorem says that the answer is positive and that the invariance with respect to small perturbation or the stability is proportional to the degree of irrationality of the rotation number r of the curve of the trajectory. This has led to a new (weakened) concept of stability which holds for the majority of the orbits; i.e. the majority of solutions (for the respective differential equations) are quasiperiodic.

What happens exactly to the relatively rare exceptional orbits which are unstable is still not enough known. This new kind of stability has a number of physical applications.

3   Question three:

Do all laws obey the superposition principle?

3.1   The Main Principle of the Linear Description of Nature

The main principle of the linear description of nature is the superposition principle. It says:

If $x_1(t)$ is a possible solution of a law of motion (of a differential equation describing motion) and if $x_2(t)$ is also a possible solution of the same law then
$$x_1(t) + x_2(t) = x_s(t)$$ is also a possible solution of this law.

---

[37] That a prize should be given for an important mathematical discovery was susggested to the King by the Swedish mathematician Mittag-Leffler. The special prize question was proposed by Weierstrass (the committee consisted of Weierstrass, Hermite and Mittag-Leffler). Cf. Moser (1978). Weierstrass himself was surprised about Poincaré's answer because he arrived (earlier) at the opposite answer. He showed that Poincaré did not in fact prove his result. Today it is knwon that for very special frequencies such series may in fact converge. Cf. ibid. p. 70f. For some of the historical questions concerning that matter cf. the letters of Weierstrass to Sofia Kowalevskaya, Weierstrass (1993) especially the letter from 15.8.1878, ibid. p. 226ff.

There are famous examples in physics of laws or of the respective physical systems which satisfy the superposition principle. Some of them are the following:

(1) Accustic waves

(2) Electromagnetic phenomena (time dependent Maxwell equations)

(3) Optical phenomena

(4) Michelson experiment (independence of the light velocity in respect to moved reference frames). This experiment works only if interference (superposition) is possible.

(5) Schrödinger's equation is a linear differential equation. Quantum phenomena are explained with probability amplitudes which can have superpositions. In fact most fundamental equations of QM are linear so far.

3.2   Non-linearity

Not all interesting laws satisfy the superposition principle. In general it holds that if the equations of motion are non-linear (cf. note 13) then the superposition principle does not hold. However there are a few exceptions: The solution of the Toda-equation which describes a chain of particles coupled together with non-linear springs. The solutions are so called solitons (found only in non-linear systems of continuum mechanics). Solitons are such solutions which describe an isolated bump moving at a constant speed.

Non-linear equations are required for describing three important groups of phenomena: Order and structure, scale-invariance and self-similarity, chaos. The first two will be described briefly in 3.2.1 and 3.2.2.

3.2.1 Order and Structure

Phenomena of order and selforganization can turn into chaotic phenomena if some of the parameters are changed only slightly. An example is the Bénard-experiment: A fluid layer heated from the bottom shows first heat conduction. After some time this state becomes unstable at a critical value (threshold) and socalled convection rolls are developed which represent a highly ordered structure. Observe that adjacent rolls turn like gear wheels and that the elements of the fluid go up (the warm ones) and go down (the cold ones) periodically. In this process different possible modes of movements are in competition. In the course of time successful ones (those modes which grow most rapidly) win over less successful ones and enslave them. The underlying princple - called the "Slaving-Principle" - says that the instable modes of slowly relaxing magnitudes enslave the relatively stable modes of quickly relaxing magnitudes. The result is a drastic reduction in the number of freedoms because the boundary conditions limit the number of rolls. An ordered hierarchy developes. If the temperature is increased further up to a certain threshold (measured by the Rayleigh number which is proportional to the increase of the temperature) then the motion begins to become chaotic.

Another example for order and structure in open dissipative systems is Karman's water turbulence which has an often observed characteristic picture emerging if a stick is put into a stream of water. Observe that the respective systems are *open* systems. The self organization of order and structure does not violate the law of entropy because this law holds in closed systems only.

Still another interesting example of a prototype of system which satisfies the "Slaving Principle" is the laser. Self-organizing structures of that sort have been studied intensively by Hermann Haken and his school.[38]

3.2.2 Scale-invariance and Self-similarity

Scale-invariance means that there is no natural scale; or in other words that one cannot distinguish a system from its (by some factor) enlarged copy just by its inner properties (laws). A map can be enlarged or reduced in order to be suitable. But atoms cannot be enlarged or reduced and thus laws in which basic physical constants play a role are not scale-invariant (cf. chapter 6). As it was said above scale-invariance and self-similarity can only be described by non-linear equations, i.e. the superposition principle does not hold. If $x(t)$ is the solution of a non-linear differential equation describing a process which leads to self-similarity then there are values which approximate the solution fairly well but which do not depend anymore from the initial condition x_0. And this means that the scales can be changed (stretched, enlarged or contracted, reduced) by the factor λ such that scale-invariance can be described by the equation $x(\lambda t) = \lambda^n x(t)$ where n is the similarity-exponent.[39] If an approximate solution is found which does not depend anymore on the initial condition x_0 then the question is whether there is some (even very small) maximum (bump) in its time evolution. If there is then scale invariance (self-similarity) of the curve leads to refined repetitions which have structural similarity (self-similarity). These structures are called "fractals". A very simple example is Koch's curve: A line with a certain length l_0 is divided into 3 equal parts and a convexity (bump) is constructed with two sides of $l_0/3$ on the middle part. If this construction is repeated the length L_n after n iterations is $L_n = l_0 \cdot 4^n/3^n$. Fractals seem to be widespread in nature like coast lines, ferns, trees etc.[40]

The discoverer of scale-invariance in phenomena of nature seems to have been L.P. Kadanoff by investigating magnetism.[41] Already in school we learned that a perfect magnet would consist of smallest elementary magnets all of which are lined up in one direction. In such a magnet (which has to be at a very low temperature close to 0) the order is complete and Kadanoff's question whether it would look different at different scales is answered with: No. That is, it is scale-invariant. The same happens when the magnet is in complete disorder, i.e. when it is at a very high temperature such that the elementary magnets all fluctuate independently of one another. At temperatures in between the magnet looks different at different scales. At a low temperature the order is not complete because some elementary magnets are out of line. Thus increasing the scale means that

---

[38] Cf. Haken (1982), Haken (1984). See also the contribution of Wunderlin (these proceedings).

[39] For further details see Großmann (1991).

[40] Cf. Mandelbrot (1982) and Peitgen and Richter (1986).

[41] Cf. Peitgen and Richter (1986), p. 130.

it is more coarse and so the little fluctuations of the few elementary magnets are unobservable. The magnet looks like on a lower temperature. It's the other way round when the magnet is observed on high temperature. By coarsening the scale the few ordered elementary magnets cannot be observed any more and the magnet looks like on a still higher temperature. The main point is that scale transformations are related to changes in temperature. The same magnet of a given temperature looks as if it were on different temperatures when viewed on different scales. In order not to produce wrong results we have to renormalize the temperature. That is, scale transformation forces a corresponding renormalization transformation (in this special case of temperature otherwise of another magnitude). Between the two attractors ($T = 0$ for low temperatures and $T = \infty$ for high temperatures) there is a boundary, the Curie-temperature. At that temperature the magnet looks the same on all scales, i.e. its temperature does not change under renormalization. And that means that the pattern of the fluctuations of the elementary magnets at the Curie-temperature is self-similar. On the other hand if the temperature of the magnet deviates only slightly from the Curie-temperature this deviation may increase by iteration and lead to either complete order and structure or complete disorder and chaos.

4   Question four:

Can symmetric (invariant) laws describe and predict asymmetric phenomena?

The general answer to this question is: Yes. Symmetric laws describe ("produce") and predict asymmetric phenomena if the initial conditions are asymmetric.

Galileo understood that the physical laws are the same in different inertial frames (i.e. in frames which are - relative to one another - at rest or are moving with uniform velocity). That is the physical laws are invariant (symmetric) in respect to inertial frames. - The following is a nice passage from his dialogue which describes his "Gedankenexperiment":

"*Salviatus*. Shut yourself up with some friend in the main cabin below decks on some large ship, and have with you there some flies, butterflies and other small flying animals. Have a large bowl water with some fish in it; hang up a bottle that empties drop by drop into a wide vessel beneath it. With the ship standing still, observe carefully how the little animals fly with equal speed to all sides of the cabin. The fish swim indifferently in all directions; the drops fall into the vessel beneath; and, in throwing something to your friend you need throw it no more strongly in one direction than another, the distances being equal; jumping with your feet together, you pass equal spaces in every direction. When you have observed all these things carefully (though there is no doubt that when the ship is standing still everything must happen in this way), have the ship proceed with any speed you like, so long as the motion is uniform and not fluctuating this way and that. You will discover not the least change in all the effects named, nor could you tell from any of them whether the ship was moving or standing still. ..."[42]

Galilei assumed that the orbits of the planets are circles. One of his reasons was probably aesthetic. The other - more important one - might have been his understanding that the laws of motion are rotationally symmetric and therefore allow circles as its simplest solutions even if circles are not required by the laws of motion. But what if he would have thought that the orbits are exclusively determined by the rotationally-symmetric laws? Then he would have been fully justified to believe in the orbits as circles. In fact *initial conditions* in addition to the laws determine the orbits. And these initial conditions may by asymmetric, may break the symmetry; i.e. determine the deviation from circles to produce ellipses.

That the orbit of a planet lies in one plane which contains the sun is determined by the laws, it follows from the conservation law of momentum. But which plane it is or better which angle this plane has in respect to, say, another star is not determined by the laws; i.e. is dependent on initial conditions. That means that the laws would allow each of the possible planes but only one as the realized one. Thus in some sense the rotationally symmetric laws produce the asymmetry (the symmetry-breaking) of just one realized plane but which one depends on the initial conditions.

Other situations created by locally asymmetric initial conditions are: that our heart is on the left side. But since there are few cases where children are born with the heart on the right side the biological laws do not seem to exclude such cases, i.e. the laws seem to be invariant in respect to right or left and the selection of "left" (for most cases) seems to be produced by initial conditions. Another example are the houses of snails which have the spiral turning in one direction within one species but with a few exceptions where the spiral turns the opposite way and the helical form of biomolecules. Similarly when polarized light is twisted to the right when passing through a sugar solution. In many such cases we do not know the exact kind of initial conditions and when and how they produced the asymmetry during the evolution of the universe.

The universe or better the distribution of masses in it is also locally asymmetric, i.e. locally unisotrop (not rotationally symmetric) and not homogenous (not translationally symmetric). This is due to asymmetric initial conditions but according to the Cosmological Principle the universe as a whole (and in large parts) is homogenous and isotrop, i.e. symmetric in respect to translation and rotation. This principle is a hypothesis which is fairly well corroborated by the following two facts: (1) The temperature of the cosmic background radiation is independent from the direction of the radiation. That indicates that at least at the time when the radiation began the universe was homogenous and isotrop. And if there were later asymmetries by asymmetric conditions they did not affect the temperature of the background radiation. (2) The velocities of the expansion of galaxies at far distance are proportional to the distance of this galaxies from

---

[42] Galilei (DWS) Second Day. Cf. the discussion in Berry (1978), p. 30ff. (The reference in Berry (p. 31) to Galileo "Dialogues concerning two new sciences" is incorrect.

the earth. The Cosmological Principle is an assertion of symmetry. If it holds then the laws of nature are symmetric in respect to translation and rotation and so is the whole universe.[43]

5   Question five:

Are the laws of nature time-symmetric?

5.1   The Oldest Invariance-Principle

The invariance with respect to displacement of time (and of place) are the oldest and perhaps most important invariance properties of physical laws and of laws of nature in general. It seems that the concept of law (of nature) is violated if these invariance conditions would not be satisfied.

"This principle can be formulated, in the language of initial conditions as the statement that the absolute position and the absolute time are never essential initial conditions. The statement that absolute time and position are never essential initial conditions is the first and perhaps the most important theorem of invariance in physics. If it were not for it, it might have been impossible for us to discover laws of nature."[44]

"The paradigm for symmetries of nature is of course the group of symmetries of space and time. These are symmetries that tell you that the laws of nature don't care about how you orient your laboratory, or where you locate your laboratory, or how you set your clocks or how fast your laboratory is moving."[45]

The following passage points especially to the fact that the relations between the events which are described by the laws depend only on the intervals but not on a point of time when the first event occurred. That means that time-symmetric laws cannot designate or select a beginning in time or a first event:

"Thus the time displacement invariance, properly formulated, reads: the correlations between events depend only on the time intervals between those events; they do not depend on the time when the first of them takes place."[46]

5.2   Philosophical Significance

The first philosopher who seems to have realized this very clearly was Thomas Aquinas. In his quarrel with Bonaventura at the university of Paris he defended the view that the beginning in time of the world (universe) cannot be proved from universal principles (laws) of (about) this world. Because universal principle which have their foundation in the essence of things (of nature) abstract from hic (place) et nunc (point of time):

---

[43] For further discussion see Genz and Decker (1991), p. 68ff.

[44] Wigner (1967), p. 4.

[45] Weinberg (1987), p. 73.

[46] Wigner (1967), p. 31.

"That the world has not always existed cannot be demonstratively proved but is held by faith alone. ... The reason is this: the world considered in itself offers no grounds for demonstrating that it was once all new. For the principle for demonstrating an object is its definition. Now the specific nature of each and every object abstracts from the here and now, which is why universals are described as being *everywhere and always*. Hence it cannot be demonstrated that man or the heavens or stone did not always exist."[47]

In this connection I want to mention that the question whether it can be demonstrated that the world has always existed or that it has a beginning in (with) time - answered differently by the two competing theories of the universe the Steady State Theory and the Standard Theory (Big Bang) - is a question about the completeness of the laws of nature. Or at least of that laws we know. A system of laws L about a certain part P of reality is complete if and only if every truth about P is provably (derivable) from L.[48] Thomas Aquinas' standpoint was that the universal laws of nature (about this world) are not complete with respect to all questions (all truths) about this world. It is not just our insufficient knowledge of the laws of nature what he has in mind, but the true laws itself are incomplete according to him with respect to some special questions. That means that there are some statements about this world which are undecidable from the laws about this world. Or in more modern terms: the laws of nature are incomplete with respect to some important initial conditions. This problem plays an important role in the Big Bang Theory of the cosmological evolution in respect to (at least) the "first three minutes".[49]

It may be interesting to cite some other passages which are connected with

---

[47] Thomas Aquinas (STh), I, 46.2.

[48] From results of Gödel and others we know that if P is the elementary theory of Boolean algebras then (the respective axiom system) L is both complete and decidable, if P is First Order Predicate Logic (including relations) (the respective) L of it is complete but not decidable, if P is the elementary theory of dense order (the respective) L is not complete but decidable and if P is the arithmetic of integers (the respective) L is neither complete nor decidable. These and other results on completeness and decidability about areas within mathematics do not automatically imply results about systems of laws of a part of natural science about a part of reality. This has many reasons. First in natural science one is usually interested only in a finite number of truths about (a finite) reality and only in a finite number of consequences of theories (systems of laws and hypothesis). Secondly it is not clear whether the presuppositions made in the case of logic and mathematics for the mentioned results like logical closure conditions, standard form or conditions for defining provability etc. can be met or transmitted in a suitable way to the laws used for example in physics. Thus only the more general intuitive idea of completeness can be applied also to the laws of natural science.

[49] Cf. Weinberg (1977). That we cannot decide concerning such initial conditions (like the question whether the universe has a certain age) was true until very recently indeed when such things as the cosmic background radiation have been discovered which is a rather strong support for the finite age of the universe - even if we could not say it is an absolute proof (demonstration).

the one above and with basic features of the theory of relativity. Thomas Aquinas says there that place and space are bound to the material world, i.e. there is no absolute space independently of the world:

"It is not enough to conceive the void as that in which nothing is; you have to define it, as Aristotle (Phys), 208b 26) does as a space capable of yet not holding a body. Our contention however is that before the world existed there was no place nor space."[50]

And again according to Aquinas there is no thing at (absolute) rest:

"For our part we assert that motion was always from the moment subjects of motion began."[51]

And finally time began with the beginning of the world, i.e. there was no time "before" the world (universe): "The phrase about things being created in the beginning of time means that the heavens and earth were created together with time; it does not suggest that the beginning of time was the measure of creation."[52]

Those are indeed interesting passages which show Thomas Aquinas to be a predecessor in some basic features of the theory of relativity. The whole questio 46 which treats the problem of the beginning and the duration of creation shows that Aquinas held already important assumptions of the Theory of Relativity.[53]

5.3   Development in the 20th Century

The view that the laws of nature are (space and) time-symmetric fits very well to the greek ideal of science mentioned in the introduction: the description of contingent and changing objects by necessary and non-changing laws. In the 20th century the space-time symmetry and the idea of the similarity between space and time coordinates was especially supported by the Theory of Relativity. And both theories, Relativity Theory and Quantum Theory - being the main theories of physics of the 20th century - are (space)-time-symmetric, i.e. they do not define an arrow of time. Nevertheless one was aware of the following two basic facts which underly both Relativity Theory and Quantum Mechanics:

(1) the fact that a (contingent) particle can move with respect to t on its world line only in one direction, i.e. in the direction of increasing t whereas it can move with respect to one of the space coordinates (x, y, z-axes) in both (positive and negative) directions. This fact separates in an important sense the time- from the space coordinates.

---

[50] Thomas Aquinas (STh), I.46.1, ad 4.

[51] Ibid. ad 5.

[52] Ibid. 46,3 ad 1. With this passage we could compare the following very modern one: "This approach is consistent with the view that there is no time at the most fundamental level, and it is necessary to understand how time emerges from a quantum theory not possessing one" (Halliwell (1994), p. 375f).

[53] It seems to me that this can be assumed in spite of the fact that there are (unrealistic but consistent) interpretations of General Relativity without matter like the De Sitter universe.

(2) the (partially epistemic) fact that it seems more natural to understand initial conditions as characterizing the state of the system (using all three space coordinates) at one definite instance of time than to understand initial conditions as characterizing the state of the system for all times but only for a single instance of one of the space coordinates.[54] Dependent on this understanding the position-momentum uncertainty is more frequently used that the time-energy uncertainty.

In spite of all this there were known processes in nature and known laws which describe irreversibility in time: Important examples of such processes belong to five categories:

(1) Thermodynamic processes (entropy increases in isolated systems)

(2) Psychological processes (remembering the past, predicting the future)

(3) Processes of radiation

(4) Processes in Quantum Mechanics (CP-non-invariance and CPT-invariance)

(5) Cosmological processes (expansion of universe and growing inhomogeneity of universe)

Boltzmann and Planck were concerned with phenomena of category (1) and Planck in addition very much with that of category (3), especially with black body radiation. Both Boltzmann and Planck had difficulties to interpret the second law of thermodynamics, the law of entropy, but thought they could make it compatible with the general idea of invariance with respect to time:

"§1. Der zweite Hauptsatz wird mechanisch durch die natürlich unbeweisbare Annahme A erklärt, daß das Universum, wenn man es als mechanisches System auffaßt, oder wenigstens ein sehr ausgedehnter, uns umgebender Teil desselben von einem sehr unwahrscheinlichen Zustande ausging und sich noch in einem solchen befindet."[55]

"Für das Universum sind also beide Richtungen der Zeit ununterscheidbar, wie es im Raum kein Oben oder Unten gibt. Aber wie wir an einer bestimmten Stelle der Erdoberfläche die Richtung gegen den Erdmittelpunkt als nach unten bezeichnen, so wird ein Lebewesen, das sich in einer bestimmten Zeitphase einer solchen Einzelwelt befindet, die Zeitrichtung gegen die unwahrscheinlicheren Zustände anders als die entgegengesetzte (erstere als die Vergangenheit, den Anfang, letztere als die Zukunft, das Ende) bezeichnen ..."[56]

Also Planck tried to give a "mechanical" interpretation of the law of entropy:

"Zermelo, however, goes farther [than I], and I think that incorrect. He believes that the second law, considered as a law of nature, is incompatible with any mechanical view of nature. The problem becomes essentially different, however, if one considers continuous matter instead of discrete mass-points like the molecules of gas theory. I believe and hope that a strict mechanical significance

---

[54] Cf. Wigner (1972).

[55] Boltzmann (1897a), p. 579.

[56] Ibid. p. 583. For more details of Boltzmann's understanding of statistical laws compared to dynamical laws and his reasons - also against Zermelo - that the former are compatible with Poincaré's recurrence theorem see Weingartner (these proceedings).

can be found for the second law along this path, but the problem is obviously extremely difficult and requires time."[57]

"The principle of energy conservation requires that all natural occurrences be analyzable ultimately into so-called conservative effects like, for example, those which take place in the motion of a system of mutually attracting or repelling material points, or also in completely elastic media, or with electromagnetic waves in insulators. ... On the other hand, the principle of the increase of entropy teaches that all changes in nature proceed in one direction. ... From this opposition arises the fundamental task of theoretical physics, the reduction of unidirectional change to conservative effects."[58]

It has to be emphasized however that time irreversibility seems not necessarily to be connected with an increase of entropy. As Popper says[59] the process when a stone is dropped into water and causes circular waves is hardly reversable and Maxwell's example in chapter 2.1 is another case in point.

Independently of this question it should also be mentioned that statistical processes as listed above can be interpreted more modestly with respect to an arrow of time: If one assumes only non-recurrency one may speak of an internal arrow of the process without assuming irreversibility in time.[60]

A new situation again developed with both, further research on the cosmological development of the universe and recent discoveries of non-linear phenomena, i.e. phenomena of order, of structure of selforganization, of self-similarity and of chaos. The two competing main theories of cosmology are a special sign for it: In the Steady State Theory of Bondi and Gold the universe expands. The basic assumption is that there is no designated place nor time in the universe, i.e. the universe is homogeneous and unchanging, neither geography nor history matters as Bondi says. The process of expanding is balanced by a permanent production of matter ($\sim$ one hydrogen atom per 6 km^3 per year) to such a degree as to keep the density and temperature unchanged (constant). According to this theory the universe has no age and does not select a special point of time although it has an arrow of time because the permanent production of matter increases the entropy with time. The Steady State Theory however is in conflict with observations which show that the deceleration parameter is positive, i.e. the expansion of the universe is slowing down, whereas according to the theory it should be negative.

The universe described by the Standard Theory (Big Bang Theory) has a certain age (about $2 \cdot 10^{10}$ years) which depends on (the most recent estimation of) Hubble's constant but does not define an arrow of time. This theory is supported by the astrophysical theory of star evolution where stars have ages (usually 10^{10} years or less) and by the decay rates of radioactive isotopes which give an age to the elements of about 10^{10} years also. The theory received great further support

---

[57] Planck in a letter to his friend Leo Graetz. Cited in Kuhn (1978), p. 27.

[58] Planck in a paper read to the Prussian Academy of Science in 1897. Cited in Kuhn (1978), p. 28.

[59] Popper (1956).

[60] Cf. ch. 5.4, Chirikov (these proceedings), ch. 2.2.

from the discovery of the cosmic background radiation by Penzias and Wilson (1965) which was predicted already by Alpher and Herman (1949).

Both the cosmic background radiation and the new phenomena of order, selforganization, selfsimilarity and chaos are - according to Prigogine - strong cases for a "new (or revised) physics" which incorporates the arrow of time.[61]

5.4   CPT Invariance

The combination of the symmetries of charge parity and time was never violated in any process investigated so for, i.e. its invriance is very well confirmed. This does not however imply the invariance of each constituent seperately.[62]

In 1956 the first violation of parity (predicted by Lee and Yang) was confirmed by C.S. Wu et al. in a cobalt-60 beta decay (decay of polarized cobalt-60 nuclei into electrons) experiment.

That means that the laws of physics are not invariant in respect to an exchange of a physical system (of the universe) with its mirror image. Further evidence is the possibility of an absolute definition of right and left by the spin of the antineutrino (which corresponds to a right hand screw in the direction of the momentum) and the neutrino (which corresponds to a left hand screw).

In 1964 and 1967 violation of charge (also conjectured by Lee and Yang in 1957) was established (by Christenson et al. and Bennett et al.) by the decay of K_2^0 mesons and K_L^0 ions. K_L^0 is unstable and may decay into e^- (electron) v (neutrino) and Π^+ (positive pion) or into e^+ (positron) v^- (antineutrino) and Π^- (negative pion). These two different kinds of decay can be distinguished by magnetic separation of electrons and positrons and it can be established that the decay rate is different: the neutral K_L^0 decays faster into positrons than into electrons. That means that the experiments (nature) are not symmetrical with respect to the plus and minus signs of electric charge. In the CP combination of charge and parity the "big" violations which occur with respect to C and P seperately are almost balanced but not completely. More accurately: all violations of C and P separately (which have not been described here) are completely neutralized in CP except one. This is the one with neutral K_L^0 mesons described above. The respective experiments have been repeated many times with great care such that the result is well corroborated. Therefore there is no complete CP invariance.

On the other hand in all processes known so far the invariance of the combination of charge, parity and time was always confirmed. This means that there is a close connection among the three symmetries C, P and T. CPT invariance - one of the most important symmetries of Quantum Field Theory - says that

---

[61] How this can be done he proposes in Prigogine and Stengers (1993), part IV and in Prigogine (1993).

[62] For a lucid discussion of C, P and T (and CPT) invariance see Lee (1988) first part. For details see Genz and Decker (1991) - a very useful book which contains a lot of interesting material -, Gell-Mann and Hartle (1994), and Particle Data Group (1990).

physical laws seem to be symmetric with respect to the complex exchange of particle - anti particle, right - left and past - future. This CPT-symmetry has remarkable consequences: It implies that the mass of the (any) particle must be the same as that of its respective antiparticle. The same holds for their lifetimes. Their electric charges must have the same magnitude but opposite signs, their magnetic moments must agree.

A most striking consequence however is that since CP invariance is violated but CPT invariance is not T has to outbalance the difference and therefore T-invariance cannot hold unrestrictedly. This is indeed a serious consequence and - if true - would have a lot of implications. On the other hand T-invariance (time-reversal) is hardly compatible with a dipol-momentum of elementary particles and it would reverse velocities and exchange initial with final states. Such a time reversal operation would be strongly non-linear in character as Wigner pointed out long ago. If it could be proved that the neutron has a dipol momentum then the laws of nature could neither be P-invariant nor T-invariant. As it has been said above we know from other experiments that they are not P-invariant. But for the violation of T-invariance no direct experiments are known so far. Since CP invariance is violated but CPT invariance is not it is important to ask for the basic assumptions which underly the CPT invariance. These are mainly the following:[63]

(1) The particles which are not composed are finite in number.

(2) The symmetries of Special Relativity hold (with respect to continuous transformations and one time and 3 space coordinates).

(3) The laws are local.

(4) Energies cannot be arbitrarily negative, i.e. there is a lowest energy level.

(5) The laws of QM hold in accordance with local relativistic quantum-field theories in four dimensions.

(6) The total probability of the quantum system is constant in time.

The relatively small violation of time symmetry in elementary particle physics described above does not seem to be sufficient however for an explanation of time irreversibility in macro-physical processes; I mean processes like in thermodynamics and in cosmology and in the recently discovered areas of selforganization and chaos.

First of all it has to be remembered that basic laws which are time-symmetric can "produce" asymmetric phenomena (states and processes) if there are asymmetric initial conditions (cf. chapter 4 above); i.e. an asymmetrical world with asymmetrical states and processes does not imply asymmetrical laws but requires asymmetrical initial conditions. Asymmetrical initial conditions of a very strong kind in fact the greatest symmetry breaking - also with respect to time - have to be assumed at the beginning of the universe in all the theories which describe a universe finite in time or with a finite age since the beginning. This amounts to an extreme and - in the sense of the law of entropy - most unprobable but most ordered and structured singularity. Also the great religions, judaism, christianity

---

[63] Cf. Wess (1989) and Genz and Decker (1991), p. 169.

and islam had in mind a singularity of order and structure when they spoke of creation.

Secondly it should be emphasized that basic laws which are time symmetric on the microscopic level are compatible with laws on the macroscopic level which are not time symmetric, but describe an arrow of time like the law of entropy (even if granted that the reverse process is not completely impossible but very highly unprobable). A nice "Gedankenexperiment" for such a compatibility is given by Lee.[64]

Assume a number of airports with flight connections in such a way that between any two of these airports the number of flights going both ways along any route is the same. This property will stand for microscopic reversibility. Some of the airports may have more than one air connection (they are connected with more than one other airport) whereas other airports have a connection only to one airport (let's call such airports dead end airports). A passenger starting from a dead end airport (or starting from any other airport) can reach any other airport and can also get back to his starting airport with the same ease. This property stands for macroscopic reversibility. In this .case we have both microscopic and macroscopic reversibility.

But suppose now we were to remove in every airport all the signs and flight informations, while maintaining exactly the same number of flights. A passenger starting from a dead end airport A will certainly reach the next airport B since that is the only airport connected with A. But then - especially when assuming that B has many flight connections - it will be very difficult to get further to his final destination, in fact it will be a matter of chance. Moreover his chance to find back to his dead end airport A will be very small indeed. Thus in this case we have microscopic reversibility maintained but macroscopic irreversibility and both are not in conflict.

A necessary condition for chaotic phenomena was the sensitive dependence on initial conditions. This exponential dependence (see chapter 2) can be used as some explanation of the non-reversibility of the process. If ergodicity could be proved for physical systems which show chaotic behaviour it would provide a more direct explanation for their time-irreversibility. Since the ergodic hypothesis is, in a good sense, the mathematical formulation of Boltzmann's idea of the irreversible time evolution of a complicated physical system. If we represent the time evolution (of positions and velocities) of all atoms in a specific phase space (with a certain energy) by the movement (time evolution) of one point then the ergodic hypothesis can be stated thus:

The point (describing the system) when moving through the phase space spends a certain time in each region (within the phase space) and this time is proportional to the volume of that region.

There is a proof for the ergodic hypothesis for certain chaotic phenomena for instance for Sinai's Billiard. For others there is no proof yet and it may be that a suitable weakening of the hypothesis would still provide a reasonable

---

[64] Lee (1988), ch. Symmetries and Asymmetries.

explanation for time-irreversibility and could be provable for interesting cases of chaotic systems.

6 Question six

Are the laws of nature invariant with respect to scales?

This invariance is also called affine-invariance or invariance with respect to an affine-group. An affine-group translates points into points, straight lines into straigth lines, planes into planes, cubes into cubes, balls into balls ... etc. Thus it forgets distances but keeps intersection properties. Thus such a transformation - also called scale transformation - changes all lengths by the same factor a, all planes by a^2 and all volumes by a^3, if $a > 1$ or $a < 1$ respectively. A modern xeroxmachine enlarges or diminishes copies, maps are diminished representations of countries or cities ... etc. (cf. chapter 3.2.2). But can we enlarge things of nature or art (technical constructions) for instance can animals or trees or ships or buildings be twice or three times or hundred times as large as they are? This was already a question for Galileo and he answered it in the negative:

"From what has already been demonstrated, you can plainly see the impossibility of increasing the size of structures to vast dimensions either in art or in nature; likewise the impossibility of building ships, palaces, or temples of enormous size in such a way that their oars, yards, beams, iron-bolts, and, in short, all their other parts will hold together; nor can nature produce trees of extraordinary size because the branches would break down under their own weight; so also it would be impossible to build up the bony structures of men, horses, or other animals so as to hold together and perform their normal functions if these animals were to be increased enormously in height; for this increase in height can be accomplished only by employing a material which is harder and stronger than ususal, or by enlarging the size of the bones, thus changing their shape until the form and appearance of the animals suggest a monstrosity."[65]

Galileo was right to understand that we cannot enlarge or diminish reality. Even if possible - within certain limits - for things made by technical construction we cannot make an enlarged or diminished copy of an atom. The radius of the hydrogen atom is something fixed, it is about $5 \cdot 10^{-9}$. Also the Avogadro Number which is $6,022 \cdot 10^{23} mol^{-1}$. On the other hand the smallest distance between elementary particles in collision is 10^{-18} cm seems to be a technical limitation depending on the energy of our accelerators. Larger systems do not contain larger atoms but just more atoms. Since the laws of nature contain such constants they cannot be scale-invariant.

6.1 Fundamental Constants

For nonrelativistic QM three constants are very important: Planck's constant h (or for many occasions more practical: $h/2\pi = \hbar = 1,055 \cdot 10^{-27} g \cdot cm^2 \cdot sec^{-1}$),

---

[65] Galilei (DNS), p. 130 [169].

the mass of the electron $m_e = 9,11 \cdot 10^{-28} g$ and the elementary charge of the electron $e = 4,81 \cdot 10^{-10} g^{1/2} cm^{3/2} s^{-1} (= 1,6 \cdot 10^{-19}$ Coulomb). With the help of these three constants one can define a length (equal to the Bohr-radius that is the radius of the hydrogen atom in its base state) and a time which is related to the energy E_e of the electron in this state by $\hbar/2t = E_e$.

For relativistic QM + Gravitation the three main constants are $\hbar$, c (light-velocity) and G (gravitational constant), where $c = 3 \cdot 10^{10} cmsec^{-1}$ and $G = 6,67 \cdot 10^{-8} cm^3 g^{-1} sec^{-2}$. With the help of these three one can define the socalled Planck's scale, the Planck length $l_{Pl} = 1,6 \cdot 10^{-33} cm$, the Planck mass $m_{Pl} = 2,176 \cdot 10^{-5} g$ and the Planck time $t_{Pl} = 5,39 \cdot 10^{-44} sec$.

In addition to that there are scale invariant dimensionless constants, the most important of them the mass proportion of proton (neutron) and electron: $m_p/m_e = 1836 (m_n/m_e = 1838, 6)$ and the fine structure constant $\alpha = e^2/\hbar \cdot c = 1/137$.[66]

There are important questions concerning these constants which suggest themselves: What is the deeper reason for these magnitudes? What is their interrelation to other important magnitudes? Why is it so that if some constants are relevant to some area of physics then just by multiplication among them new magnitudes are produced which are again relevant for that area (if not for other areas too)? Do these constants (for instance those of the Planck scale) belong to the laws i.e. are they determined by the laws or should we understand them as the most fundamental initial conditions?

6.2   Dirac's Large Numbers Hypothesis

Especially with the first two question above Dirac was concerned in an essay of 1937 and in later papers.[67] In his last paper he is especially concerned with the dimensionless constants which are independent of the system of units one uses:

"At present, we do not know why they should have the values they have, but still one feels that there must be some explanation for them and when our science is developed sufficiently, we shall be able to calculate them."[68]

Dirac found another dimensionless constant which is the ratio of the electric force e^2/r^2 between electron and proton and the gravitational force $G \cdot m_p \cdot m_e/r^2$ between electron and proton: $e^2/G \cdot m_p \cdot m_e$ which is dimensionless and of the order of about 10^{40}. This number he compared with the age of the universe (T) in terms of atomic units, for example expressed in time units $t_e = r_e/c$, i.e. t_e is the time the light needs for the distance of the diameter of the electron. The age of the universe (as known today) in time units t_e is also about 10^{40}. Thus he proposed the equation $e^2/G \cdot m_e \cdot m_p = T/t_e$ as a fundamental equation expressing his socalled "Large Numbers Hypothesis". This Hypothesis states that very large numbers (numerical coefficients) cannot occur without reason in the basic laws of physics:

---

[66] For details see Genz and Decker (1991), p. 302ff.
[67] Cf. Dirac (1973).
[68] Ibid. p. 45.

"It involves the fundamental assumption that these enormous numbers are connected with each other. The assumption should be extended to assert that, whenever we have an enormous number turning up in nature, it should be connected to the epoch and should, therefore, vary as t varies. I will call this the Large Numbers Hypothesis."[69]

This hypotheses and more specifically the above mentioned equation which connects the gravitational constant with the age of the universe has severe consequences:

(1) If this equation is true then the laws of nature are not time-symmetric. According to Dirac $\dot{G} \neq 0$ and G should decrease with time. A further consequence would be that the law of the conservation of energy would no more hold.

(2) As Dirac points out the Big Bang Theory when developed in accordance with the Large Numbers Hypothesis implies continuous creation of matter which violates the law of conservation of energy.

A consequence of (1), i.e. $\dot{G} \neq 0$, would be that the moon should depart from the earth in the course of time and that some of the constants in the above equation - or respectively in the following transformation of it - would not be really constant: $T = e^4/(m_e^2 \cdot m_p \cdot G \cdot c^3)$. The most exact measurements concerning a departure of the moon did not show a significant effect or, more accurately, are compatible with both $\dot{G} = 0$ and the expected deviation.[70] Furthermore there are - so far - no serious indications that e, m_e, m_p or c change with time. Specifically no respective effects are known concerning spectra of far galaxies.

Concerning the constancy of c an interesting theory has been proposed by August Meessen.[71] According to this theory of space-time-quantization $c = 2a \cdot E_U/h$, i.e. c is dependent on the amount of energy E_U of the whole universe (i.e. the positive energy represented by the matter of the universe) and thus c will be constant as long as the numerical value of E_U is. Here a is the ultimate limit for the smallest measurable distance - called quantum-length - and not necessarily identical with Planck's length.

There have been tests for other constants mentioned above. The fine-structure of spectral lines has been compared to that of its red shift, but there was no indication that the fine-structure constant would change with time. Also tests have been performed for h by measuring energy E and wave length λ of photons coming from "old" and "new" astronomical sources of light. There was no indication that $h = E \cdot \lambda/c$ would depend on the age of the source of light.[72]

The discussion of the constancy of the fundamental constants of nature is a good example of an important question which is open to further revision. Although it seems so - acording to the best tests performed so far - that G, h and α (and other constants) do not change with time the question is not settled.

---

[69] Ibid. p. 46.

[70] Cf. Genz and Decker (1991), p.94, Irvine (1983) and Damour et al. (1988).

[71] Cf. Meessen (1989).

[72] For further details see Dyson (1972), Irvine (1983) and Norman (1986).

Future experiments may indeed show a dependence on time. Since such constants enter fundamental laws this would mean that the laws of nature are not time-invariant (cf. chapter 5).

6.3 Local Scale Invariance

Though the laws of nature are not scale-invariant because some important constants always enter a law there are local areas for which scale invariance holds. These are areas where those fundamental constants mentioned above, including atomic distances and proportions, do not enter. An interesting example is the Principle of Archimedes. Within resonable limits this principle is scale invariant. We may enlarge the things (rigid bodies like stones or ships) to be put into water (or another fluid) arbitrarily. No fundamental constant enters this principle. The importance of local scale invariance for self-similarity and fractal structures has been pointed out already in chapter 3.2.2. The example of Kadanoff discussed there shows that in the two extreme cases of a magnet being in complete order (temperature = 0) and in complete discorder or chaos (temperature = ∞) this physical system (magnet) - or better: the laws govering it - are scale-invariant.

7 Question seven

Are the laws of nature valid also in other universes which differ from our universe only with respect to initial conditions?

Popper (1959) proposed the following definition of natural (physical) necessity as an essential feature of laws of nature:

"A statement may be said to be naturally or physically necessary iff it is deducible from a statement function which is satisfied in all worlds that differ from our world (if at all) only with respect to initial conditions."[73]

Let us call the set of all the changes which do not change the laws (of nature) the symmetry group of nature. About this symmetry group Weinberg says:

"It is increasingly clear that the symmetry group of nature is the deepest thing that we understand about nature today ... Specifying the symmetry group of nature may be all we need to say about the physical world beyond the principles of quantum mechanics."[74]

But how can we determine the set of all changes which leave the laws invariant? This would mean to know the line of demarcation between contingent initial conditions and necessary and invariant laws. It would mean to know which

---

[73] Popper (1959), p. 433.

[74] Weinberg (1987), p. 73. Weinberg's definition of "symmetry group of nature" is a little bit different and has some subjective element in it: "point of view" and "way you look at nature". "The set of all these changes in point of view is called the symmetry group of nature." cf. ibid. p. 72 and 73. I shall try to avoid this subjective element and use as a preliminary version: the set of all changes which do not change the laws.

constants when changed do not affect the laws and which do; and which initial conditions and boundary conditions would affect the laws when changed and which would not. Are the laws of nature invariant with respect to a slight (drastical) change of the amount of energy (mass) of the whole universe (which is constant by the law of conservation of energy)? Or could we change the ratio of electron and proton mass slightly without changing laws? Very probably not! Could there be a constellation of the planets (or some star-systems) which differs from the one realized now (at a certain time) or in other words: Are different constellations of systems of stars (of our planetary system) at a certain time after the Big Bang compatible with the laws?

7.1   Groups of Symmetries

Before I am trying to give an answer to some of these questions I shall give a brief list of the four most important symmetry groups.

(1) Permutational Symmetry

Permutational Symmetry means that "different individual" particles of the "same sort" are treated identical. Thus the laws and the respective physical world (universe) described by these laws remain the same if we interchange any two electrons. The same holds for protons, neutrons, neutrinos and μ-mesons (according to the Fermi-Dirac statistics) and also for photons, π-mesons, κ-mesons and gravitational quanta (according to the Bose-Einstein statistics).

(2) Continuous Space-Time Symmetries

(a) Translation-symmetry in space. This leads to three conservation principles of momentum. Unobservable: absolute place.

(b) Rotation-symmetry in space. This leads to three conservation principles of angular momentum. Unobservable: absolute direction.

(c) Translation-symmetry of time (i.e. delay in time makes no difference). This leads to the principle of conservation of energy. Unobservable: absolute (point of) time. This symmetry is violated to some extend because of the expanding universe and the decreasing of temperature for instance in the cosmic background radiation.

(d) Translation with uniform velocity (small in relation to c) in a straight line relative to the reference frame. Unobservable: who is moving.

(a) - (d) plus the assumption of Euclidean Space give the full Galilean invariance (symmetry) which includes relativity in respect to inertial frames (d). This symmetry group (a) - (d) underlies Newton's Theory.

Observe that condition (c) may be violated if the gravitational constant G (or other important constants) is not really constant but changes with time (cf. chapter 6.2). (d) was already discovered by Galileo (cf. chapter 4) although fully understood only by Newton. It is the important principle of relativity entering the Theory of Special Relativity.

(e) Lorentz-invariance = Full Galilean invariance + the principle that the speed of light is the same in all inertial reference frames. This is the invariance of Special Relativity.

(f) Invariance of General Relativity: Physical laws are the same in all "free falling" (with acceleration moving) - but not rotating - systems. This is the so-called weak equivalence-principle though the Theory of General Relativity satisfies also the strong equivalence-principle. This principle includes experiments dependent on gravitation like Cavendish-experiments and says that in all of them inertial mass and gravitational mass are the same.

(3) Discrete symmetries

(a) Charge-symmetry or particle-antiparticle symmetry. This symmetry leads to the conservation of charge conjugation. Unobservable: absolute sign of electric charge.

This symmetry is in fact not completely valid, i.e. experiments with neutral kaons differentiate between + and - of electric charge (cf. chapter 5.4).

(b) Parity or Right-Left-symmetry or Mirror-image-symmetry. Conservation: Parity. Unobservable: absolute right (or left).

This symmetry is satisfied for electro-magnetic effects but not completely fulfilled for radioactive phenomena (decay). Cf. chapter 5.4.

(c) Time Reversal-symmetry or Past-Future-symmetry. Conservation: Time reversal. Unobservable: Absolute sign (+, -) of time.

All fundamental laws of physics (of quantum mechanics and of the theory of relativity) are invariant with respect to time reversal. According to Prigogine this is a sign that the laws of physics are still incomplete since many processes are irreversible in time.[75] But time-reversal symmetry may even not hold on the micro level as the experiments prove CP violation (cf. chapter 5.4)

(d) CPT-symmetry. This symmetry seems not to be violated by any processes known so far (cf. chapter 5.4)

(4) Gauge symmetries

Gauge symmetry means that the physical world remains the same if ψ (representing the wave function) is multiplied by some phase factor. A consequence of this symmetry is the conservation of electric charge and - when applied to other phases - the conservation of hypercharge baryon number and lepton number. Unobservable: the phase difference between two states of different charge.

(a) U1-symmetries. They lead to conservation laws of baryon number, lepton number, electric charge and hypercharge.

(b) SU2-symmetry: Isospin symmetry. This symmetry which means an interchange of proton and neutron is not completely satisfied because of the slight mass-difference (0,14%) of proton and neutron.

(c) SU3-symmetry: Colour and flavour symmetry. This symmetry is the basis of quantum chromodynamics.

7.2   The Laws Satisfy More than One Universe

Thesis 1: The laws of nature (known laws of nature) are valid just in our universe only if the following conditions are satisfied. Or in other words: If the

---

[75] Cf. his book (together with I. Stengers) "Das Paradox der Zeit". An English version is in preparation. See further Prigogine (1993).

laws of nature (known laws of nature) are valid just in our universe then the following conditions are necessary:

(1) The laws (known laws of nature) are complete with respect to our universe.

(2) All laws of nature (known laws of nature) are deterministic.

(3) Permutation change (interchange) of elementary particles of the same kind (see chapter 7.1 (1)) does not change the world (universe).

(4) All sets of initial conditions compatible with the laws of nature (known laws of nature) occur as states (are played through) during the life time of our universe.

(5) All fundamental constants (cf. chapter 6.1) are ruled by laws of nature (known laws of nature).

Thesis 1 says in other words that all five conditions above are satisfied if the symmetry group of nature - i.e. the set of all changes which do not change the laws of nature (the known laws of nature) - is the empty set. Or in other words: All five conditions above are satisfied if the set of all models which are satisfied by the laws of nature is the unit set, i.e. if there is just one model and this is our universe. Einstein's question was more general: Not whether the laws of nature allow more than one world as a model but whether God was free to create another world (even perhaps with other laws).

Thesis 2: The above five conditions are not (all) satisfied, i.e. the laws of nature (the known laws of nature) are valid also in other universes which differ from ours. In order to support thesis 2 we have to show that at least one of the above five conditions is not satisfied.

7.2.1 Condition (1): Completeness

As said in 5.2 a system L of laws about a certain part of reality, in this case about the whole universe U, is complete if every truth[76] about U is derivable from L. Let L be the class of the laws of nature (the known laws of nature). First of all and more trivially L cannot be complete concerning particular contingent truths about U since L does not contain initial conditions. Thus let us ask whether $L+$

---

[76] In order not to run into some logical difficulties (paradoxical situations) I do not permit here a set with the usual logical closure. That means that from a certain specific truth about U (say that proton and neutron have the same spin or that the fine structure constant is 1/273) not every consequence which is allowed by logic is permitted in this set of truths about the world. Since logic allows a lot of redundant truths to be derived from one true sentence like "p or q" as a logical consequence of p (where q is anything whatsoever). Therefore I restrict the logical consequences by some suitable relevance criterion which has been applied successfully to many different areas like explanation and confirmation theory, verisimilitude (theory of approximation to truth) quantum logic and still other areas like epistemic and deontic logic. The criterion eliminates redundancies and permits only most informative and compact consequence-elements. For more information see Weingartner (1993), Weingartner (1994) and Schurz and Weingartner (1987). Cf. note 45. For the general question of the completeness of physical laws in general see Weingartner (1996).

all initial conditions determining particular states of U in the past ('I_p' for short) is complete.

First I think there will be considerable agreement that if "laws of nature" means the laws of nature known so far then these laws are (even together with I_p) very probably not complete in the above sense.

The doubts of Einstein concerning the completeness of the laws of physics known so far are expressed in his EPR-Gedankenexperiment[77] and Prigogine has also doubts concerning completeness because the arrow of time is not incorporated in the most important laws of physics of today.[78]

Despite the incompleteness with respect to laws also I_p is incomplete. This is so since there is an important subset of I_p, namely the set of initial conditions determined at the moment of the beginning of the universe which is unknown to great extend and very difficult to approach scientifically (cf. condition (4) below).

Second if we understand by "laws of nature" a complete set by definition then the answer is of course trivial.

Third, since every truth about U includes also truths about the future of U, the answer of this question depends very much on the kind of the laws. If condition (2) is satisfied i.e. all laws are deterministic, one might think it possible that there can be a complete set $L + I_p$. Since (2) is probably not satisfied the question is open and has probably to be answered in the negative.

7.2.2 Condition (2): Deterministic Laws

As the phenomena of thermodynamics, quantum mechanics, radiation and the new discovered processes of cosmological evolution, of selforganization and of (certain kinds of) chaos suggest condition (2) is not satisfied. Of course this holds under the assumption which I accept here that not all randomness comes ultimately from inherent and hidden deterministic laws. That means that statistical laws in the realistic interpretation, i.e. describing nature, not just degrees of our ignorance, can be genuine laws of nature.[79] But if this is true there are degrees of freedom for the development of the universe in the future. And this means that more than one universe is a model for these laws.

7.2.3 Condition (3): Permutation Invariance

Permutation change, i.e. interchange of elementary particles of the same sort does not change laws but also - according to the usual understanding of a physical system (this system may be the whole universe) - does not change this system. Cf. 7.1(1). That means that elementary particles of the same kind are treated as indistinguishable although numerically different.

The point that permutation change does neither change the laws nor the physical system (the world) has a bearing on the definition of "symmetry group of nature". Because this was defined as the set of all changes which do not

---

[77] Cf. Einstein et al. (1935).

[78] Cf. footnote 66.

[79] Cf. Weingartner (these proceedings).

change laws. But under "nature" we understand usually "our" nature i.e our universe. Therefore in order to make sure that the "symmetry group of nature" selects just our universe one would have to define this symmetry group as the set of all changes which neither change laws nor our universe. But this restriction seems not to reflect adequately the character of a law of nature and so in this case "symmetry group of nature" could not be used to characterize the most important properties of laws of nature.

Philosophically this may seem controversial if by some principle of individuation numerically different particles are viewed as different individuals. If Leibniz's definition of identity (agreement in *all* properties) is taken literally, then such particles are not identical. However taken literally this definition of identity is hardly applicable at all. The identity expressed by physical equations does not always satisfy this definition. For instance if on the right sight there are observable magnitudes (for example masses, lenghts, times) and on the left side not (forces) - like in Newton's Second Law of Motion. That is "for all properties" has to be restricted in a suitable way - for example for all physical magnitudes representable by real numbers. Thus a stronger principle of individuation may not be helpful here. The distinctions and differences go as far as discovered inner properties go; if new such inner properties would be discovered which hold only for a part of the particles belonging to one kind, then new differences will appear. Thus an interchange of two particles (of the same kind) does not lead to a different world (universe). And so thesis 2 cannot be supported with permutation change.[80]

7.2.4 Condition (4):

In order to speak of initial conditions at all a basic assumption taken here is that the distinction beween initial conditions and laws of nature represents at least some true kernel.[81] That means that not everything is ruled by laws. Only if this is true question 7 makes sense.

Can we imagine now that all possible initial conditions (that is all those initial conditions which are compatible with all the laws of nature) are or will be realized during the life time of the universe? I think that this is not very probable. Even if the choice of initial conditions at the beginning of the universe would have been rather small - a question which is hardly decidable - as soon as we take statistical laws seriously (cf. condition (2)) there will be a great number of states which have not been realized because of the degrees of freedom which allow different states by chance. But even if taking just deterministic laws, with asymmetric initial conditions asymmetric effects are produced: this special plane of the orbit of a planet (not that it is a plane which follows from the rotationally

---

[80] For a discussion of the problem of identity of particles see Van Fraassen (1991), chapters 11 and 12.

[81] Cf. the introduction and the passages cited from Wigner. The notion "initial condition" is taken here in a rather wide sense. It includes also socalled boundary conditions but not the fundamental constants of chapter 6 which make up a seperate condition (5).

symmetric laws) is due to initial conditions and the plane could lie in a different angle to the one realized (recall chapter 4).

Slight changes in the constellations of stars (and planets) seem not to violate laws because such changes occur since planetary motion (and probably this is similar with systems of stars) is to some extend chaotic.[82]

Or take charge symmetry-violation. The ratio of the decay rate could be slightly different (due to some change in the initial conditions) from the one observed (cf. chapter 5.4). Take parity: The ratio of the rates of snails having left screw houses to those having right screw houses (or the respective ratio of heart on the left side and on the right side) could be different without affecting biological laws. In radioactive decay phenomena parity violation could be more frequently than observed.

Such examples (which could be continued) suggest that it is highly unprobable that all possible changes of initial conditions will be realized some time in this universe. That means that if not all initial conditions are played through during the life time of this universe then there are also other universes satisfying the laws of nature and having some of those initial conditions which are not realized in our universe.

7.2.5 No Boundary

There is the theory of Hawking which tries to dispense with certain initial and boundary conditions especially with the focus on the initial conditions at the start of the universe: "The boundary condition of the universe is that it has no boundary."[83] Clearly the theory of General Relativity is incomplete with respect to the start of the universe. It does not tell us how the universe began. And that it began or that it has a certain age it tells us only when we add certain parameters (for the expansion, for the density of matter etc.) of which it is very difficult to have exact knowledge. According to a result of Penrose and Hawking: If General Relativity is correct and the universe contains as much matter as we observe, then there must have been a Big Bang singularity. But this first singularity is burdened with a lot of guesswork and ignorance. It is therefore natural (and known from many examples in the history) to choose one of the following two strategies: (1) to throw out the unconvenient entity or (2) to introduce some new entity which could give an explanation for the problematic case(s). Examples for the latter are Plato (ideas), religion (creator), Leibniz (monads), Newton (forces) and Cantor (sets) etc. Examples for the former are strategies of elimination: Hume (causation; in fact he replaced it by a psycholgical entity: habit), Vienna Circle (metaphysics) etc. Hawking - it seems - has chosen strategy (1): Elimination of boundaries and in a sense elimination of "real" time by replacing it by imaginary time. It appears as a consequence of a theory of

---

[82] Cf. Laskar (1994).

[83] Hawking (1988), p. 144. To avoid misunderstandings it should be mentioned that the "no-boundary" approach of Hawking does not mean no boundary at all. For example the geometries needed presuppose certain regularities which are determined by boundary conditions concerning the radius of the 3-sphere and the scalar field.

quantum gravity which uses Feynman's idea of sum over histories (or paths) of particles in space time and - in order to avoid certain difficulties with this idea - introduces imaginary time. In contradistinction to the usual difference between the space coordinates and the time coordinate the distinction between time and space disappears completely if one measures time by using (instead of real) imaginary numbers: τ (imaginary time) $= itc$. Thus $c^2 dt^2$ becomes $-d\tau^2$ and the difference between space and time disappears within ds^2. This suggests that the important description of nature is then a geometrical ("Euclidean") one and that the reason for the difficulties with the first singularity would be due to a wrong or improper concept of time which has to be replaced by the concept of imaginary time as the proper concept of time.

In spite of the general doubts concerning eliminations of inconvenient entities and more specific ones like the somewhat unnatural disappearance of the difference between space and time coordinates (with all its consequences) and the mathematical trick with introducing imaginary time, Hawking's theory will have to be judged by tests in the future.[84] As he says himself:

"I'd like to emphasize that this idea that time and space should be finite without boundary is just a *proposal*: it cannot be deduced from some other principle. Like any other scientific theory, it may initially be put forward for aesthetic or metaphysical reasons, but the real test is whether it makes predictions that agree with observation."[85]

After this critical passage it is the more astonishing that Hawking claims that "the idea that space and time may form a closed surface without boundary also has profound implications for the role of God in the affairs of the universe." And further: "So long as the universe had a beginning [B], we could suppose it had a creator [C]. But if the universe is really completely self-contained, having no boundary or edge [S], it would have neither beginning nor end: it would simply be [¬ B]. What place, then, for a creator?" [C? or ¬C?][86]

As a short comment to that I want to say two things:

(1) If the last passage is supposed to be an argument with the conclusion ¬C it would be fallacious. Since B → C, S → ¬B, S ⊢ ¬C is a logical fallacy. From the premisses B → C, S → ¬B and S one cannot draw any conclusion about C (or ¬ C). The conclusion ¬C would follow if we had instead of B → C C → B; but then - despite the question of the truth of S - this premiss C → B is questionable too because a creator is compatible also with a creation which does not have a certain age or beginning.

---

[84] A strange consequence of an earlier proposal of Hawking (1985) was that the arrow of time would reverse at the maximum of expansion of the universe. However some of his pupils (Lyons, Page, Laflamme) found solutions which avoid that consequence such that the contracting phase has no reversal of time and entropy and irregularities increase in the expanding and contracting phases. Cf. Hawking (1994) and Laflamme (1994). There are also other recent theories of the inflationary universe without a beginning in time which do not use imaginary time. Cf. for example Linde (1990).

[85] Hawking (1988), p. 144.

[86] Ibid. p. 149. The letters in square brackets are mine.

(2) Some philosophers at the end of antiquity tried to attack the idea of God as a first cause by proposing a cyclic universe, self-contained in the sense of a cyclic process (with only internal causation). Of such a process it does not make sense to cut it up somewhere (it would destroy the process) and to call one piece next to the cut the first cause. To this Basilius (3rd century) replied that even if this were the right description of the universe it is always allowed (consistent) to ask for the explanation of the whole, i.e. to ask who instituted this circle.

Also a very strong claim is made at the end of the book: "However, if we do discover a complete theory ... then we shall ... be able to take part in the discussion of the question of why it is that we and the universe exist. If we find the answer to that, it would be the ultimate triumph of human reason - for then we would know the mind of God."[87]

This claim is entirely different from the one cited above. In fact both claims are even somewhat inconsistent in the sense that the latter presupposes a God (with mind) whereas the former does not only not presuppose one but suggests not to have a "place" for him in a universe without boundaries (forgetting the fact that in the main religions God is transcendent with respect to the universe). In spite of the above inconsistency: What if some Christian philosophers were right in saying that since the creation (the universe) is an action of God's free will and not a necesary outcome of his essence, knowing the creation (the universe) does not mean knowing his essence, which is impossible in this life (even if revealed texts accessible to religious belief tell us some aspects of his essence by analogy). Knowledge of the universe could mean then knowing him as a most powerful thinker and cause but would not reveal or exhaust the structure of his mind.

7.2.6 Condition (5): Fundamental Constants

Important questions about the fundamental constants have been stated already in chapter 6. The one which is relevant here is whether these constants are all ruled and interconnected by laws. Can we change one of them without changing laws? This is connected with the question whether all of these constants are really constant or whether they change (slightly) with time (with the evolution of the universe, recall ch. 6.2). If Dirac would be right with his Large Numbers Hypothesis or in general if the fundamental constants of nature change with time then one can imagine that by some change of initial conditions (say the amount of energy of the universe is higher) the change could be faster or slowlier. In this case again the laws would be satified by other universes which differ from ours just by the speed of change of those constants. From all what we know so far however it seems very unlikely that these fundamental constants are in some sense independent from the laws because they enter at least the fundamental laws. In this case they do not allow different universes from ours to satisfy the laws of nature.

Summarizing the discussion of the 5 conditions stated above I think it is more likely that the laws of nature are valid also in other universes which differ

---

[87] Ibid. p. 185

from ours. They may differ first of all because the laws of nature known to us so far are not complete and so allow to be satisfied of more than one universe. But even if we take a set of laws of nature which is together with the set I_p of initial conditions complete the falsity of conditions (2) and (4) allow universes different from ours to satisfy the laws.

References

Alpher, R. A., Herman, R. C. (1949): Remarks on the Evolution of the Expanding Universe. Physical Review 75, 1089–1095

Aristotle (Met): Metaphysics. In *The Complete Works of Aristotle*,The revised Oxford Translation, Vol. 2, Barnes J. (ed.), Princeton University Press, Princeton 1985, 1552–1728

Aristotle (Phys): Physics. In *The Complete Works of Aristotle*, The revised Oxford Translation, Vol. 1, Barnes J. (ed.), Princeton University Press, Princeton 1985, 315–446

Aristotle (Heav): On the Heavens. In *The Complete Works of Aristotle*, The revised Oxford Translation, Vol. 1, Barnes J. (ed.), Princeton University Press, Princeton 1985, 447–511

Benedicks, M., Carleson, L. (1991): The Dynamics of the Hénon Map. *Annals of Mathematics* **133**, 73–169

Berry, M. (1978): Regular and Irregular Motion. In: Jorna, S. (ed.) *Topics in Nonlinear Dynamics*, Amer. Inst. of Physics (New York) 16–120

Berry, M. V., Percival, I. C. - Weiss, N.O. (eds.) (1987): *Dynamical Chaos*, Proceedings of the Royal Society of London. A. Mathematical and Physical Sciences 413, 1844

Boltzmann, L. (1896): Entgegnung auf die wärmetheoretischen Betrachtungen des Herrn E. Zermelo. In: Boltzmann, L. *Wissenschaftliche Abhandlungen, Vol. III*, §119

Boltzmann, L. (1897a): Zu Herrn Zermelos Abhandlung 'Über die mechanische Erklärung irreversibler Vorgänge'. In: Boltzmann, L. *Wissenschaftliche Abhandlungen, Vol. III*, §120

Boltzmann, L. (1897b): Über einen mechanischen Satz Poincaré's. In: Boltzmann, L. *Wissenschaftliche Abhandlungen, Vol. III*, §121

Boltzmann, L. (18 - 1905): *Wissenschaftliche Abhandlungen, Vols. I-III*, F. Hasenöhrl (ed.), Leipzig 1909. Reprinted by Chelsea Publ. Comp., New York 1968

Brudno, A.A. (1983): Entropy and the Complexity of the Trajectories of a Dynamical System. Transactions of the Moscow Mathematical Society 2, 127–151

Casati, G., Chirikov, B. V. (1994): The Legacy of Chaos in Quantum Mechanics. In: Casati, G., Chirikov, B.V. (eds.) *Quantum Chaos Between Order and Disorder* (Cambridge University Press)

Chirikov, B. V. (1979): A Universal Instability of Many-Dimensional Oszillator Systems. Phys. Rep. **52**, 463

Chirikov, B. V. (1991a): Time-Dependent Quantum Systems In: Giannoni, M.J., Voros, A., Zinn-Justin, J. (eds.) *Chaos and Quantum Physics (Les Houches 1989)*, Elsevier, Amsterdam, 443–545

Chirikov, B. V. (1991b): Patterns in Chaos. Chaos Solitons and Fractals 1(1), 79–103

Chirikov, B. V. (1992): Linear Chaos. In: *Springer Proceedings in Physics* 67, 3–13

Chirikov, B. V. (these proceedings): Natural Laws and Human Prediction.

Damour, T. et al. (1988): Limits on the variability of G using binary-pulsar data. Phys. Rev. Lett. **61**, 1151ff

Dirac, P. A. M. (1937): Cosmological Constants. Nature **139**, 323ff

Dirac, P. A. M. (1972): The Fundamental Constants and Their Time Variation. In Salam and Wigner (1972), 213-236

Dirac, P. A. M. (1973): Fundamental Constants and Their Development in Time. In: Mehra, J. (ed.). *The Physicist's Conception of Nature*, Reidel, Dordrecht, 45-54

Dyson, F. J. (1972): The Fundamental Constants and Their Time Variation In Salam and Wigner (1972), 213-236

Einstein, A., Podolski, B., Rosen, N. (1935): Can Quantum-mechanical Description of Physical Reality be Considered Complete. Physical Review **47** 777-780. Reprinted in: Wheeler, J.A. - Zurek, W.H. (eds.) *Quantum Theory and Measurement*, Princeton University Press (1983)

Farmer, J. D. (1982): Information Dimension and the Probabilistic Structure of Chaos. Zeitschrift für Naturforschung 37a, p. 1304

Feynman, R. P. (1967): *The Character of Physical Law* (MIT Press)

Galileo Galilei (DNS): *Dialogues Concerning Two New Sciences*, transl. by H. Crew and A. de Salvio (Dover, New York)

Galileo Galilei (DWS): *Dialogue Concerning the Two Chief World Systems - Ptolemaic & Copernican*, transl. St. Drake (University of California Press, Berkeley)

Galton, F. (1889): *Natural Inheritance* (London)

Gell-Mann, M., Hartle, J.B. (1994): Time Symmetry and Asymmetry in Quantum Mechanics and Quantum Cosmology. In: Halliwell, J.J. (et al.), 311-345

Genz, H., Decker, R. (1991): *Symmetrie und Symmetriebrechung in der Physik* (Vieweg, Braunschweig)

Großmann, S. (1989): Selbstähnlichkeit. Das Strukturgesetz im und vor dem Chaos. In: Gerok, W. (ed.) *Ordnung und Chaos* (Hirzel, Stuttgart), 101-122

Haken, H. (1982): *Synergetics. An Introduction* (Springer, Berlin)

Haken, H. (1984): *Advanced Synergetics* (Springer, Berlin)

Halliwell, J.J., Perez-Mercader, J., Zurek, W.H. (eds.) (1994): *Origins of Time Asymmetry* (Cambridge University Press)

Halliwell, J.J. (1994): Quantum Cosmology and Time Asymmetry. In Halliwell et al. (1994), 369-389

Hawking, St. (1985): The Arrow of Time in Cosmology. Physical Review D **32**, 2489

Hawking, St. (1988): *A Brief History of Time* (London, Bantam)

Hawking, St. (1994): The No Boundary Condition and the Arrow of Time. In Halliwell et al. (1994), 346-357

Hénon, M. (1976): A Two Dimensional Map with a Strange Attractor. Commun. Math. Phys. **50**, 69-77

Holt, D.L., Holt, R.G. (1993): Regularity in Nonlinear Dynamical Systems. British Journal for the Philosophy of Science **44**, 711-727

Irvine, J.M. (1983): The Constancy of the Laws of Physics in the Light of Prehistoric Nuclear Reactors. Contemporary Physics **24**(5), 427-437

Kuhn, T. (1978): *Black-Body Theroy and the Quantum Theory 1894-1912* (Oxford)

Laflamme, R. (1994): The Arrow of Time and the No Boundary Proposal. In Halliwell et al. (1994), 358-368

Laplace, P.S. (1814): *Essai philosophique sur les probabilités* (Paris, Courcier)

Laskar, J. (1994): Large-Scale Chaos in the Solar System. Astron.Astrophys. **287** L9

Lee, T.D. (1988): *Symmetries, Asymmetries and the world of Particles* (Univ. of Washington Press, Seattle)

Lighthill, J. (1986): The Recently Recognized Failure of Predictability in Newtonian Dynamics. Proceedings of the Royal Society London A **407**, 35–50

Linde, A. (1990): *Inflation and Quantum Cosmology* (Academic Press)

Lorenz, E.N. (1963): Deterministic Nonperiodic Flow. *J. Atmos. Sci.* **20**, 130ff

Mandelbrot, B.B. (1982): *The Fractual Geometry of Nature* (Freeman, San Francisco)

Maxwell, J. C. (MaM): *Matter and Motion* (Dover, New York 1992)

Meessen, A. (1989): Is It Logical Possible to Generalize Physics through Space-Time Quantization? In: Weingartner, P., Schurz, G. (eds.) *Philosophy of the Natural Sciences. Proceedings of the 13th Int. Wittgenstein Symposium* Vienna, Hölder-Pichler-Tempsky, 19–47

Miles, J. (1984): Resonant Motion of a Spherical Pendulum. Physica **11D**, 309–323

Moser, J. (1978): Is the Solar System Stable? Mathematical Intelligencer **1**, 65–71.

Norman, E.B. (1986): Are Fundamental Constants Really Constant? Amer. J. of Physics **54**, 317ff

Particle Data Group (1990): Review of Particle Properties. Physics Letters **B 204**, 1ff

Peitgen, H.-O., Richter, P.H. (1986): *The Beauty of Fractals* (Springer, Berlin)

Peirce, Ch.S. (1935, 1960): *Collected Papers of Ch.S. Peirce* ed. by Ch. Hartshorne and P. Weiss (Harvard U.P., Cambridge - Mass)

Penzias, A.A., Wilson, R.W. (1965): A Measurement of Excess Antenna Temperature at 4080 Mc/s. Astrophysical Journal **142**, 419–421

Pesin, Ya.B. (1977): Characteristic Lyapunov Exponents and Smooth Ergodic Theory. Russian Math. Surveys **32**(4), 55–114

Poincaré, H. (1892): *Les Méthodes Nouvelles de la Méchanique Céleste* Vol. 1 (Paris)

Popper, K.R. (1956): The Arrow of Time. Nature **177**, 538

Popper, K.R. (1959): *The Logic of Scientific Discovery* (London, Hutchinson)

Popper, K.R. (1965): Of Clouds and Clocks. In Popper (1972), 206–255. Originally published as the second A.H. Compton Memorial lecture (Washington)

Popper, K.R. (1972): *Objective Knowledge* (Oxford)

Prigogine, I. (1993): Time, Dynamics and Chaos: Integrating Poincaré's 'Non-Integrable Systems. In: Holte, J. (ed.) *Nobel Conference XXVI. Chaos: The New Science* (London, Univ. Press of America)

Prigogine, I., Stengers, I. (1993): *Das Paradox der Zeit. Zeit, Chaos und Quanten* (Munich, Piper)

Ruelle, D. (1980): Strange Attractors. Mathematical Intelligencer **2**, 126–137

Ruelle, D. (1990): Deterministic Chaos: The Science and the Fiction. Proceedings of the Royal Society London **A 427**, 241–248

Salam, A., Wigner, E.P., Eds., (1972): *Aspects of Quantum Theory* (Cambridge University Press, Cambridge)

Schuster, H.G. (1989): *Deterministic Chaos* (VCH, Weinheim)

Schurz, G., Weingartner, P. (1987): Verisimilitude Defined by Relevant Consequence Elements. In: Kuipers, Th. (ed.) *What is Closer-to-the-Truth?* (Amsterdam, Rodopi), 47–77

Shannon, C.E., Weaver, W. (1949): *The Mathematical Theory of Information* (Univ. of Illinois Press, Urbana)

Thomas Aquinas (STh): *Summa Theologica*, transl. by Fathers of the English Dominican Province (Christian Classics, Westminster, Maryland 1981)

Van Fraassen, B. (1991): *Quantum Mechanics. An Empiricist View* (Oxford)

Weierstrass, K. (1993): *Briefwechsel zwischen Karl Weierstrass und Sofja Kowalewskaya*, ed. R. Bölling (Akademie Verlag, Berlin)

Weinberg, St. (1977): *The First Three Minutes* (London, Deutsch)

Weinberg, St. (1987): Towards the Final Laws of Physics. In: *Elementary Particles and the Laws of Physics. The 1986 Dirac Memorial Lectures*, (Cambridge), 61–110

Weingartner, P. (1993): A Logic for QM Based on Classical Logic. In: de la Luiz Garcia Alonso, M., Moutsopoulos, E., Seel, G. (eds.) *L'art, la science et la métaphysique. Essays in honour of A. Mercier* (Bern, P. Lang), 439–458

Weingartner, P. (1994): Can there be Reasons for Putting Limitations on Classical Logic? In: Humphreys, P. (ed.) *Patrick Suppes: Scientific Philosopher Vol. 3* (Dordrecht, Kluwer), 89–124

Weingartner, P. (1995): Are Statistical Laws Genuine Laws? A concern of Poincaré and Boltzmann. Grazer Philosophische Studien (to appear)

Weingartner, P. (1996): Can the Laws of Nature (of Physics) be Complete? in: Dalla Chiara, M.L. et al. (eds.) *Proceedings of the 10th International Congress of Logic, Methodology and Philosophy of Science*, North Holland Publ. Comp., Amsterdam.

Wess, J. (1989): The CPT-Theorem and its Significance for Fundamental Physics. Nuclear Physics B (Proc. Suppl.) **8**, 461ff

Wigner, E. P. (1967): *Symmetries and Reflections. Scientific Essays of Eugene P. Wigner* ed. by Moore, W.J. and Scriven, M. (Indiana Univ. Press, Bloomington)

Wigner, E.P. (1972): On the Time-Energy Uncertainty Relation. In: Salam, A., Wigner, E.P., Eds., 237–247

Wunderlin, A. (these proceedings): On the Foundations of Synergetics

Zermelo, E. (1896a): Über einen Satz der Dynamik und die mechanische Wärmetheorie. Wied. Annalen **57**, 485ff

Zermelo, E. (1896b): Über die mechanische Erklärung irreversibler Vorgänge. Wied. Annalen **59**, 793ff

Contingency and Fundamental Laws
Comment on Paul Weingartner
"Under What Transformations
Are Laws Invariant?"

M. Stöckler

Universität Bremen, Germany

In the beginning of his paper Paul Weingartner reminds us of a central idea of the Greek conception of science. He refers to the distinction between - universal laws, which do not change and are necessary in the sense that they are universal patterns in such worlds we want to take into account, and - particular events, states of the world which are contingent in the sense that they could be different without changing the laws. Laws of nature determine the evolution of processes in time if some initial conditions are given. Since most of the events can be explained in this way, only the initial conditions are contingent in a narrower sense. All other events could not be different given the initial conditions and the validity of the laws. Weingartner discusses the problem of necessity and contingency in the world using the distinction between laws and initial conditions. This approach fits well to the logical form of explanations as they are reconstructed in modern philosophy of science.

Weingartner focuses on the relation between laws and initial conditions mainly in the light of new discoveries of physics. At the end of his stimulating paper he investigates arguments which make it plausible that not everything in the world is necessary. That means we can think about changes in the states of the world which would not change the laws. Our world is not the only model of the laws of nature, the laws of nature are valid also in other universes which differ from ours. The thesis that there are other universes satisfying the laws of nature and having some of those initial conditions which are not realized in our universe (7.2.4) should not be understood as maintaining the material existence of such worlds, as it is postulated by the many-worlds interpretation of quantum mechanics (in order to avoid the reduction of the wave function) or by some astrophysicists (in order to explain the fine tuning of constants and initial conditions in cosmology). The other universes exist as our fictions or, more objectively, as solutions of the equations describing the laws of nature. I do agree with the conclusion of Weingartner developed in 7.2. Nevertheless it could be profitable to think about slightly different ways to arrive at the same conclusion. So, I am going to comment on the five conditions which are valid if it is not possible to change our world without changing the laws or, for short, if we live in a Leibnizian world.

Condition (1): The criterion of completeness can be read in different ways. I am not so happy with epistemic variants of incompleteness. It might be the case that we never shall know the complete set of laws and the special initial conditions of the universe. But despite this lack of knowledge it could be possible that every change of our world would include a change of the laws. And the other way round, we could know the initial conditions of the universe very well (for example by relying on some Anthropic Arguments) without being able to explain them and without having arguments that they could not be different. By the way, it seems to me (and to many other people) that quantum mechanics is not incomplete in the sense explicated by A. Einstein. In any case, this form of incompleteness is slightly different from the notion of incompleteness used by Weingartner. I agree that indeterministic laws make science incomplete in the sense explicated in 7.2.1. Leaving this aspect to condition (3), I would say that, except for indeterminism, the present fundamental theories (quantum field theory and relativity) seem to be complete. I think that most scientists would agree that there are no special arguments for thesis 2 of 7.2 which refer to incompleteness and which are not cases of indeterminism (condition (2)) or of the choice of initial conditions (condition (5)) at the same time.

Condition (2): Weingartner explicitly states that epistemic limits to predictability do not give reasons for the suggestion that condition (2) is not satisfied. But then only the measurement process in quantum mechanics is an example of an indeterministic process. In a different way of speaking, we could say that the dynamic law (Schroedinger's equation) is deterministic but incomplete in the sense of condition (1). I propose to weaken condition (2), which should be restricted to fundamental laws. Even if the world is deterministic there could be some laws on higher levels which are indeterministic because of lack of knowledge. A complete set of deterministic fundamental laws of nature could be valid just in our world but nevertheless we could have some special laws which are indeterministic for epistemic reasons. At the end of my comment I will come back to the concept of 'fundamental law'.

Condition (3): This condition requires that an interchange of elementary particles of the same kind does not change the world. The idea of the argument seems to be the following: if the exchange of the place of two individuals in the world would lead to a different state of the world, we could create a twin universe in this way which satisfies the same laws as our universe. Consider the class of theories which allow different distributions of the individuals to different states like the distribution of similar horses to different starting places in a race. In all these theories, a special choice of the distribution is a part of the initial conditions. So far the argument of Weingartner is correct and a special case of condition (4). On the other side, the permutation invariance of quantum field theory is a rather tricky problem. I prefer the following description of the situation: The permutations in quantum field theory refer to the number of some special basic states. So the permutation invariance requires that a special convential way of bookkeeping should not influence the observables of the theory. If we follow quantum field theory, the elementary particles are not individuals like the individual described by constants in predicate logic. I prefer to represent them as excitation of some basic field. In this view particles are special states of the world and not individuals. The permutation invariance refers to a transformation which does not change the world but which is a redescription of the same situation. For finding out whether the laws of nature are satisfied just in our universe, we should restrict ourselves to real changes of the world and skip all transformations which just change the form of the description (like transformations of the coordinate system)[1]. Of course, condition (3) can be accepted, if the permutation invariance is not understood in the technical sense of quantum field theory. Nevertheless, I would prefer to discuss this general form of permutation invariance as a special case of choosing initial conditions (cf. condition (4)).

Condition (4): This condition requires that all sets of initial conditions compatible with the laws of nature occur (are played through) as states during the life time of a Leibnizian world. But this formulation could be misunderstood. Consider Kepler's laws of planetary motion. We could imagine a huge set of initial conditions which are compatible with Kepler's laws. But some of the initial conditions (or even a considerable part of the whole set) could be ruled out, for example by other laws describing the origin of planetary systems. So even in a Leibnizian world there could be some initial conditions of a special law which are not realized because they are forbidden by other laws. In any case condition (4) has to be restricted to such initial conditions which are compatible with all the laws in the world. However, such a condition is not very useful because it is very hard to judge whether a special initial condition is contingently not realized (but could be realized in another universe) or is forbidden by some removed law. It sounds a little bit strange that all sets of initial conditions compatible with the laws of nature must occur as states during the life time of a Leibnizian world. In some sense that is too weak. We could have many universes which all play through all possible sets of initial conditions but which do so in different orders. I agree that condition (4) is necessary for a Leibnizian world. But the interesting point about the contingency of the initial conditions is that two universes differ in the initial conditions at the same instant. Universes having the same set of events but differing in their time order are different too.

Condition (5): I agree with this condition. But again, the choice of the fundamental constants, as far as they are not part of fundamental laws, is an element of the set of initial conditions. Nevertheless, condition (5) is necessary for a Leibnizian world.

Conclusion: It seems to me that, first of all, a Leibnizian world has to be deterministic and to exclude the existence of genuine initial conditions. If the fundamental laws are indeterministic there are different universes which coincide at some starting point. If there are genuine initial conditions there are universes which differ at the starting point (especially the deterministic ones) or branch at some instant (if the laws are indeterministic).

I propose to keep (2) and (4) as necessary conditions for a Leibniz world: (1') The laws are deterministic. (2') There are not any genuine initial conditions. I suppose that these two conditions together are even sufficient.

Finally, I have to add some remarks on fundamental laws. Quite different

---

[1] Cf. Redhead (1975).

things are called 'laws' in natural science. Some of them can be deduced from other, more fundamental laws if some special conditions are specified (for example Kepler's laws can be deduced approximatively from Newton's law of gravitation). Very often the derived laws have less symmetries than the fundamental laws. That is the reason why the symmetry group of the world should be restricted to the fundamental laws (like that of quantum field theory and relativity). In a sense, such derived laws are necessary because they can be explained. In another sense, they are contingent because they are valid only under special conditions. In the order of explanation there are some laws which cannot be explained by other laws and which I called fundamental laws. They are contingent in a special way because their is no explanation why we have these laws and not some others. We have also different sorts of initial conditions. Some of them are genuinely contingent: they cannot be explained. Other initial conditions (for example the orbit of the earth) are consequences of former processes. In a deterministic world, at least in principle, they could be explained by genuine initial conditions and other laws. That is the reason why I proposed to discuss the problem whether we live in a Leibnizian world restricting ourselves to genuine initial conditions and fundamental laws. In this way one gets into some trouble with the idea that initial conditions are contingent and laws are necessary. I don't see a way of solving this problem without a clear conception of 'law' and unfortunately I don't see any people having a convincing conception of 'law of nature'.

References

Redhead, M. L. G. (1975): Symmetry in Intertheory Relations. Synthese **32**, 77–112

Discussion of Paul Weingartner's Paper

Chirikov, Miller, Noyes, Schurz, Suppes, Weingartner, Wunderlin

Wunderlin: I want to ask you why you spoke only of symmetries. For many examples you gave "symmetry-breaking" would be more appropriate I think.

Weingartner: I think it depends very much on the level of generality whether you speak of symmetry or of symmetry breaking. Most houses of snails are like right hand screws, very few turn the other way. If the law is that all are right hand screws we may speak of a symmetry breaking in very few exceptional cases. But the more plausible way to see it is that there is a hidden symmetry: The law is symmetric in respect to the right or left spirale and todays preference for right was caused by (asymmetrical) initial conditions during the evolution. A less relative example is the Avogadro-number which breaks scale-symmetry. But take a gas on the other hand. Macroscopically the gas is a good example of symmetry. But imagine we have one film picture (snapshot) of the state (positions of all molecules) at a certain point of time. This picture is of course neither rotationally symmetric nor (with respect to parts of it) translationally symmetric. But for the physicist the symmetry obtained by averaging over positions in the course of time is usually more interesting than these asymmetries.

Suppes: I'd like to ask you a question about invariance and the problem you mentioned about the initial conditions, the boundary conditions. Of course I agree with your remarks but it also seems to me that we can think of the matter this way. We simply have the notion of models of a theory and the concept of the group of transformations that leave models of a theory invariant. The requirement on the group of transformations is that it carries any model of the theory into another model. You may want to require something else, something more restrictive. So, for example, the way the Galilean transformations carry any mechanical model into another mechanical model, but even more, carry particles with inertial paths into particles with inertial paths. And this gets around the problem of invariance of worrying about the initial conditions because the initial conditions are given only implicitly in the particular model. Each model will have some implicit boundary conditions. I only mentioned this because it's a way of characterizing the group of transformations, because they leave - as we would say intuitively - the theory invariant, which means more precisely carrying models into models.

Weingartner: When you say "that gets around the problem of worrying about the initial conditions because they are given in the particular model" then this doesn't seem to me to be a solution of the problem how to divide the "world" into contingent initial conditions and symmetric and invariant laws (Wigner's problem). Putting the initial conditons into the model makes fixed and hidden assumptions out of them. But the very question is which of those assumptions can be changed without changing the laws. There are of course relatively easy cases where your idea does not make problems. For example in the case of inertial systems as models we know what can be changed without changing laws. But in case of fundamental laws of QM and Relativity (or the fundamental constants respectively) the very problem is whether this existing universe is the only model. In this case all "initial conditions" would be hidden and determined by laws. I argued that this is not so because I accept statistical laws as genuine laws and because - assuming that the world is finite in time - it would be not time enough to play through all possible initial conditions. But the question is a difficult open problem.

Miller: Can I follow up that what Suppes said. I was surprised by his response to the question of your title because I did not understand that question as an empirical one particularly. I did not imagine that you were asking for the transformations under which our present laws are invariant, but something more philosophical and general, namely can we characterize laws in general, not knowing what they are, by requiring them to be invariant under certain groups of transformations? It is surely not an answer to say: if it is classical mechanics then Galilean invariance characterizes the laws, and if it is relativistic mechanics then Lorentz invariance characterizes them. That sounds to me like a question of physics, not of philosophy.

Weingartner: The distinction, I think, which is important here is whether I interpret my question in a normative sense or in a factual sense and I am not against the view that such questions can be interpreted in a normative sense. But my view is that if you think so then we have first to look at the type of the symmetry and we have first to compare what symmetry we have. One has certainly to take into account, the status of generality of the symmetry and its main consequences, and also the presuppositions of these kinds of symmetries in order to get a more suitable "normative" concept of what a law is. But I am not too quickly proposing any normative concept here because I would fear it will become too idealistic. Then it's too far away from being (at least partially) "discovered" by science and to some extend immune with respect to criticism coming from new experimental results.

Miller: I agree that somebody who insisted that our theories of the heavens must be spherically symmetric with respect to the centre of the earth would perhaps be regarded as obscurantist with regard to the development of celestial mechanics. But would you be prepared for the eventual conclusion that there aren't any laws of nature. Is that a possibility?

Weingartner: I would at least say that this is very unprobable. And to the other question, I mean to the question of Galileo, I see it in a much more relative way. I mean the main point is this: It is just astonishing that no one, neither the church nor Galileo had the idea that this (the point of reference of movement) could be relative. This is the more astonishing since Galileo understood already the invariance with respect to inertial reference frames as his "Gedankenexperimente" on the ship show (in his dialogue concerning two new World Systems). The earth as a frame of reference is the most practical arrangement for most of our practical purposes. Would you make a time table of a train, by relating it to the sun?

Miller: Not to the earth either!

Weingartner: You see, even still a more concrete point of reference (middle European time area) we need here. I only want to point out this fact because it is something often forgotten. Most of the practical things, concerning which we need measurement must be related to the earth as a point of reference. Of course if you make astronomical investigations other points of reference prove to be more suitable, the sun, the milky way or the cosmic background radiation. In fact the Greeks were much closer to the idea that there is no absolute point of reference. And also the great philosophers of the middle ages, partially because of religious reasons. For only God is absolute, and nothing can be absolute in this universe. There cannot be an absolute space or time since space is only - already according to Aristotle - made up of the distances between material bodies and time is bound to moving bodies, since time is the measure of moving things in respect to earlier and later. Thus space and time are bound to the material moving bodies. These views hold also for Aquinas who writes in detail about it in his questio 46 of his Summa Theologica.

Miller: There is some importance given in part of your talk to the distinction between laws and boundary conditions or initial conditions. Some people would like everything to be lawlike, with no boundary conditions, and there are others who would want to take the opposite view and say that in a sense everything is boundary conditions, though there may be local regularities.

Suppes: Of course, I want to make a comment on that since you're so anti these standards. For example, let's take the most interesting, perhaps the most standard philosophical example of these transformations, namely an old result of Tarski's that you can of course characterize if you have a collection of models of objects, structures of any set theoretical type. You can characterize standard logic by the statements, i.e., logical laws, that remain invariant under 1 - 1 transformations. Now, let us take up your view. If you take the extreme view of course, you would not even accept that. But a sceptical empiricist might still accept that there were logical laws in that sense. These are invariant. But that is as far as you can get with the general proofs that depend on 1 - 1 transformations.

Schurz: There is an argument of Charles Sanders Peirce. It concerns the justification of induction. It says that there must always be some statistical uniformities, some statistical laws in the universe. For, even if the statistical distribution of the possible states of the universe were completely symmetric - if there were, in other words, no statistical dependencies - this would also be some kind of statistical law. Peirce concludes that there must be always some uniformity in the universe and hence, the inductive method is justified. - I am not sure whether Peirce's argument is correct. But at least it is extremely difficult to imagine how a completely irregular universe could look like. There must be arbitrary fluctuations all time ...

Miller: I guess that it would be a universe where the frequencies do not converge.

Schurz: But assume the universe is finite in time. Then trivially every kind of event has a limiting frequency.

Noyes: With regard to the relativity of rotational motion at the time of Galileo, the issue was raised by the Tychonic system in which the planets rotated around the sun but the sun rotated around the stationary earth. For circular orbits, the two systems are kinematically indistinguishable. And the Jesuit astronomers were willing to sacrifice Ptolemy to this geometrical monster in order to avoid conflict with the biblical passages that require a stationary earth. According to de Santillana's analysis in *The Crime of Galileo*, Galileo was genuinely afraid that this compromise would temporarily succeed and hence prevent the application of physical reasoning to the heavens. He therefore refused to even mention the Tychonic system in his *Dialogue Concerning the Two World Systems*. We now know from the analysis of his observational notes by Stillman Drake in *Galileo, Pioneer Scientist*, Toronto, 1990, p. 153, that even at the geometric-kinematic level it is possible to *prove* from his observational records of the eclipses of the satellites of Jupiter that *both* the earth and Jupiter go around the sun. But this was after the edict of 1616 which prohibited him from publishing anything that was not hypothetical on this subject. This could, literally, have lighted the faggots at his feet if he had made it public! He came close enough to that fate as it was. To quote Drake: "As was said at the beginning of the chapter, Galileo wrote of satellite eclipses only in an appendix to his sunspot letters. There alone did he ever assert unequivocally that motion of the earth must be taken account of in astronomy. After 1616, no Catholic was permitted to make such an assertion; hence it is no wonder that one crucial pro-Copernican argument was not included in Galileo's famous *Dialogue* of 1632."

Weingartner: I agree on that. My point was that the idea of the relativity of the point of reference didn't occur clearly to Galileo although he understood it with respect to experiments on the ship. And may be if he would have been acquainted better with the philosophical tradition in this respect (he criticized correctly Aristotle in a number of other points) it could have helped him. Like Mach's challenging ideas in his history of mechanics were important for the basic considerations of Einstein concerning his theory of General Relativity.

Noyes: I want to make a point about Einstein's opinion on Mach's Principle, which I learned of c. 1952 thanks to correspondence with Bergmann, and through him with Einstein. Unfortunately the original is lost, but both Bergmann and Grünbaum saw the original and agree with this paraphrase of Einstein's remarks: "As you know the general theory is a field theory defined by differential equations, and any such theory must be supplied with boundary conditions. In the early days it was believed that the only solutions of the field equations far from gravitating matter were believed to be the flat space of special relativity, or an overall cosmological curvature; the uniqueness of these boundary conditions was believed to meet this problem. Since the discovery (Gödel, Taub) of solutions of the field equations with non-vanishing curvature everywhere in the absence of gravitating matter, this argument from uniqueness no longer applies. In a sense this is a violation of Mach's principle. But now that we have come to believe that space is no less real than matter, Mach's principle has lost its force." So the question of Mach's Principle and whether the rotation is given once or twice,

I would say, is still important and Einstein rejected it. I think many theorists would like to see Mach's principle and its global implications destroyed; because of the evidence for non-locality in quantum mechanics I think this might be a mistake.

Weingartner: I agree that this was a development of Einstein's thoughts later. But Einstein himself says (in his biography in the Schilpp Volume) that Mach's proposal does not fit in a field theory but he points out at length that Mach's criticism was sound and important.

Noyes: Oh yes, sure.

Miller: I should like to come back to the principle that similar causes lead to similar effects. Part of the problem is that what counts as similar to what is usually quite unclear. When are two initial states similar? If similarity is indicated by numerical distance, then the principle fails in typical chaotic processes, such as the continued multiplication of a number by 10 and removal of the integral part. Here rational starting points lead to periodic activity, while irrational do not. Still, we could easily think that rational numbers are more similar to each other than irrational numbers, however close.

Weingartner: I agree of course that this makes a difference. And moreover there are degrees of irrationality which play an important role here. Non-chaotic, stable and predictable motion with integrability of the system has a rational rotation number (i.e. motion frequency through perturbation). However already in the case of KAM integrability where the system survives weak perturbation but is not yet fully chaotic the rotation number has to be sufficiently irrational.

Miller: Is the principle that similar causes lead to similar effects meant to be testable?

Weingartner: Of course it's meant to be testable. Already Maxwell pointed out that in many cases it's satisfied but there are also cases in which it is violated and he gives the counterexample of the railway. But today we can add all cases of chaotic motion as counterexamples because a positive Lyapunov exponent measures the exponential separation of adjacent conjugate points (cf. chapter 2.2 of my paper).

Suppes: David's various questions call to my mind a distinction. For example when we talk about causes, we think immediately of two or more than two. One is when we are observing and testing by empirical observation, another concerns models of the theory. These two are often mixed up. We have some theoretical conceptions defined in classical mechanics and so we talk about causes and we ask: do similar causes produce similar effects? And we are referring now not to things we observe but to the models of the theory of mechanics. Then we go on to a notion of similarity in this model-theoretic sense. On the other hand, we may believe that we have no laws that are absolutely true empirically because we can always discover exceptions. But we still have this notion of law relative to a certain model or theory. And that gives us a different sense. There we have a very natural environment to study invariance. We must distinguish between models of a theory in the ordinary model-theoretic sense and the empirical results of experiments testing the theory.

Chirikov: My remark is very general, not related to chaos only. I think that the primary law is a fundamental equation, the motion equations or whatever related to the model as general as possible. And then you can derive the specific phenomenon like chaos which is a secondary law. But it would not help you to understand the fundamental law unless you find some very sharp contradiction in the results.

Weingartner: I agree, Newton thought that his laws are completely invariant in respect to changing the distances of the planets. Kepler thought that this is not so because these distances express very special mathematical proportions which are responsible for the harmony ("Weltharmonie") of the system, for the harmonic and stable motion. Thus if these proportions would be changed irregular (chaotic) motion would be the consequence.

Chirikov: This is an old story, now we know that it is not a fundamental question. You fix the equations of the model, for example the motion equations. Whether these equations are correct or not is not a problem in the studies of chaos.

Weingartner: This is not my claim.

Chirikov: But, for example, what is in this equation the law and what are the initial conditions. This is, I agree, a fundamental question. But when we study, for example the motion stability all these questions are fixed. There is no choice.

Weingartner: We do not know what would happen (with the laws) if we would change fundamental constants like h, c or G.

Chirikov: No, we know but principally. Technically it is difficult. You should distinguish between technical difficulties and the principle.

Weingartner: Of course. Think also of the purely numerical constants like the proportion of electron and proton mass or the fine structure constant.

Chirikov: We know what happens if some constant would have got a different value. But we don't know, for example, whether the gravitational constant is a constant or is time dependent.

Suppes: I've then another question: you know of course the big dispute about geometry between Frege and Hilbert. It reminds me of this discussion about law. Frege, was very firm that the only axioms of geometry are those that are true. Hilbert had a very different attitude in terms of building a theory that has models, it doesn't necessarily mean that the axioms of the theory are true. The axioms are only true in the models of the theory, but not true simpliciter. Now my question. I'm not sure what your position is when you ask about laws. Are you thinking à la Frege or à la Hilbert?

Weingartner: Now, this is a very difficult question. I would like to say first that both views, Frege's and Hilbert's, have been defended for mathematics not for natural science. But the laws I was dealing with were those of nature or of natural science. Within mathematics it seems to me that Hilbert's view fits more to the area which is constructed by mathematicians, he himself said from Cantor's paradise nobody should expel us. Despite the fact that Hilbert required finitistic methods for the metatheory (metamathematics) with which one should prove the consistency of the theory. Frege's view on the other hand fits more to the area of natural numbers and arithmetic to which his main work in mathematics belongs and to elementary geometry. But when applied to natural science the situation is more complicated as you pointed out already. Then the Hilbert view that a law is only true in a model fits it seems to me better to the working scientist who constructs models to mirror reality and changes these contructs if necessary because of new results. But for a realist this picture is not sufficient. He thinks that by revising models and laws because of controlling and testing them constantly with the help of new experimental results one can find fundamental laws which are in fact approximations to the true laws of nature.

Part II

Determinism and Chaos

The Status of Determinism in an Uncontrollable World

D. Miller

University of Warwick, UK

Abstract. The modern theory of non-linear dynamical systems [chaos] has certainly taught us a salutary lesson about unpredictability in classical physics. It is less obvious what it should have taught us about determinism. While many regard deterministic metaphysics as triumphantly vindicated, Popper (1982), for example, sees its only real support pushed unceremoniously aside. It is the burden of this paper to outline this problem (section 1); to challenge both the usual connection between determinism and predictability and the relevance of the argument leading from deterministic chaos to determinism (section 2); and finally to suggest that, for those who see prediction as an essentially probabilistic enterprise, unpredictability may promise an even more severe headache than is currently envisaged (section 3). It is only fair to say in advance, perhaps, that I am an indeterminist. This seems once more to be an unfashionable position, but I hope that it is still a sustainable one.

1   Determinism & Predictability

In *The Open Universe. An Argument for Indeterminism*, which was written in the 1950s, well before the current enthusiasm for chaos began, Popper (1982) distinguished several varieties of determinism, in particular what he called *metaphysical determinism* and *scientific determinism*, together with their opposites, as many indeterminisms. The difference between these doctrines lies partly in their methodological character — metaphysical determinism is metaphysical, and scientific determinism is supposed to be scientific — but also in their content. Metaphysical determinism is the simple doctrine, surely unfalsifiable by empirical methods, that the future of the world is as unalterably fixed as is the past (p.8). Scientific determinism — which will be spelt out in more detail shortly — is a doctrine asserting the unrestricted predictability of the state of the world by scientific or rational methods — that is, by deduction from universal theories and initial conditions (pp.29-40). Popper claimed that metaphysical determinism, though venerable, originating as it does in religious views (p.5), has little to recommend it — it is plainly in conflict with common sense — except that it is a consequence of scientific determinism (pp.8, 27). But scientific determinism, he maintained, can be shown to be false, even in a world constrained by classical physics. It follows that metaphysical determinism, though not refuted, deserves to be rejected. These arguments make no allusion to statistical theories such as quantum mechanics. In this I follow them.

Everyone now agrees that scientific determinism is untenable. But that is almost the limit of the concordance. Earman (1986), p.9, for example, criticizes the very appellation 'scientific determinism', on the grounds that determinism always was, is now, and ever will be a metaphysical doctrine solely about the world, and that questions of predictability are not relevant to its truth or falsity. (We can agree with this at least to the extent of understanding the unqualified terms 'determinism' and 'indeterminism' to stand for the metaphysical doctrines.) Earman nonetheless maintains, on other grounds, that a world governed by classical physics is nothing like as incontestably deterministic as it has traditionally been taken to be. Hunt (1987) too concedes that the connection between the two doctrines is tenuous, but extracts from the humbling of scientific determinism — which he calls epistemic determinism — what looks like a verdict in favour of metaphysical determinism. His argument, shortly stated, is that the predictive failures of classical physics can now be understood and forgiven, and there is no need to suppose that Laplace's demon would share them. As I say, Popper draws very nearly the opposite conclusion. According to him the predictive successes of classical physics — though marvellous human achievements — can now be recognised as special cases, successes at uncharacteristically straightforward prediction tasks (op.cit., section 13, and also Popper (1957), section 27). No conclusions about the world should be drawn from these successes, and metaphysical determinism stands quite unsupported, a doctrine at odds with empirical science, rather than at home with it.

What is not in dispute is that both metaphysical determinism and metaphysical indeterminism are doctrines that are logically and empirically unfalsifiable, and that neither is refuted by the failure of scientific determinism. It is just this that makes them metaphysical, and so difficult to discuss (see also Miller (1995), section 1). As I understand it, Popper's motive in developing the doctrine of scientific determinism was not to banish metaphysical determinism from rational discussion, but to initiate a more pointed discussion, by bringing it within the scope of empirical investigation. The logical and methodological situation may be put like this. Metaphysical determinism is not in its own right a component of empirical science, and empirical evidence cannot properly be cited in its favour. But a metaphysical theory may well be a component (perhaps an idle component, perhaps an essential one) of a theory that is empirically falsifiable. It does not follow from its being metaphysical that determinism cannot be a part of some greater scientific theory. There are numerous consequences of scientific theories that are themselves not empirical — the most obvious (though not the most interesting) examples are tautologies, existential statements, and vague uniformity principles. How is an empiricist to approach such unempirical elements in science? The only decent answer (Popper (1974), p. 1038, Miller (1994), pp. 10f) is that unfalsifiable consequences of scientific theories must be eliminated from science as soon as the theories from which they are derived are eliminated (though, of course, they can be reinstated as consequences of some successor theory), despite their not having themselves been subjected to empirical scrutiny. They receive a kind of courtesy status on entry similar to that bestowed on the families of diplomats. Should a diplomat be ordered home, his family will lose their status, and their right to remain in the country in which they are living, even if not themselves guilty of any undiplomatic activity. So it is for scientifically incorporated metaphysics. And if metaphysical determinism is derivable from some empirical scientific theory, either from classical physics or from some claim that is inspired by classical physics, then it has to be rejected as soon as those claims are refuted. It appears to have been Popper's intention to promote scientific determinism as such an empirical theory. Although it is doubtful that this could succeed — as noted below — it is at least true that scientific determinism, unlike metaphysical determinism, is inconsistent with many real scientific theories. At one remove, that is to say, it is open to empirical investigation.

Following Popper, we may call a theory *prima facie* deterministic if, given a completely precise specification at some instant of all the initial or boundary conditions pertaining to a closed system within the domain of the theory, then for each future instant there is exactly one state description compatible with the theory. Classical mechanics, and other theories of classical physics are apparently *prima facie* deterministic (but see Earman *op.cit.*, Chapter iii). And there can hardly be much doubt that metaphysical determinism, however it is formulated, would follow from a comprehensive *prima facie* deterministic theory. So metaphysical determinism would in such a case be rightly regarded as a genuine metaphysical component of science. But none of the theories of classical physics, nor even the collection of all of them together, implies metaphysical indeterminism, simply because none of them ever actually encompassed all phenomena, or even seriously showed itself likely to be able to do so. There are many aspects of the universe, as we know, that are not explicable in classical terms. The plain fact is that metaphysical determinism was itself never a consequence of classical physics, which is not comprehensive (Popper (1982), p. 38).

To the extent that it is empirically backed, metaphysical determinism has, according to Popper, relied exclusively on the doctrine that he calls scientific determinism. An outline is overdue. Scientific determinism is fundamentally a Laplacean demon constrained to testable size — not, as for Laplace, a superhuman scientist causally detached from the physical universe yet able to predict in exact detail its future evolution (in particular, the longterm prospects for the solar system) — but an idealized scientist, interacting physically with the universe, able to predict future events not necessarily precisely but to arbitrarily severe standards of precision (this limitation is essential if we are to admit, as we must admit, that some equations defy exact solution) — and, of course, accuracy. More carefully formulated, the doctrine of scientific determinism says that any sharply and clearly formulated prediction task concerning a closed system can be carried out to any required degree of accuracy and precision by a combination of available scientific theory and sufficiently precisely stated initial or boundary conditions. Prediction tasks may be one of two types: those that ask for a prediction of the state of the system at some definite future instant, and those that ask for a (longterm or asymptotic) prediction of whether or not the system will ever be in a certain state (*op.cit.*, pp.36f.).

Now since scientific determinism is supposed to be a scientific doctrine, not just another foray into metaphysics, some explanation must be given of what are to count within any given task as 'sufficiently precisely stated initial conditions'. For without this, the failure of any prediction could without embarrassment be blamed on a slackness in the initial evaluations. The obvious way forward is to ask that once the required precision of the task is given, there should be a uniform way of calculating (using our theories) what degree of precision would be required in the initial conditions. More strongly, and more satisfactorily from the point of view of transforming scientific determinism into a scientific theory, we should require that what can be calculated is the degree of precision required of the measurements that we might have to make in order to be able to evaluate the initial conditions. Popper calls this requirement on measurements the strong form of *the principle of accountability* (Popper (1982), p. 13). Note further that if scientific determinism is to be an empirical thesis it by no means suffices for it to assert only that the required level of precision of the measurements is open to calculation. It is necessary too that it should be physically possible to perform measurements of the required precision. For the postulation only of a method of calculation would make scientific determinism little more than a mathematical theory (and one that might be false on purely mathematical grounds). I shall say more in the next section about this aspect of the matter — that scientific determinism must be understood to assert that the needed measurements can practically be accomplished.

This doctrine of scientific determinism is not a part of classical physics, but an independent theory conjectured in the light of classical physics. Indeed, it is generally agreed that scientific determinism can be rejected without requiring any rejection of any part of classical physics. Popper's own arguments in *The Open Universe* are of two kinds: there are those that appeal to results in non-linear dynamics, and those that appeal to the difficulties of self-prediction. Writing in the 1950s Popper did not of course take advantage of any of the now familiar results of the theory of chaos, such as the evolution of the time-dependent logistic function $f(t + 1) = \lambda f(t)(1 - f(t))$[1], but instead referred to a theorem published in 1898 by Hadamard concerning geodesics on surfaces of negative curvature (Popper (1982), section 14). In each case we are presented with a simple example of a process whose asymptotic behaviour, though governed by a *prima facie* deterministic equation, cannot be predicted with any precision unless the initial conditions are given with absolute precision. Such processes — if physically realized — plainly contradict scientific determinism — in the form in which it refers to asymptotic behaviour — since it is accepted by all interested parties that there is no possibility of measuring values of continuous variables with absolute precision. It is worth noting that this objection to scientific determinism is not a purely formal one, though it may well appear to be such; for there is nothing in pure mathematics to say that any unstable processes of the required type exist in nature. Nor does the objection constitute

---

[1] When $\lambda = 4$ the function f evolves constantly from $f(0) = .75$, but haphazardly from any initial value arbitrarily close to .75. See Miller (1994), pp. 153-155.

an empirical refutation of scientific determinism, for the existence of unstable processes is not a fact of observational experience either. All that is true is that scientific determinism turns out to clash with science. It must accordingly be rejected. Arguments of this kind are now so commonplace, and so widely accepted, that I need not expatiate on them further at this point.

Popper's other arguments against scientific determinism are rather different, and rather harder to assess. They make no mention of precision, and if valid they presumably show that even absolutely precise initial measurements are insufficient to enable the generation of absolutely accurate predictions. The main idea is clear enough — the problem is 'whether a predictor can predict changes in those parts of its own environment with which it strongly interacts' (*op. cit.*, p.73, emphasis removed). Popper at one point states that the conclusion is that 'there cannot be a scientist able to predict all the results of all his own predictions' (Popper (1982), p. 63, emphasis removed). The arguments rest in the main on a demonstration (*op. cit.*, section 22; see also Popper (1950)) that no physical device, however carefully prepared, can invariably predict in detail its own future behaviour, though it may be able to do it on some occasions, and though it may perhaps be able to announce at any time t what it is doing at time t. (For a recent discussion of similar issues, see Breuer (1995).) It is required that the device should issue its predictions in some standard and explicit format that demands mechanical time-consuming output (so that, for example, the starting state of the device is not taken as an implicit statement of the desired prediction); and there are a number of other conditions imposed to ensure that the device is a genuine predicting machine, and not just imaginatively interpretable as one. It is then argued that the best that the predictor can do in the way of issuing a statement of its own state at some future time is to issue it at exactly that time, not before. But while I am sympathetic to the idea that every step in the calculation will have to take account not only of the original task, but also of the changes in the conditions induced by previous steps in the calculation, I am afraid that I do not find the general argument in Popper (1982) entirely convincing. It seems to claim that as there is, for each device, some self-prediction task that cannot be undertaken successfully, it follows that unqualified prediction is impossible. But a predictor that had the occasional blind spot, otherwise performing competently, would hardly suffice to demolish determinism as a doctrine based in science. For nothing in that doctrine says that the same method of prediction must be used for every task, only that the methods used should be general methods rather than acts of opportunism. It may be that there is some predictor equal to any task, though none is equal to every task. (Everything of macroscopic dimensions is, no doubt, visible to some eye; but no eye can see its own retina.) I grant that if the whole universe is taken as a predictor attempting to predict its own future behaviour, then it will fail in this task. But I am not persuaded that the general impossibility of self-prediction — unlike the argument from non-linear dynamics — shows that there is any real physical prediction task that cannot be carried out successfully.

2   Two Doubtful Connections

In this section I wish to challenge two lines of argument that are implicit in what has gone before. One is the validity of the argument from predictability — that is, scientific determinism — to metaphysical determinism; the other is the relevance (the invalidity is plain) of the argument from chaos — that is, scientific indeterminism — to metaphysical determinism. My challenge is in each instance rather obvious, and is heavily dependent on well known ideas. I should not be surprised if each complaint had been made in more or less the same way before.

It is usually supposed that scientific determinism logically implies metaphysical determinism. Popper himself states categorically that 'metaphysical determinism is, because of its weakness, entailed by "scientific" determinism' (*op.cit.*, p.8). Watkins (1974), pp. 373f explains: 'Scientific determinism superimposes on metaphysical determinism the epistemological claim that there is, in principle, no limit to the extent to which the already fixed and determined future may be scientifically foreknown from our knowledge of present conditions and of the laws of nature. ... Scientific determinism is, of course, a stronger doctrine than metaphysical determinism, which it incorporates and stiffens.' The logical connection must be conceded if, as here, metaphysical determinism is taken to be a simple conjunct (or axiom) in the formulation of scientific determinism. But this construal seems to impose on scientific determinism exactly the kind of idle metaphysical baggage that a purportedly scientific theory ought never to be asked to carry (see Watkins (1984), section 5.32, especially p. 205). In any case, the question still arises whether scientific determinism — understood merely as unconstrained predictability — implies metaphysical determinism. It would do so, of course, if it were an inconsistent doctrine, as it is sometimes claimed to be. But as noted in the previous section, there is nothing in mathematics alone that requires that there should be any non-linear processes in the world — a block universe in which nothing happens is not impossible — and therefore nothing there that requires that there be any failure of scientific determinism. The proof, moreover, of the impossibility of self-prediction, is definitely not free of contingent assumptions. I accept that there may be quite weak truths about the world that suffice to ensure that there exists somewhere, some time, some element of non-linearity and of unpredictability. All I am interested in here is whether scientific determinism alone, without further contingent assumptions, really implies the metaphysical unalterability of the world.

It is easy to see why the connection is thought to be there. If we can systematically make true predictions of the behaviour of any closed system then, to those who are realists, the theories that sustain those predictions must be true, or very nearly so; and since it seems inescapable that the theories in question are *prima facie* deterministic, metaphysical determinism appears to follow. (See Hunt *op.cit.*, p.132. This is not the old argument that, because future contingent statements are devoid of truth value, we cannot make true predictions about an unfixed future. The role of theoretical statements in the generation of the predictions is essential.) But this way of thinking strangely neglects the fact that the activity of prediction is a physical activity; and that even if, unrealistically, the making of calculations is regarded as a purely cerebral business, the making of measurements required for the initial conditions is not. In other words, the appearance of fixedness of the future of the system under study may be an artifact of the carrying out of the prediction task, a version of the Oedipus effect (Popper (1957), section 5; Popper (1976), section 26). It may be that by tampering with the world so as to obtain the values of the initial conditions needed for our predictions (not to mention making the calculations) we introduce a determinacy that was not there before — just as some interpreters of quantum mechanics say that by making measurements we introduce a determinateness that was not there before. I do not see why this should not be the general rule, why closed systems should not generally be tamable by the measurement of sufficiently many initial conditions sufficiently precisely. To be sure, this supposition can become trivial if we impose no restriction on what qualifies as an act of measurement: I don't want to allow as a measurement the transformation of a playful kitten into a predictable system by fixing its initial position with an anaesthetic needle. It has to be granted too that the system that we cast into a predictable state cannot be expected to behave exactly as it would have behaved had we not intervened. But this is true of almost all measurements. In any case, scientific determinism hardly resembles a testable thesis if it asserts only the derivability of statements of what would have happened had the system not been prepared, not of statements of what will happen when it has been.

Another point that needs to be made is that there is no (implausible, but not impossible) suggestion here that making a series of measurements on an unpredictable system would in every case convert it permanently into a fully predictable system. Recall that scientific determinism requires only measurements that allow prediction at a prescribed level of precision; and many systems measured exactly enough for a certain task may rapidly return to being unpredictable. It has to be admitted that measurements sufficient for making asymptotic predictions may perhaps have a quasi-permanent effect on the system predicted. But this is by no means obvious. Remember that we are assuming at this point that scientific determinism is true, so that there exist no unstable chaotic processes. It may be that accurate asymptotic predictions, at least of a qualitative kind, may be achieved without binding the system in perpetuity.

I stress that I do not take very seriously any of these possibilities. To my mind neither scientific determinism nor metaphysical determinism is true. But even if scientific determinism were true, it would lend much less support to metaphysical determinism than is commonly supposed.

One thing that should by now be becoming plain is the extent to which the claims of scientific determinism require investigators to be able freely to prepare and measure initial conditions of a precision appropriate to the intended predictions. Something like this, though something weaker, is indeed the case for all empirical activity when seen from a critical or falsificationist perspective: if we are unable to meddle in parts of the world — those parts that contain items of scientific apparatus — more or less at will, then we shall unable seriously to put our theories through their paces. Substantial freedom to intervene is quite simply a presupposition of rational science. In other words, only if metaphysical determinism is false is science possible. This point may be compared with the argument attributed by Popper (*op.cit.*, section 24, and Popper & Eccles (1977), section 20) to Epicurus, Augustine, Descartes, and Haldane to the effect that determinism, if true, is not susceptible to rational discussion. It cannot be too strongly emphasized that these considerations do not show that metaphysical determinism is false; but — like a thoroughgoing anti-realism — they do render the success of science inexplicable.

Now consider again the argument from the existence of chaos. What the existence of processes such as the logistic function demonstrates is that we are unable in general to prepare a deterministic system sufficiently precisely to be able to predict its longterm evolution. We live in a partly uncontrollable world. The metaphysical determinist concludes that unpredictability is no bar to determinism. What has to be noted, however, is that if metaphysical determinism is true, then the intermediate conclusion is already inescapable: it would be miraculous if we, mere puppets in a deterministic world, were routinely able to adjust our initial measurements to ensure a preordained level of precision in all our prediction tasks. The kind of predictability required by scientific determinism, that is, is already excluded by metaphysical determinism, and the existence of chaotic systems seems not to be relevant. Metaphysical determinism implies scientific indeterminism when that doctrine is understood to insist on our actual ability to undertake measurements. If metaphysical indeterminism too implied scientific indeterminism, then scientific indeterminism would be a logical truth. I have tried above to repudiate this latter implication. If I am right, we are at liberty to hold scientific determinism to be consistent, true only if the world is metaphysically indeterministic.

What the determinist's argument invoking chaos amounts to is in essentials this: first, a demonstration that if determinism is true then failures in predictability are to be expected; and further, the assertion that therefore there is nothing in our experience that obliges us to surrender determinism. The demonstrated step may be admitted, but it can be established with invoking chaos. The further consequence is doubtful, given how much there is in our experience that seems to contradict determinism. But let us ignore this point. My interest here is in why the existence of unpredictable chaotic processes has been thought so gloriously to enrich the determinist's case. One reason is that chaotic processes are able to simulate random processes; and random processes have for long been phenomena without a clear deterministic explanation (see Miller (1995), section 3). Yet the pseudorandomness of the outcomes of some chaotic processes must not be confused with their unpredictability. The simulation of randomness is a distinct issue, and whether it speaks for determinism or against it depends on how expert the simulation is.

The existence of unpredictable non-linear processes has been welcomed by determinists, I think, only because there is an implicit presumption that in a fully linear world unpredictability would disappear. This, I have argued, involves a presumption of metaphysical indeterminism. I am not insinuating that there is any circularity. Indeed, it is not a ground for complaint if someone assumes the negation of a thesis that he is trying to establish. The metaphysical determinist may, if he wishes, assume metaphysical indeterminism in order to refute it — provided that he does not assume determinism at the same time. Unfortunately, it is just this that seems to be happening.

What we really need to learn is not only that metaphysical determinism is compatible with unpredictability — which is true enough — but that it implies it; not only is a thoroughgoing predictability compatible with metaphysical indeterminism, it implies it. Once this connection is properly seen, the temptation to succumb to deterministic ways of thinking should surely fade.

3   Chaotic Epistemology

In the previous two sections I have stressed that scientific forecasting of the behaviour of a physical system is a complex physical process, beginning with the measurement of initial conditions — and the inevitable disturbance on the system being investigated — continuing with the expenditure of time and energy in the calculation — nearly always using approximation methods — and ending with the publication of a prediction. Little of this way of describing the process would make much sense in a deterministic world, and henceforth I shall assume that indeterminism is correct. What we learn from non-linear dynamics is that even so, and even if the second phase of the prediction process is regarded more abstractly — as an exact mathematical transformation of externally provided initial values into predicted values — the prospects of unconstrained predictability are not good. For in the measurement of most physical quantities complete precision is unattainable, and the values obtained (and the predictions, where they can be made) are usually announced in the form $\vartheta = \mu \pm \delta$ — here ϑ stands for a physical quantity, and μ and δ are numerals — meaning at best that ϑ lies somewhere in the given interval, δ being a characteristic of the measurement process. Even this oversimplifies considerably. For one of the few things that we can definitively learn from experiment and experience is that few pieces of measurement apparatus yield the same value on each occasion that they are used. All, or almost all, real measurements are infected with random errors or inaccuracies. It is therefore standard to interpret the equation $\vartheta = \mu \pm \delta$ to indicate the mean and standard deviation of the actual measurements of ϑ. This again is not really satisfactory, since the actual measurements themselves remain imprecise. It is scarcely helpful to confound the imprecision in the measurement process with the inaccuracy of the actual measurements. (We can have imprecision without inaccuracy, as when we repeatedly ask a computer to state the value of π, and inaccuracy without imprecision, as when we count votes.) But that is not my concern here. What I want to suggest is that new possibilities for chaotic unpredictability emerge once we introduce a probability distribution over the measurements, even if the measurement process itself is taken to be perfectly precise. This is especially so if — as in Bayesianism — the probability distribution is modified (via Bayes's theorem) as the evidence accumulates.

In Chapter 8 of Miller (1994) I gave examples of distributions — strange distributions, to be sure, but authentic ones — that appeared to evolve in a chaotic manner. The idea used there was to characterize a distribution by a single parameter, and to show — admittedly only by computational methods — that arbitrarily small variations in that parameter occasioned wild variations in the distribution after successive conditionalizations. For example, suppose that h is a hypothesis of interest, and that $P_0(h)$ is its initial probability. Suppose further that $e_0, \cdots$ are successive items of evidence, which may occur positively (e_t holds) or negatively ($\neg e_t$ holds). After each item of evidence is received, the probability $P_t(h)$ of h is amended by Bayesian conditionalization:

$$P_t(h \mid e) = P_t(e \mid h)P_t(h)/P_t(e) \tag{1}$$

$$P_{t+1}(h) = P_t(h \mid e) \tag{2}$$

where $e = \pm e_t$ is either e_t or $\neg e_t$. Equation (1) here is Bayes's theorem, which gives the relative probability (at stage t) of the hypothesis h on the evidence e in terms of the absolute probabilities $P_i(h)$ and $P_i(e)$ and the *likelihood* $P_i(e \mid h)$. Equation (2) says that, after receipt of evidence $\pm e_i$ at time t, the absolute probability $P(h)$ should change from $P_t(h)$ to $P_i(h \mid e)$. We suppose that at each stage t the likelihood function $P(e \mid h)$ is given by the following relative of the logistic function.

$$P_t(e_t \mid h) = 4P_t(h)(1 - P_t(h)) \tag{3}$$

$$P_t(e_t \mid \neg h) = 1 - 4P_t(h)(1 - P_t(h)) \tag{4}$$

Note how similar the right-hand side of (3) is to the right-hand side of the equation on p. 4. It is possible to calculate from (3) and (4) the terms $P_i(e)$ and $P_i(\neg e)$, and hence to calculate $P_i(h \mid e)$ and $P_i(h \mid \neg e)$ by means of (1). In other words, the specification (3) and (4) determines (via (1) and (2)) the value of $P_{i+1}(h)$ once $P_i(h)$ is given. Using this method of updating we find that — provided that the distribution of positive and negative items of evidence is more or less evenly balanced — the longterm behaviour of $P_t(h)$, the probability of h at t, is most sensitively dependent on the initial value $P_0(h)$. For $P_0(h) = (2 \pm \sqrt{2})/4$ it is constant, but for initial values about 10^{-8} from these fixed points it is highly agitated even after a few hundred items of evidence. Some pictorial illustrations are provided on pp.164f. of Miller (1994).[2]

---

[2] I should like to take this opportunity to correct an unaccountable slip in my treatment in Miller (1994) of this likelihood function. Noting there (p.161) that it follows from the law of total probability (a consequence of (1)) and (2), (3), and (4) that

$$P_t(e_t) = 1 - 5P_t(h) + 12P_t(h)^2 - 8P_t(h)^3 \tag{5}$$

I added (pp.167f.) that $P_t(e_t)$ clearly does not take the constant value 0.5, which is embarrassing given that positive and negative items of evidence are supposed to be equally distributed. But this arithmetical assertion is just false when

One moral drawn from these speculations (*op.cit.*, pp.170f.) was that we cannot in general rest content with probability distributions given to us approximately rather than exactly. The exact starting point of the function $P_t(h)$ in the example above is crucial if we are to have any idea of its value after a reasonable amount of evidence has been absorbed. This is as serious a drawback for Bayesians — who claim that probabilities may be elicited by the crude mechanism of betting — as it is for objectivists who rely on the accumulation of statistics. In fact, it is much more serious, since objectivists do not care to update their probability assignments in response to every new item of evidence (Popper (1983), section 12, Part ii). It might be retorted, of course, that the function $P_t(h)$ must have an exact initial value, and that is all that matters for its exact evolution; in other words, the value of $P_{1000}(h)$ may not be predictable, but it will be there when it is needed. But given how insistently probabilistic considerations underwrite most epistemology — though not falsificationism — this amounts to admitting that prediction is sometimes impossible even in the presence of exactly specified empirical data. Even if the physical phenomena under study are stable and well behaved, the prediction process may introduce uncontrollable fluctuations. Of course, not all likelihood functions are as irrepressible as is the one defined by (3) and (4). No one suggests that all efforts at systematic prediction are doomed.

References

Breuer, Th. (1995): The Impossibility of Accurate State Self-Measurements. Philosophy of Science **62**, 197–214

Earman, J. (1986): A Primer on Determinism. *The University of Western Ontario Series in Philosophy of Science*, volume 32 (D. Reidel Publishing Company [now Kluwer Academic Publishers], Dordrecht)

Hunt, G. M. K. (1987): Determinism, Predictability & Chaos. Analysis **47**, 129–133

Miller, D. (1994): Critical Rationalism. A Restatement & Defence. (Open Court Publishing Company, Chicago & La Salle)

Miller, D. (1995): Propensities & Indeterminism. Anthony O'Hear, editor, *The Work of Karl Popper*, Royal Institute of Philosophy Lectures (Cambridge University Press, Cambridge, forthcoming), 113–139

Popper, K. R. (1950): Indeterminism in Quantum Physics & in Classical Physics. The British Journal for the Philosophy of Science 1, 117–133, 173–195

Popper, K. R. (1957): *The Poverty of Historicism* (Routledge & Kegan Paul [now Routledge], London)

Popper, K. R. (1974): Replies to My Critics, 961–1197 of Schilpp (1974)

Popper, K. R. (1976): *Unended Quest* (Fonata, London 1976 [new edition, Routledge, London 1993]); originally *Intellectual Autobiography*, 3–181 of Schilpp (1974)

$P_0(h) = (2 \pm \sqrt{2})/4$: it is easily checked that $P_t(e_t)$ begins at 0.5 and remains there. And since the function given by (5) is continuous, it remains very close to 0.5 when $P_0(h)$ is close enough to $(2\pm\sqrt{2})/4$. The distribution given by (3) and (4) is therefore a good deal more satisfactory than I thought it was.

Popper, K. R. (1982): *The Open Universe. An Argument for Indeterminism*, Volume ii of *The Postscript* (Hutchinson, London [now published by Routledge, London])

Popper, K. R. (1983): *Realism & the Aim of Science*, Volume i of *The Postscript* (Hutchinson, London [now published by Routledge, London])

Popper, K. R., Eccles, J. C. (1977): *The Self & Its Brain* (Springer International, Berlin [now published by Routledge, London])

Schilpp, P. A., editor (1974): *The Philosophy of Karl Popper* (Open Court Publishing Company, La Salle)

Watkins, J. W. N. (1974): The Unity of Popper's Thought, 317-412 of Schilpp (1974)

Watkins, J. W. N. (1984): *Science & Scepticism* (Hutchinson, London)

Discussion of David Miller's Paper

Batterman, Miller, Noyes, Schurz, Schuster, Strack*, Suppes, Weingartner, Wunderlin

Suppes: I share lots of your sceptical views. There are three points I would like to comment on. The very first point is simply a point of which I think I'm right by a bare example - that Popper had it wrong in emphasizing prediction so heavily in the sense that if we use the classical differential equation definition of determinism, namely with sufficiently smooth forces, sufficiently smooth conditions and no collisions, that then the differential equation for a system is deterministic ... This is the particular example now. I want to give an example. There exists this beautiful example which I referred to earlier and Gerhard (Schurz) mentioned in his lecture of the three-body problem - this marvellous theorem starting with Sitnikov and then Alekseev, and there is also a good account given by Moser. This example, in which you can construct a random sequence within the three-body problem, does not depend upon any lack of definiteness in the initial conditions. There exists an initial condition that will produce any random sequence. The precisely determined initial condition is in fact what produces it. It has nothing to do with the problem of measurement and it is very important to recognize the existence of such problems in the heartland of mechanics.

Miller: But I think that's true with the logistic function too. If you start with any value one of the fixed points, you will get an evolution ...

Suppes: I think you recognize the theoretical point, but in the discussion of mechanical examples you can emphasize the precision of measurement of initial conditions. And my point is that in your mechanical example qualitatively you did emphasize that. It isn't a big point, because it has been discussed before. But the point is you could know the initial conditions and boundary conditions precisely in this example and still we have no hope of prediction, a long-term prediction. You could predict locally if you had measured, but you certainly can't predict the long run, because you can't predict a random sequence which you are not able to produce.

Miller: That is true.

Suppes: So, that's my first point, I don't think that's a big point, you probably agree with this.

Miller: Yes, I did not discuss the question. I did not actually say anything about the problem of non-solvability.

Suppes: Non-computability?

Miller: Yes, non-computability.

Suppes: I mean the point is that you have exact initial conditions and you have non-computable trajectories.

---

* Hans-Bernd Strack, Institut für Chemie und Biochemie, Salzburg, Austria.

Miller: This is a prima facie deterministic theory, nonetheless.

Suppes: Well, I think what we are doing here is separating the existence and the uniqueness theorems of differential equations. Any undergraduate in physics can derive a differential equation for this three-body-problem, but the solution involves quite difficult mathematics. I mean you can prove something about it. I think what's important is to separate the existence and uniqueness of the solution from the question: is the solution computable?

Miller: Yes.

Suppes: So I don't think you really disagree much about that. My second comment is historical and I am disappointed that you didn't discuss Kant a bit more on this metaphysical determinism, given his entanglement with it and the many useful things he has to say about it. That's simply a historical remark, because in a way if I had to choose my philosophers - though I have great affection for Karl -, I think Kant got deeper into the problem in many respects. My third remark is about the Bayesian example. I just want to illustrate. Let's take the case of a coin with unknown probability: So we want to know the true probability - it's maybe biased. In the real Bayesian approach we could take on the zero-one interval a β-distribution for the prior. You take a distribution, a smooth function - a β-distribution will be fine, on the zero-one-interval, - which expresses my prior as to what the true probability distribution of the coin is. Let's say that all unindoctrinated Bayesians take such a prior, you are pretty confident. But if you say, well I am doctrinated, you still have smooth functions on the interval, then it is a good question: can you construct such a thing as in your example of diverging distributions? It is a matter of study, it is not clear in this kind of case that you could. I am not suggesting by the way that you are not right about this. I think you are absolutely correct that it is a mistake if you think you can get divergence in the kind of example you gave. I just want to point out that there are ways of safeguarding convergence in many cases in a standard approach.

Miller: I don't want to deny that at all. I think it's quite clear that there are cases where convergence is a fact and is observed.

Suppes: But all you have is this smoothness assumption of a distribution rather than a point prior.

Miller: All I am trying to do is to provoke a certain scepticism.

Suppes: Oh, no, no. I do not disagree with this. I do want to protect my Bayesian.

Batterman: I've got a question about Bayes' theorem too. I did not understand the way you set up the logistic map. What role did Bayes' theorem play? How was it used for up-dating probabilities in this case?

Miller: The updating is done by Bayes' theorem. There was an initial probability for the hypothesis. And then I gave a rather bizarre likelihood function that was unchanged throughout entire evolution. The question was: Can I actually make this sequence behave like the logistic function and it turned out: yes I can.

Batterman: But didn't de Finetti prove a theorem to the effect that if one's prior probability assignments are coherent, then if you have two different people who assign different initial priors to the same hypothesis, there will be convergence in probability assignments as evidence is accumulated?

Schurz: De Finetti proved it under the condition that the probability function is symmetric and regular.

Suppes: Savage also discusses that. I mean, I agree with him. This example does not satisfy the hypothesis under which de Finetti proved his theorem.

Batterman: I mean exactly what's not satisfied? Is it just the fact that there are not symmetric or the fact that there are different priors in this case?

Suppes: I stick to a very simple example, a real trivial example that does not satisfy de Finetti and there is no convergence. Suppose you believe that Elvis Presley is still alive - you assign probability one, that he is alive, I assign probability zero. Whatever evidence is produced, we are not going to converge.

Miller: That's not regular.

Suppes: That's not regular. He [Miller] gave a more sophisticated example.

Schurz: But under certain conditions it will converge - that has been proved by de Finetti.

Miller: The problem is that most Bayesian philosophers think - or write as if they think - that de Finetti or somebody proved that in all interesting cases you have convergence.

Suppes: Remember, de Finetti's representation theorem has very strong assumptions in terms of exchangeable events. For the study of most events in time you don't have exchangeability.

Schurz: That was what I meant with symmetric.

Strack: If I understood your summary correctly you wanted to exclude prediction by accident - you wanted to have deterministic theories just with a rational means of prediction. Now I wonder where the borderline is for you, if just for the sake of the argument you would say that theories are schemes which link perceptions with predictions some kind. And then if you also believe that theories are selected - basically selected - for the partial success they have had then you select theories which repeatedly have been successful in this limited sense. Now, that may just be an accident or at first it certainly is an accident. Now, where is the borderline for these theories to become rational in your sense, basically at first they are all accidental.

Miller: Where is the borderline? I shall try to draw a border for you. Prediction has to be systematic. Whatever theory we use, the theory has to be something that is stated in advance. It is not easy to make useful predictions from a theory in a systematic way - unless you take an inconsistent theory, from which you can predict everything. Actually getting a theory that will produce any predictions at all that is no easy matter. What I wanted to rule out was the sort of case where we might think that a prediction is made successfully, but it is really just pulled out of the hat.

Wunderlin: A technical question: You have written this equation for the probabilities - one writes such equations of probabilities which are each time normalizable. Now I don't see how this fact is guaranteed when you use the logistic equation for the time development of such probability.

Miller: No, Bayes's theorem controls the time development.

Wunderlin: But on the other hand this equation is still restrictive. In physics we call it detailed balance. It is a special condition to the probabilities if time is involved - this means, if you put this $p(e)$ to the left side, the left side is the time forward part, and the right side is the time backward part. If both are equal then this is a very special condition, so I am not sure whether this holds generally.

Schurz: May I mention that in this equation there is no time development involved. These (e and h) are propositions, just propositions without time dependence involved.

Suppes: You start at a given time. Then you simply get new additional evidence updating your posterior $p(h/e)$. [to Miller]: I have some more comments, I think you are too quick on accidental prediction. I want to make a couple of remarks. First of course if we took your horizon of rational thought literally, then of course the Bayesian is doing accidental predicting because the Bayesian has the following kind of scheme: he uses a theory T in making a prediction. But then it is not quite clear what the circumstances are. So he uses his judgement - as the Bayesian say - which is not theoretically given, but is based on experience, given in a non-cognitive way to make the final prediction. This is not just an imaginary example we are talking about. This is the way a real weather forecaster can make predictions. They use the information and the theory, but then they make a final adjustment in terms of their experience. I would say Sir Karl is too rationalistic. There is too much of experience that we really count upon digesting in a non-theoretical way. I mean, we cannot give an explicit verbal account of the way in which we digest all this experience. But that means that if you make your scientific predictions in that framework, in that Bayesian framework, then there is a clear accidental component in your sense, namely, the predictions are tempered by judgement and not simply by pure calculation. So, for what I would call a typical Bayesian, there is a mixture of calculation that is done by theory and then there is a final component of judgement that corresponds to your idea of accidental. We talked about weather forecasting - that is a practical matter. Let's take a different kind of example. Take any really complicated experiment in physics. Then there is an enormous amount of apparatus involved, and the standard comment of the technicians and engineers is: "God help us. The physicists don't understand this equipment." They simply trust what they were told by the engineers and how it's going to work. It is an enormous matter of judgment. So if you are experienced to work in these fields, you have a nonverbal evaluation of the trustworthiness of your technicians. This also applies to computer work. You don't have anything like a rational analysis in the ordinary sense of ratio. The meaning of rational here being that you could give a theoretically based verbal account of the basis of your prediction. Now, in that sense even in scientific work, you don't make rational predictions. There are too many unverbalized adjustments made in the experiment.s. But it's even more rational, in the sense of practical success. So that may be what you might call the 'extended' or 'weak' sense of rational.

Miller: But that seems to me a doctrine, that is very artificial indeed. The success of a prediction depends on very many different ways and circumstances.

Suppes: That is also what the Bayesian would say.

Noyes: I just want to support Pat. Actually in my own experience in physics, it is very important to develop a sense of judgment as to what laboratory results you are going to trust and which you are not. I've been in the data analysis business for a long time. For several years I was engaged in proton-proton scattering analysis aimed at obtaining theoretical meaningful and trustworthy parameters. I found that you cannot just uncritically accept what is published in the main stream, refereed literature. You learn that results from some laboratories, and some experimental teams hold up and that others do not. Of course you cannot simply throw out the points in a data set which are so far from the mean that they make no sense. What you have to do is to throw out a whole experiment, or preferably simply ignore some sets of published data because you distrust the team, or the laboratory or the method. This is often difficult or embarrassing to defend publicly.

Miller: But I think that is a rational process.

Noyes: Well, a matter of judgment. It is not a calculated process.

Schuster: I feel a little bit uncomfortable in this discussion now, because I think we assume and we strictly assume that the good laboratories give us the results we trust. They work on a rational basis. So we just have some experience. So we better not trust some of them because they do not work in a rational way. And we do not assume that the better laboratories just have the better intuition and the better guessing people, or do you?

Miller: No, they do their error correction before they submit their results for publication rather than afterwards.

Suppes: I want to comment on that. I have a marvellous example in my first experiment. Recently I worked with a very smart guy from China. He had his Ph.D. in mechanical engineering. - I have a long experience, 40 years of people working for me who have a lot of technical experience. This guy exemplified rationality a 100% and he had experience. He could read the manuals, could understand them, etc. What he lacked was to challenge data, to explicate what is meant by them. His judgement was based upon experience as a technician who cannot tell you what is the basis of his judgement. But because he has had a great deal of experience, he could actually make good judgements and I knew that I could trust them. But he could not give a verbal account. And I want to make this comment. It seems to me that though we can use the word 'rational' in a nonstandard way - if you talk about such non-verbal judgment as rational - but the real problem is to explicate what its components are. What is in fact a rational laboratory technician in this sense?

Schuster: I have an argument for that, and that is reproducibility. I have a marvellous example. There was one laboratory with one technician and they were able to do an experiment that did not work in any other laboratory for a very long time. So you could still argue, that this particular technician does not know what she does - but she is fading somehow. But then it turned out that the intuition to trust this particular technician was correct because after all one learned how to do it and then it was reproducible.

Suppes: Reproducible by other people?

Schuster: Yes. After one learned what she did and she did it correctly. But others just did it not correctly.

Suppes: I want to comment on your example which I agree with but I also want to make the following point. You can have criteria of rationality in terms of success. But I want to emphasize that there is a long tradition in analyzing rationality that tried to go deeper than this. And what we don't have is - to start with your example of reproducibility - we do not have a really sophisticated body of knowledge about these laboratory experiments or about similar experience, that constitutes what one would call a general theory of rationality. What we have are important and well documented observations of the concept. But what I am saying to David, what is missing is: what are the characteristics of rational law, of rational behavior?

Miller: My answer is that I think that there is no such thing. The idea of rationality is somehow a procedure, an appropriate procedure ...

Suppes: In this case you are more sceptical than I am. A very good example is this: There is a great deal known about efficient ways of training animals. There is a lot known about how to do a better job of training nurses or training doctors.

Miller: But that is making it irrational. Training here seems to mean doing things without being constantly aware of what the effect of what one is doing is, and how one may be incorrect, without looking for mistakes.

Suppes: Oh, that is really nonsense. I really must disagree with that in an extreme way. I want to give you another example. We are running, as I told you, an educational program for very gifted students. We are teaching them physics at an early age. We have a home laboratory developed for them. Some of these very bright students - because we did not give them earlier training - could not understand how to use a soldering iron. They did all the problems with great elan, but they could not touch soldering. If you think that we don't know something about how to train people, you are wrong. Training is a rational procedure. That is what I mean by rational. And the problem is to have a theory of practice, where the theory of practice is, for example, a theory of how to do the job of training technicians or training doctors. I mean the thing about doctors is that they cannot verbalize accurately what they can do well. But they can get very good under proper regimes at actual diagnosis. And your verbal criterion of rationality - the doctors don't satisfy. But they are very good on the basis of the training that can be provided.

Schurz: I think the rationality problem is a difficult problem where you have to weigh several aspects - difficult to decide. Maybe we go on to Paul's remark.

Weingartner: Determinism can be characterized in a twofold way. Either we characterize laws in a certain way and call them deterministic, because they have this and that property. Or we apply determinism to nature and ask whether the structure of nature is deterministic. I prefer the first sense because I think it is difficult to define such a thing as the structure of nature, but I think it has good sense to say that if laws have certain features and if the laws are successful, then we call these laws the deterministic laws. Now, you spoke about refutation of determinism. I would say what the research on chaos shows us is that we have to revise our earlier view on laws. We have to revise this view, because this view was that we have differential equations and if there was some slight change in the initial conditions or some perturbation, the view was that these can be handled by the calculus of mistakes or by perturbation theory to any suitable degree of accuracy. And this I think is now challenged, this is really new. For very small deviations in the initial conditions may lead to chaotic behaviour which is no more predictable with the help of these laws. We have to revise the concept of law. Would you agree?

Miller: I do not quite understand what the sensitivity to initial conditions has to do with the concept of law. Which kind of consequences do you want to draw for the concept of law?

Weingartner: If you ask what essential features do laws have, fundamental laws, then you have the problem of either still upholding the dynamic deterministic laws as the necessary fundamental underlying laws and explain chaotic motion and its unpredictability with the contingency and chance-like behaviour of initial conditions - determinism wouldn't be refuted in this case - or to try to find a revised concept of law which is even at the fundamental level not as strict as we usually understand dynamical deterministic laws and allows so to speak a statistical law as a part of a dynamical law.

Miller: I should prefer to say that we have the same concept of law as before. We just discovered new things about its extension.

Weingartner: It seems to me, that means that we get either to a new concept of law, to a revised concept, or we put the "new things" into the range of initial conditions.

Miller: No, we discover new things about the same old concept, and eliminate some of the mistaken ideas we have about it. One of these mistaken ideas may perhaps be the belief in statistical - or even probabilistic - laws, which may have to give way to deterministic laws governing propensities.

Schurz [to Miller]: I agree with you, but I want to connect it with the debate on determinism and indeterminism - to connect it with some practically relevant aspects concerning the controllability of the world. In the debate in the 1920ies on determinism and indeterminism in quantum mechanics, people debated on this question with the idea that only if the world is indeterministic then there is a place of free will in it. Now we know - and this is one of Pat Suppes' favourite topics - that the question of determinism versus indeterminism is really not relevant, not really basic for the question of how controllable the world is. It can be uncontrollable or although it is deterministic. These old concepts maybe will become more and more irrelevant and a new concept of unpredictability versus predictability in some sense or other will become more relevant for the future. Do you agree?

Miller: The question of whether the world is in part controllable - I find that question very difficult. There must be something between chance and absolute necessity, and this gap must, I think, be filled by propensities. Although they are not determined events, events under the sway of propensities are not chance events either (though these may exist as well). Somewhere in here must lie the clue to control.

Kinds of Unpredictability in Deterministic Systems

G. Schurz

Universität Salzburg, Austria

1 Introduction

Deterministic chaos is often explained in terms of the *unpredictability* of certain deterministic systems. For the traditional viewpoint of determinism, as expressed in the idea of Laplace (Weingartner (these proceedings), ch. 1.1), this is rather strange: how can a system obey deterministic laws and still be unpredictable? So there is an obvious lession which philosophy of science has to learn from chaos theory: that determinism does *not* imply predictability. But besides this obvious point, there is much conceptual unclarity in chaos theory. There are several different and nonequivalent concepts of predictability, and accordingly, different concepts of chaos. In this paper I will differentiate several different concepts of predictability. I will discuss their significance and their interrelations and finally, their relation to the concept of chaos. I will focus on very simple examples of graphically displayed trajectories of particles (usually in a one-dimensional space), without going into the mathematical details of the dynamical generation of these trajectories from fundamental equations. I think that all major conceptual problems can already be discussed at hand of these simple examples.

A dynamical system is defined by its state space (or phase space) S. The state space of systems of classical mechanics has to contain, for each particle, its position and its velocity — or in the Hamiltonian description its position and its momentum — in the 3-dimensional Euclidean space (cf. Batterman (1993), p. 46). Classical dynamics is interested in determining the temporal development of such a system from some *fundamental* equations — usually differential equations — describing the forces which act on the particle(s) of that system. The solutions of these differential equations are functions $s : T \to S$ which describe the possible movements of particles in S in dependence on time and are called (time-dependent) *trajectories*. In the example of planetary systems, the possible trajectories are either periodic orbits or collapse trajectories where a planet collapses into the sun, or escape trajectories where a planet escapes the attraction of the sun. Note that the understanding of notion of trajectory among physicists is not unique: besides the time-dependent notion of trajectory there is also a time-independent notion of trajectory (cf. fig. 9).[1] In the normal case we under-

---

[1] Chirikov understands "trajectory" in a time-dependent sense (Chirikov (these proceedings), ch. 2.1), Batterman in a time-independent sense (Batterman (1993), p. 47)

stand the time t and position s in S to be *continuous* real-valued parameters; exceptions where a discrete modeling is assumed will be explicitly mentioned.

2 The Concept of Determinism

Usually a mechanical system is called *deterministic* if its state $s(t_o)$ at some arbitrarily chosen intial time t_o determines its state $s(t)$ for each future time point $t > t_o$ — this is *forward determinism* — and also, for each past time point $t < t_o$ — this is *backward determinism*. Of course, the fundamental equations governing the system do not determine the actual trajectory of the particle of the system (assume it is a one-particle system). But if the particle's state $s(t_o)$ is given for only one time point t_o, then its entire trajectory is determined. In other words, a mechanical system is deterministic if its trajectories never branch, neither forward nor backward in time.[2] For practical purposes, forward determinism is more important. But classical dynamical systems are forward as well as backward deterministic — something which is obvious from their well-known time-reversibility . If the time is assumed to be discrete instead of real-valued, there exists an equivalent incremental definition of a deterministic system requiring that for any given time-point t_n, the system's state $s(t_n)$ determines the immediate successor state $s(t_{n+1})$ — forward determinism — and its immediate predecessor state $s(t_{n-1})$ (backward determinism). By induction on n, this definition implies the earlier one.

A central feature of all these characterizations of a deterministic system is the (implicit) *modal* or *counterfactual* element which they involve. For the definition of determinism says that if the given system is in state s_o at time t_o, it *must* be in state s_1 at time $t_1 > t_o$ — in other words, if $s(t_1) \neq s_1$ *were* true, then $s(t_o) \neq s_o$ *would* have been true. Also the physicist's talk of trajectories is modal talk, since a trajectory is a *possible* movement of the system's particle — possible relative to the fundamental equations of the underlying *theory*. Actually a particular system moves along only *one* trajectory; the other trajectories are movements under possible but non-actual circumstances. When Russell (1953), p. 398 suggested his functional definition of determinism — a system is deterministic if there exists a *function* of the mentioned kind $s : T \to S$ specifying the system's state in dependence on time — he seemingly tried to escape the modal element in the definition of determinism, but without success: he himself noticed that this definition is too weak, because the movement of *every* particle — be it deterministic or random movement — can be described as *some* function of time, though sometimes a rather complex one.[3] So the definition of a deterministic system has always to contain the modal concept of possible trajectories, where in physics, this notion of possibility is of course not understood as primitive but as relative to the underlying theory.

---

and Haken in a more complicated sense (Haken (1983), p. 124). The understanding of trajectory used here coincides with that of Chirikov.

[2] This was added as a result of my discussion with Robert Batterman.

[3] Cf. the discussion in Stone (1989), p. 124.

How can we empirically *test* whether a given system is deterministic? As always, via empirical *generality* — we have to realize a situation of the *same kind* more than once. One possibility is that the underlying theory describes systems of a given kind A, e.g. two particle gravitational systems, and we are able to prepare several systems x_i of kind A in a way that they all are in the same state $s(x_i, t_o)$ at time t_o. Then we just have to look whether their future development is the same; if so, we have confirmed the hypothesis that systems of kind A are deterministic (w.r.t. the parameters of their state space S). Of course, it is impossible to definitely verify this hypothesis by finitely many observations (which has been stressed in Popper's *Logik der Forschung*). But note that *if* (and only if) the parameters describing the system's state space are continuous, the hypothesis of determinism can also never be falsified by finitely many observations. The reason is our *limited measurement accuracy*: we can measure the states only up to a certain *acccuracy level*, say ε. If we observe that two states $s(x, t)$ and $s(y, t)$ are the same up to ε, $s(x, t) =_\varepsilon s(y, t)$, but they cause different future developments, then this does not necessarily imply that the system is indeterministic, because the true state of x and y at time t may be different, and the system may exhibit what is called *sensitive dependence on initial states*: small and unmeasurable differences in intitial states may cause great differences in future states (see figure 1 below). So for continuous systems the hypothesis of determinism is neither definitively verifiable nor definitively falsifiable, but of course, it is confirmable or disconfirmable via (dis)confirming the global physical background theory.

In cosmology one is unable to prepare systems and it almost never happens that two different systems are in the same state at the same time. What happens is that one or two systems of the same kind are in the same state at two *different* times. In this case the hypothesis of determinism *seems* to imply that the future development starting from t_1 has to be the same as that starting from t_2. This means that the fundamental laws describing the possible trajectories have to be invariant with respect to (w.r.t.) time: time is 'causally inefficient', nothing changes if the entire system is shifted in time. If this is true, then the notion of determinism would be intimately connected with the well-known symmetry principle of *invariance w.r.t translation in time*. Is it true?

In my talk I suggested a positive answer.[4] Based on what I have learned from the discussion I want to defend here a more differentiated point of view. Classical mechanical systems which are not invariant w.r.t. translation in time are those where their fundamental differential equation — their Hamilton operator — involves an explicit time dependence. An example is given if the gravitational constant γ would change in time (which was Dirac's conjecture). Can also such a time-dependent dynamical system be regarded as deterministic? I am inclined to think that this depends on the *complexity* of the function describing the dependence of γ on time (which now is understood as a *fundamental* law, not being

---

[4] This paragraph was added to the original version presented in my talk.

derivable from some 'super'-differential equation). Not just any function can be admitted here, for then even if the gravitational constant would change in a random way, the corresponding dynamical system would count as deterministic. If there were some fundamental time-dependent laws of cosmology, they must exhibit a strong regularity to count as deterministic, e.g. a continuous periodic oscillation. What would count as a resonable 'minimal' condition for such a regularity? I do not know. One suggestion would be to require that such a function has to be *analytic* in the mathematical sense, which implies that if the value of the function and all of its derivatives are given for just one time point, then its values are determined for all other time points.[5] If this is the case, invariance w.r.t. translation in time will hold again, provided we include the time-dependent parameter(s) in the description of our state space. If we shift the actual value of γ and its derivatives in time, then the physical behaviour of the system will still remain unchanged. So on a deeper level there still seems to be a connection between determinism and translation invariance in time.

The following considerations will be independent from that question. I will consider systems without explicit time-dependence; they are deterministic in an unproblematic sense. My main question will be what it means for such a system to be unpredictable. The second question will be how being chaotic is related to being unpredictable. I will be only concerned with *deterministic* predictions - that is, predictions of the actual trajectory up to some accuracy level (cf. Schurz (1989) for a definition of this notion). I will not be concerned with *statistical* predictions of trajectory distributions. That one cannot make deterministic predictions does not imply than one cannot predict something about the statistical distributions of trajectories.[6]

Given a deterministic system, two things are necessary for making predictions. *First*, one must be able to calculate the function determining future states $s(t)$ from intitial states $s(t_o)$ in some reasonable time, and *second*, one must be able to measure the initial state $s(t_o)$ with some reasonable accuracy, sufficient for keeping the error in the prediction small. Consequently, there are two main approaches to unpredictability, one where the first condition is not met (ch. 3-4) and one where the second condition is not met (ch. 5-6). I will end up with the conclusion that in all of the different notions of unpredictability, limited measurement accuracy and sensitive dependency on initial conditions play a (if not: the) key role.

---

[5] This follows from the fact that all analytic functions can be expanded by a Taylor series with vanishing residual (cf. Zachmann (1973), p. 399, p. 262). Strictly speaking, the notion of an analytic function is defined only for complex valued functions. For real valued functions, we must require expandability by a Taylor series directly.

[6] The last two sentences where added after the talk, in reaction to a critical comment of Patrick Suppes.

3   The Open Form Concept of Unpredictability

The following conception of unpredictability has been suggested, among others, by Stone (1989). It is concerned with the complexity of the computation which calculates the future state $s(t)$ from a given initial state $s(t_o)$ — more precisely, with the dependency of this complexity on the *prediction distance* $t - t_o$, the temporal distance between the initial and the predicted state. Some of the differential equations describing dynamical systems are *integrable*: they admit so-called *closed form* solutions. For instance, the differential equation $ds/dt = k.s$ has the class of closed form solutions $s(t) = s_o.e^{kt}$ which describe exponential growth or decay (depending on whether k is positive or negative). Several differential equations, for instance the three body problem of mechanics, are not integrable. Some of them may be solved in an approximate way, but some others admit even not an approximative closed form solution. They can only be solved *pointwise* - which means that there is an algorithm which calculates $s(t_{n+1})$ from $s(t_n)$, for a given partition of the continuous time into discrete time intervals. Pictorially speaking, such an algorithm *simulates* the evolution of the system by moving incrementally along its trajectories. Such a pointwise solution is always an approximation in the case of a continuous time, but it may be exact in the case of functions depending on a discrete (time) parameter. An example is the well-known logistic function $s_{n+1} = 4\lambda s_n(1 - s_n)$, where $s_n := s(t_n)$, which describes the so-called "Verhulst-dynamics" of the growth of a population with a dampered growth rate.

The crucial difference between closed form and open form solutions is not adequately captured by saying that the former admit solutions of the form $s = f(s_0, t)$, while the latter only admit point-to-point solutions of the form $s_{n+1} = f(s_n)$. From a mathematical viewpoint one can define also in the latter case a function g such that $s_n = g(s_o, n)$, just by defining $g(s_o, n) = f^n(s_o)$, where f^n means f n times iterated. The crucial difference is that in closed form solutions the complexity of the computation of the function $s = f(t)$ is *independent* or at least almost independent from the prediction distance $t - t_o$. In contrast, in open form solutions the complexity of the computation increases proportionally, and significantly, with the prediction distance. Hence I suggest to define a solution as having a *closed form* if there exists an algorithm for its computation for which the time of computation is almost independent from the prediction time and significantly shorter than it; otherwise it has an open form.

Assume a solution has an open form. If we observe the system's initial state at time t_o and then start our predictive algorithm, we will never be able to *predict* the system's future state at times $t > t_o$ because the algorithm is not faster than the system itself and hence will terminate not earlier than t. This effect has motivated several authors, like Stone (1989), to see here the crux of unpredictability. Therefore I call this the *open form concept* of unpredictability. Let us ask: under which condition do open form solutions really lead to unpredictability? In order to eliminate other sources of unpredictability we assume that the system does not exhibit sensitive dependency on initial conditions: the error of the predicted state shall be comparable in magnitude with that of the initial state. Then we can draw the following distinction. If we cannot *control* the initial state of the system but merely observe it — as for example in the prediction of the movement of a comet — then an open form solution will indeed lead to unpredictability, for we can start the point-to-point computation procedure not earlier than we have observed the initial state. On the other hand, if we can control the initial state and hence are able to prepare it at any given time — for example a technological operation in a primitive environment — then we may run the simulating procedure and wait for its termination before we prepare the initial state of the real system. In this situation the fact that the system's solution is open form does not at all prevent us from predicting the system because we can simulate the system's evolution in advance. We can summarize this insight as follows: *open form solution and uncontrollability implies unpredictability.* There is an objection to this line of reasoning coming from chaos in computer simulations. Continuous real-valued parameters cause errors of two kinds: the first results from the inaccuracy of their measurement, and the second results from the inaccuracy of their mathematical representation by finite strings of digits. The second kind of error is made in each iteration of an open form solution. On this reason, some authors have argued that iterative computer simulations may exhibit chaotic behaviour because the small rounding errors which are made in each iterative computation step may amplify each other in an exponential way (cf. Peitgen et al. (1992), ch. 1.5). However, I think that *if* this is the case, it results from sensitive dependence on intitial states but *not* from the mere fact of an open form solution. If there is no such sensitive dependency, the expected error of an open form solution will be not greater than that of a closed form solution. Let us demonstrate this more carefully at hand of two figures. Figure 1 shows a system that exhibits sensitive dependence on intitial states: trajectories (s_1, s_2) which start from *almost* the same initial point diverge from each other very rapidly — in the typical case with exponential divergence rate (cf. ch. 5-6). As a result unmeasurable differences in initial values get amplified and cause great differences in pedicted values. The point b is called a *bifurcation point*. Note that this is just the simplest case of diverging trajectories; typically chaotic systems have a great number of bifurcation points (they are indicated by the dotted curves, but note that trajectories with many bifurcation points are better displayed in time-independent or in a more-that-two-dimensional manner). The amplification of error in the closed form solution, where the computational algorithm directly moves from $s(t_o)$ to $s(t_p)$, is indicated by asterisks $*$; the amplification of error in the open form solution, where the algorithm moves from $s(t_o)$ to $s(t_p)$ via several intermediate steps, is indicated by the boxes ($\square$). As fig. 1 shows, there is no significant difference in error amplification, for in both cases the error is already 'catastrophic'; moreover, the errors due to rounding go in both directions (up or down) and hence do not sum up but cancel each other out (although, because of the exponential divergence, they don't cancel out to zero).

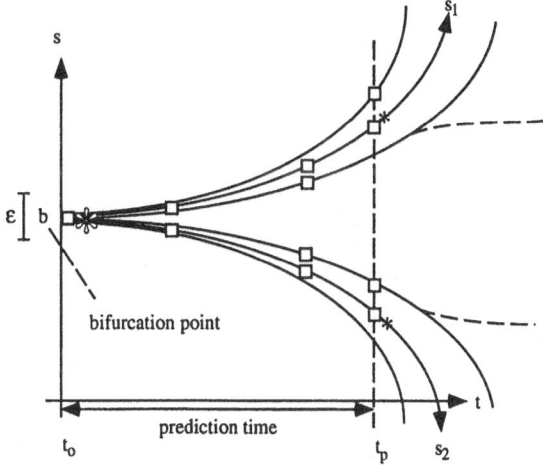

Fig. 1: Sensitive dependence on initial conditions because of exponentially diverging trajectories. ε = measurement accuracy level. $*$ = closed form, $\square$ = open form.

Figure 2 shows a system with *stable* trajectories — they do not diverge from (nor converge to) each other but keep in an almost constant distance (cf. Haken (1983), p. 132). Here neither the closed form solution $*$ nor the open form solution ($\square$) will produce an amplification of error. The expected error of the open form solution will be the same as that of the closed form solution, because (assuming the errors as randomly distributed), they cancel out to zero.

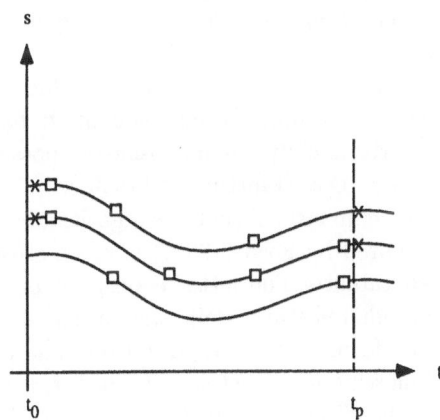

Fig. 2: A system with stable trajectories. $*$ = closed form, $\square$ = open form.

If the system is discrete but infinite, the situation concerning open form solutions is similar. If the system is discrete and finite, the situation changes because then we may be able to run the simulation starting from all possible states in S. Thereby we obtain a finite list which specifies for each initial state $s_o \epsilon S$ and each n the state s_n. By using this list, we are able to calculate the future state of the real system much quicker than it occurs in reality, even if we have no control over it. I conclude that open form solutions are not a fundamental obstacle to predictability — only in the case of systems with uncontrollable continuous or infinitely many states.

4   The Algorithmic Randomness Definition of Unpredictability

This concept of unpredictability is based on the algorithmic complexity theory as developed by Kolmogorov, Chaitin and Solomonoff (cf. Fine (1973), ch. V; see also Batterman (1993), ch. IV). Prima facie this concept applies to *discrete sequences* — trajectories in a finite state space with discrete time. The algorithmic complexity $K(SQ/I)$ of a finite sequence SQ given information I is defined as the length of the *shortest computer program* which generates (computes) the sequence SQ when combined with input I. To avoid this definition being relative to the underlying computer, one assumes this computer to be the universal Turing machine. To obtain the intended definition of randomness, one assumes the information I just to be the length n_{SQ} of the sequence SQ, and considers the behaviour of $K(SQ/n_{SQ})$ with increasing n_{SQ}. If $K(SQ/n_{SQ})$ increases with n_{SQ} in an unbounded way, hence if the limit of the quotient $K(SQ/n_{SQ})/n_{SQ}$ for $n \rightarrow \infty$ is positive, then the sequence SQ is called algorithmically random (cf. Batterman (1993), p. 57; Ford (1989)). The intuitive idea behind this definition is that if a sequence is random in the algorithmic sense then the shortest program which can generate it will be as long as the sequence itself — and hence will go ad infinitum if n goes ad infinitum. No lawlike redundancy whatsoever is embodied in an algorithmically random sequence which would allow to predict the entire sequence from a finite intial segment. It has been proved that sequences which are algorithmically random tend to possess all the standard statistical features of randomness (Martin-Löf (1966)).

First of all it has to be emphasized that the algorithmic randomness concept of unpredictability (promoted by Ford (1989)) is much stronger than the open form concept of unpredictability. The latter concept is based on the length of the *computation* of the predicted state, while the former concept is based on the length of the *program* performing this computation. The program of an open form solution has the following form: *set $s_o = k$; for $0 < n \leq p$ compute $s_{n+1} = f(s_n)$; halt* (where p is the discrete prediction time). If f is a function with a small computational complexity, which is independent from n, the length of this program will be rather short. For example, this is the case for the logistic function, where $s_{n+1} = 4s_n(1 - s_n)$. In contrast, the time which the program needs to compute s_n from s_o (via n steps) - in other words, its *halting time* — may be very long, (simply because recursive commands of unit length may cause iterations of any number). In other words, that a sequence SQ is algorithmically random does not only imply that there exists no closed form solution, but also that there exists no open form solution generated by an iterative function $f(s_n)$ with a complexity which does not increase with n. Many open form functions do not create algorithmically random sequences. For instance, the logistic function does not produce random numbers, if it is projected on a partition of the unit interval into three subintervals of equal length. In this case, certain combinations of the numbers 1, 2 and 3 will never be produced by the logistic function (cf. Peitgen et al. (1992), p. 396), and hence the sequence is not random, although the function is essentially of open form, not equivalent with any closed form function.

The interesting question is how algorithmic randomness is related to sensitive dependence on initial conditons. To answer this question one first has to find a way to apply the concept of algorithmic complexity, which is defined for discrete sequences, to continuous trajectories. This is done by partitioning the continuous state space S as well as the continuous time T into a finite number of 'cells', and by considering the projection of the trajectory on this partition — this projection is called a *symbolic* trajectory (this notion was first introduced by Hadamard; cf. Chirikov (these proceedings), ch. 2.2). Of course the algorithmic complexity of such a symbolic trajectory will depend on the underlying partition, but by making the partition finer and finer one obtains a limit algorithmic complexity which is partition-independent. It follows from theorems proved by Brudno (1983), White (1993) and Pesin (1977) that for almost all trajectories satisfying certain mathematical preconditions, their algorithmic complexity equals their metric entropy which in turn equals the sum of their positive Lyapunov exponents.[7] Thereby the Lyapunov exponents of a trajectory express the mean exponential rate of divergence of its nearby trajectories. If they are positive (negative), the trajecories are exponentially diverging (converging).

The Brudno-White-Pesin theorems establish a mathematical connection between algorithmic randomness and sensitive dependence on initial conditions. These theorems as well as their preconditions are highly complicated. Therefore it seems appropriate to try to explain the relation between algorithmic randomness (AR) and (exponentially) diverging trajectories (DT) from a purely qualitative point of view. The one direction, from DT to AR, seems to have an easy qualitative explanation. For given the trajectories are everywhere rapidly diverging from each other, then for every finite partiton $P(S)$ of the state space S, the information that the position of the particle at given discrete times $t_1, \ldots, t_n$ falls into certain cells $c_1, \ldots, c_n$ of $P(S)$ will *not* be sufficient to determine its

---

[7] Cf. Batterman (these proceedings), ch. 2; Chirikov (these proceedings), ch. 2.4. I have simplified matters: the discrete partitioning of S is needed for the definition of metric entropy, while for the definition of algorithmic complexity it suffices to consider a discrete open cover of S.

position at the next time point t_{n+1} - two trajectories coinciding in all these cells may spread between t_n and t_{n+1} into two distinct cells. Therefore the shortest computer program which can generate the symbolic trajectory $c_1, \ldots, c_n$ will always have to contain explicit information about all the discrete position values and hence will increase with increasing n in an unbounded way. So it will be algorithmically random.

The other direction, however, cannot be explained in such a straightforward qualitative manner. It is clear that if we are allowed to take as our trajectory any function, then AR will not imply DT. As an example, assume the trajectories to be functions mapping integers (Int) into themselves - hence both state space and time are discrete and infinite. Take any noncomputable function $f(x)$ and define the set of trajecories to be the set of all functions differing from $f(x)$ by an integer, $\{g(x) \mid g(x) = f(x) + m$ for some integer $m\}$. These trajectories will be algorithmically complex, because they are not computable, although their trajectories are stable. See fig. 3.

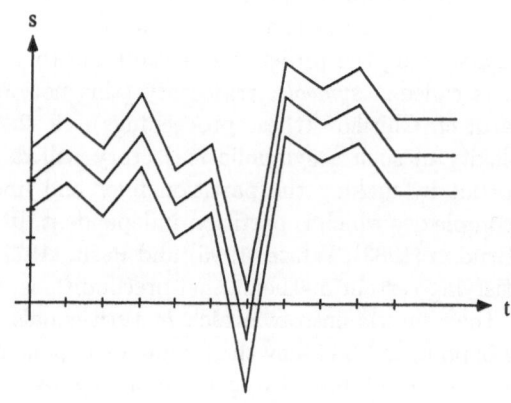

Fig. 3: Algorithmically complex but stable trajectories with discrete S and T.

But recall what was said in ch. 2 concerning time-dependent Hamilton operators. Not just any function can count as deterministic - only one which is sufficiently 'regular'. The functions of fig. 3 are totally irregular and hence cannot constitute a violation of the claim that every *deterministic* function which is algorithmically random must have diverging trajectories. Indeed, the preconditions of the Brudno-White-Pesin-theorems seem to be rather strong - they require the trajectories to be a continuous, inversely continuous, twice differentiable and measure-preserving mapping of a compact metrizable space into itself (cf. Batterman (these proceedings), ch. 2). I conjecture that these conditions are similar to the conditions for analytic functions, or functions representable by a Taylor series, which have been mentioned in ch. 2. For such a function f the

Brudno-White-Pesin-theorem has an easy qualitative explanation: if we let the values $f^n(t_0)$ of all the finitely many derivatives of f be the initial conditions, then these values determine the entire function. Hence if these values are given with some finite accuracy ε, then given the trajectories are stable, the values of f for all times t will be determined up to some given ε; and so, the expression $KSQ/n_{SQ}/n_{SQ}$ will go to zero for increasing n_{SQ}.

Let us finally consider the question of algorithmic randomness for systems with a discrete state space . If the state space is discrete *and* finite, algorithmic randomness is *impossible*. For then there exists a finite list $\subseteq S \times S$ specifying for each state in S which next state is determined by it. This list gives us a finite program which together with the given initial condition will generate the correct temporal evolution for a prediction time of arbitary length. So with increasing length of this temporal evolution (coded as a sequence), the algorithmic complexity will converge towards zero. Since every computer program is a finite and discrete system, it follows that no computer simulation can generate algorithmically random sequences, and hence that computer generated random numbers do not exist. This is important because there exist several computer algorithms for producing 'random' numbers. These computer generated 'random' numbers will always be pseudo-random, but not really random in the algorithmic sense. For similar reasons, computer generated chaos will always be pseudo-chaos in the sense explained by Chirikov (these proceedings), ch. 3.4.

Of course, if the state space is infinite, then algorithmic randomness is possible via uncomputable functions. But as was argued above, such functions can hardly be called 'deterministic'. So it seems that the only way for deterministic systems to produce random trajectories is via sensitive dependence on initial conditions because of diverging trajectories. If this is true, then deterministic unpredictability is a typical feature of *continuous* systems, because diverging trajectories lead to unpredictability only on the condition of *limited measurement accuracy*, and this condition is typical for continuous systems. In the following sections I turn to this latter concept of unpredictability. Also here we will have to face the problem that there are several different and nonequivalent concepts of trajectory divergence and, hence, of unpredictability. In ch. 5 I discuss some standard explications of trajectory divergence. Their common feature is that they are based on limit considerations. In ch. 6 I will suggest some pragmatic definitions of unpredictability in order to overcome certain problems of the limit conceptions.

5   Limiting Trajectory Divergence Concepts of Unpredictability

These concepts consider the *limit* behaviour of trajectories - their behaviour when the difference in initial conditions goes to zero and the prediction distance goes ad infinitum.

5.1   The Simple Trajectory Divergence Concept

This concept has been suggested by Suppes (1985), p. 192, in terms of Lyapunov-stability. A system is Lyapunov stable if two different trajectories keep arbitrarily close together, ad infinitum, provided their difference in initial conditions is sufficiently small. Let **S** denote the space of all possible trajectories, i.e. functions $s : S \rightarrow T$. Then the formal definiton of Lyapunov-stability and that of its logical contrary - (simple) trajecory divergence - are as follows. Note that the negation of (1) is logically equivalent with a condition, call it (2'), which is like (2) except that "$\forall \varepsilon$" is replaced by "$\exists \varepsilon$"; but diverging trajectories will always satisfy the stronger condition (2).

(1) Lyapunov-stability:
$$\forall \varepsilon > 0 \ \exists \delta > 0 \ \forall s, s' \in \mathbf{S} \ \forall t \geq t_o : \ |s_o - s'_o| < \delta \ \rightarrow \ |s(t) - s'(t)| < \varepsilon.$$

(2) Trajectory-divergence:
$$\forall \varepsilon > 0 \ \forall \delta > 0 \ \exists s, s' \in \mathbf{S} \ \exists t \geq t_o : \ |s_o - s'_o| < \delta \ \wedge \ |s(t) - s'(t)| \geq \varepsilon.$$

The Lyapunov stability concept of predictability is very *strong* - it implies the possibility of a prediction with a given accuracy level ε for the entire future, i.e. for *all* prediction times. Vice versa, the corresponding concept of unpredictability is very weak. It is violated whenever the trajectories of a system *diverge* from each other - not only if this divergence increases exponentially with time, but also if it only increases linearly in time. To give an example, assume I sling a ball fixed on a rope in a circle and let it go at some time. The angle of the line along which the ball will fly away will depend on the position where it was released. So the trajectories will diverge linearly from each other - formally they are functions of the form $s(t) = k.t + a$. Also in this case the trajectories are Lyapunov unstable (see fig. 4).

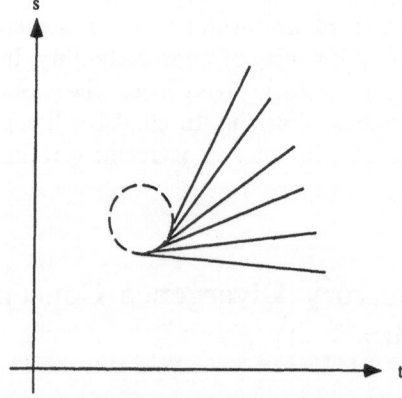

Fig. 4: Linearly diverging trajectories (produced by a slingshot experiment).

The trajectories of fig. 4 are Lyapunov unstable and hence unpredictable according to the trajectory divergence concept. Since measurement of the release time has finite accuracy, it will not be possible to predict the ball's position with a certain accuracy ε for *all* future times. *Intuitively*, however, I think one would not speak in this case of an unpredictable system. It seems that the condition is too weak.

5.2   The Exponential Divergence Concept

This is a stronger concept which identifies unpredictability with *exponential* divergence of trajectories. It is connected with the Lyapunov coefficients which have been mentioned in ch. 4 and express the mean exponential rate of divergence of two nearby trajectories. If the trajectories diverge exponentially from each other, the Lyapunov coeffficients are positive (if they converge exponentially, the Lyapunov exponents are negative, and otherwise they are zero). Hence, this concept of unpredictability is equivalent with that of positive Lyapunov exponents. It is the concept of unpredictability which underlies the Brudno-White-Pesin theorems of explained in ch. 4.

The simplest example of exponentially diverging trajectories is *exponential growth* $(s = s_0.e^{kt})$, as shown in fig. 5. (The more complicated example of fig. 1 is discussed soon.) Is this concept sufficient to explicate our intuitions of an unpredictable or even a chaotic szenario? I doubt that. Usually we are not interested in predicting the infinite future but only some finite future. As has been remarked by Batterman (1993), p. 52f, exponentially diverging trajectories do not prevent us from making finite predictions with arbitrary accuracy. It is obvious from fig. 5 that for each future time point t we may predict the systems state $s(t)$ with arbitrary accuracy provided we make the accuracy in the initial conditions sufficiently small. In other words, the following will hold in the case of exponential growth:

(3) $\forall t \geq t_0 \ \forall \varepsilon > 0 \ \exists \delta > 0 \ \forall s, s' \in \mathbf{S} : \ |s_0 - s_0'| < \delta \ \rightarrow \ |s(t) - s'(t)| < \varepsilon$
(continuous dependence on initial conditions)

Condition (3) is Hadamard's condition of continuous dependence on initial conditions (cf. Batterman (1993), p. 53). Batterman (1993), p. 53f tells us that Hadamard's condition fails for some systems which are governed not by ordinary but by partial differential equations. Such systems admit immediate growth of trajectory divergence, that is, they satisfy the following

(4) $\forall t \geq t_0 \ \forall \varepsilon > 0 \ \forall \delta > 0 \ \exists s, s' \in \mathbf{S} : \ |s_0 - s_0'| < \delta \ \wedge \ |s(t) - s'(t)| \geq \varepsilon$
(immediate growth of trajectory divergence)

A trajectory space with immediately growing trajectories is shown in fig. 6.

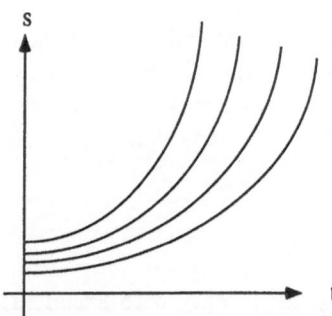

Fig. 5: Exponential growth.

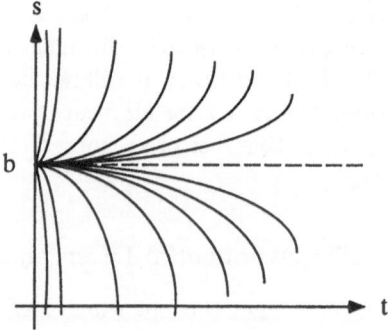

Fig. 6: Immediately growing
trajectories - condition (4).

Note that (4) is much stronger than the negation of (3), which is

(5) $\exists t \geq t_0 \; \exists \varepsilon > 0 \; \forall \delta > 0 \; \exists s, s' \in \mathbf{S} : \; \mid s_0 - s'_0 \mid < \delta \; \wedge \; \mid s(t) - s'(t) \mid \geq \varepsilon$
(negation of Hadamard's condition)

As Batterman remarks, Hadamard's condition follows as a corollary from the existence and uniqueness of solutions, and he seems to presuppose that all ordinary differential equations have unique solutions (Batterman (1993), p. 54). But this is not generally true — many ordinary differential equations have solution manifolds containing some *singular* points — (initial) points which are shared by many or even all trajectories (cf. Bronstein and Semendjajew (1973), p. 381; Haken (1983), p. 126). Consider again the trajectories of fig. 1 with one bifurcation point. If many or even all trajectories of fig. 1 would start from the point b (or go through the point b), then b is a singular point and Hadamard's condition is violated. This situation is shown in fig. 7, and the following condition will hold in such a case:

(6) $\forall t \geq t_0 \; \exists \varepsilon > 0 \; \forall \delta > 0 \; \exists s, s' \in \mathbf{S} : \; \mid s_0 - s'_0 \mid < \delta \; \wedge \mid s(t) - s'(t) \mid \geq \varepsilon$

Note that (6) is stronger than the negation of Hadamard's condition (5) but weaker than immediate growth of trajectories (4).

The situation of fig. 7 contains *branching* trajectories. So, according to what was said in ch. 2, the underlying system is no longer a *deterministic* one — for initial point b, the future is 'undetermined'. On the other hand, if the trajectory space contains no trajectory branching, then the initial values of the trajectories in fig. 1 will only come arbitrarily close to the point b without crossing it — the only trajectory which goes through point b will be the bifurcation line. This situation is shown in fig. 8, and for this situation, Hadamard's condition is again satisfied. For here, the trajectories above and below the bifurcation line may come arbitrarily close together provided their initial values are sufficiently close to the bifurcation point b.

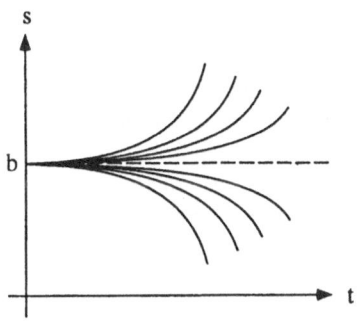   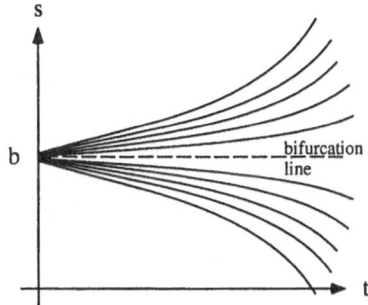

Fig. 7: A singular point without   Fig. 8: Bifurcation without
unique solutions - condition (6).   violation of condition (3).

Hunt (1987), p. 130, has argued that whenever the trajectory space contains a bifurcation point, Hadamard's condition will be violated. I do not know whether there exists a unique definition of the notion of a bifurcation point and what it is. However, if both figures 7 and 8 are examples of bifurcations, then Hunt's claim cannot be generally true - it holds only for fig. 7 but not for fig. 8.

It seems to me that Batterman's claim that Hadamard's condition is generally satisfied for all ordinary differential equations is true if it is restricted to those which produce *deterministic* trajectory spaces. However that may be, many systems usually subsumed under the rubric 'chaotic systems' satisfy Hadamard's continuity condition. Therefore, as a general condition for unpredictability the negation of Hadamard's continuity condition seems to be too strong. Hence we are confronted with a kind of dilemma. We have three explications of unpredictability in the sense of sensitive dependency on initial conditions. The first two are too weak and the third is too strong. Is there something in between which fits our intuitions better? I think that every limit concept of unpredictability will be unsatisfactory in this respect, because our intuitions of unpredictability and chaos are essentially pragmatic. We are not interested in predictions about the limit behaviour but about a finite prediction distance, and we cannot measure the initial conditions with arbitrarily high precision but only with finite precision. Also, the error in our predictions need not be made arbitrarily small, but only if it exceeds a certain relevant degree, it will be practically harmful. Based on these considerations I will suggest in the next chapter some pragmatic definitions of unpredictability.

6   Pragmatic Trajectory Divergence Concepts of Unpredictability

I make three pragmatic assumptions. *First*, there exists a smallest measurement accuracy ε_0 of the initial conditions relative to the given background of theoretical knowledge as well as the given technical possibilities. Let me emphasize that it follows from quantum theory that there are not only technical reasons but also reasons in principle for the assumption of a finite lower bound of measurement accuracy — for if ε reaches quantum dimensions then Heisenberg's uncertainty relation will imply that a further decrease of ε is impossible. *Second,* there exists a smallest level δ_0 of differences in our predictions which count as practically relevant, relative to the given background of practical circumstances and goals. *Third,* I assume there exists a smallest prediction time t_p which counts as interesting; hence we are interested in making predictions only for prediction times $t \geq t_p$.

Now I call a system pragmatically unpredictable iff the following holds:

(7) $\forall t \geq t_p \; \exists s, s' \in \mathbf{S} : |s_o - s'_o| < \varepsilon_0 \; \wedge \; |s(t) - s'(t)| \geq \delta_0$
(Pragmatic unpredictability)

In words, for all practically relevant prediction distances there exist trajectories with unmeasurable difference in their initial points but practically relevant difference of their predicted state. The negation of (7) is written down in (8). It may be called weak pragmatic predictability, because it says that there is at least *some* future time for which all practically relevant differences in the prediction outcome will be caused by measurable differences of the initial states. In contrast, strong pragmatic predictability says that this holds for *all* future times.

(8) $\exists t \geq t_p \; \forall s, s' \in \mathbf{S} : |s(t) - s'(t)| \geq \delta_0 \; \rightarrow \; |s_o - s'_o| \geq \varepsilon_0$
(weak pragmatic predictability)

(9) $\forall t \geq t_p \; \forall s, s' \in \mathbf{S} : |s(t) - s'(t)| \geq \delta_0 \; \rightarrow \; |s_o - s'_o| \geq \varepsilon_0$
(strong pragmatic predictability)

Let me finally discuss Batterman's example of the roulette wheel in the light of this definition. Batterman's point is that despite the fact that the differential equations describing the roulette wheel are linear and integrable, which is the typical situation of regular and nonchaotic behaviour, the roulette wheel is used in practice to produce a completely unpredictable random process (Batterman (1993), pp. 63–65). Assuming a small friction, the spinner's trajectories will even be Lyapunov-stable: since the movement of the wheel comes to an end after some finite time, it will be possible to keep the trajectories of the spinner arbitrarily close together for *all* future times if we specify the initial condition — the initial momentum of the wheel — with a sufficient accuracy. Hence the roulette wheel is predictable according to the limit concept, and thus it is not algorithmically random (according to the Brudno-White-Pesin theorems). But still we use the roulette wheel to produce a random process. How is this possible?

With our pragmatic concept of unpredictability we have an easy explanation for that. Consider the *time-independent* trajectories of the roulette wheel in the state space with the angular momentum Θ and the position s of the spinner in fig. 9. The circular path of the spinner is projected on a linear axis s with marks for the periodic circles — $2\pi r$ is the length of one period, where r is the radius of the circle (the third axis of time lies perpendicular to the plane of the paper.) We assume an almost linear relation between angular momentum and position. Although the trajectories are stable linear functions, we have pragmatic unpredictability: because of the *small* friction, the gradient of the trajectories is so small that unmeasurable differences in the initial momentum of the spinner will cause 'hughe' differences in the position where the spinner stops: differences which are greater than one period of the spinner and hence which are much greater than the practically relevant difference of the outcome, which is $2r\pi/37$, if the circle is divided into 37 numbered intervals. Hence it will be completely impossible to predict the 'number' at which the spinner will stop from any measurement of the initial momentum. The moral of these considerations is that pragmatic unpredictability is very different from the various kinds of dynamic irregularity or algorithmic complexity.

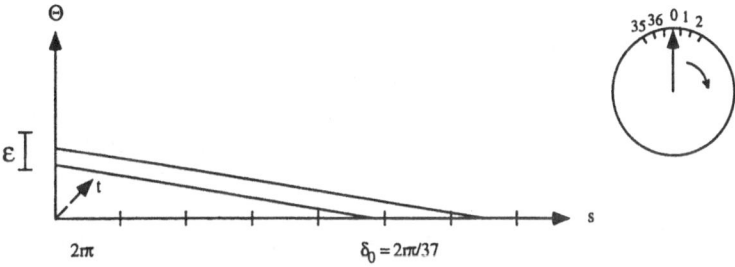

Fig. 9: Time-independent trajectories of a roulette wheel.

7 From Unpredictability to Chaos

Chaoticity is certainly something stronger than unpredictability. Though this is not the main focus of this paper, let us finally ask in what this stronger property consists. There are several suggestions to characterize chaos in terms of the mathematical features of the differential equations describing the system. For instance, Batterman (1993), p. 62, suggests as a necessary condition for a chaotic system that it must not be integrable, i.e., its differential equation must not admit of an exact closed form solution. I am not sure whether this condition is really necessary. On the other hand, it is certainly not sufficient because there might be open form solutions which behave quite regular. Another frequently mentioned characterization of chaos often mentioned is that the differential equations have to be nonlinear. Nonlinearity gives often raise to bifurcations in the trajectory space; however, it does not necessarily lead into chaotic behaviour (cf. Chirikov (these proceedings), ch. 2.4).

I want to propose that there are three conditions which are necessary and taken together sufficient for chaos. *One* condition is pragmatic unpredictability. For according to the limit concept of unpredictability, i.e. the trajectory divergence concept, even our solar system is unstable and chaotic (cf. Chirikov (these proceedings), ch. 2.5), but the time after which it becomes unstable is of the same dimension as the cosmological time and hence without practical relevance. I think it makes no sense to call our solar system chaotic if we want to avaoid making the concept of chaos almost empty; therefore I think that pragmatic unpredictability is a necessary condition for chaos. But pragmatic unpredictability is not enough for chaos, as is seen from the roulette wheel: though it is pragmatically unpredictable, its trajectories are regular and stable and thus not at all chaotic. So I think that a *second* condition for chaos is *exponential trajectory divergence*, and thus, by the Brudno-White-Pesin theorems, *algorithmic randomness*. But also this is not enough, for intuitively we want to distinguish chaotic behaviour from exponential growth (or exponential 'explosion') — and the trajectories describing exponential growth satisfy the second condition, and with a suitably chosen relevant outcome interval δ_0 also the first condition. As emphasized by Chirikov (these proceedings), ch. 2.4, the important *third* condition of chaos is the *boundedness* of the trajectories: in distinction to exponential growth, the (time-independent) trajectories remain within a finite region of the state space. This third condition explains several further characteristic features of chaotic trajectory spaces. First, the boundedness of trajectories is usually produced by adding a nonlinear term to a linear differential equation; hence the importance of nonlinearity. Second, boundedness together with algorithmic randomness implies that the (time-independent) trajectories will oscillate in a finite region of the state space without being periodic, i.e. recurrent in time (cf. Weingartner 1995, ch. 1.3.4); for if they were periodic, they could not be exponentially diverging from each other. This implies, third, that the 'symbolic' trajectories mentioned in ch. 4 will contain all possible sequences (Chirikov (these proceedings), ch. 2.2) and thus simulate a statistical random experiment; and fourth, that the set of trajectories starting from one finite cell will, after some time, have filled the entire state space (cf. Weingartner 1995, ch. 1.3.4).

Let me conclude with a conceptual problem. It seems to me that the third condition of boundedness is - not in conflict with the first condition of pragmatic unpredictability, but - in *conflict* with the second condition of trajectory divergence. For if the trajectories remain within a *finite* region S_f of the state space S, then it is impossible that the mean distance of neighbouring trajectories increases - linearly or exponentially - with time in an unrestricted way: the distance will never exceed the 'diameter' $| S_f |$ of the finite region S_f. So strictly speaking, the *limit* definition (2) of trajectory divergence cannot be satisfied if the trajectories are bounded.

I do not know what the best solution of this problem will be. Maybe we should drop the second condition and only work with the first and the third. Alternatively, we could restrict the second condition of exponential divergence to some finite initial segment of time. However, these considerations lie beyond the scope of this paper.

References

Batterman, R. (1993): Defining Chaos. *Philosophy of Science* **60**, 43 - 66

Batterman, R. (these proceedings): Chaos: Algorithmic Complexity versus Dynamical Instability

Bronstein, I. N., Semendjajew, K. A. (1973): *Taschenbuch der Mathematik* (Verlag Harri Deutsch, Zürich)

Brudno, A. A. (1983): Entropy and the Complexity of the Trajectories of a Dynamical System. *Transactions of the Moscow Mathematical Society* **2**, 127-151

Chirikov, B. (these proceedings): Natural Laws and Human Prediction

Fine, T. (1973): *Theories of Probability* (New York, Academic Press)

Ford, J. (1989): What is Chaos, That We Should be Mindful of It? In: P. Davies (ed.), *The New Physics*, Cambridge Univ. Press, Cambridge, pp. 348-371

Haken, H. (1983): *Synergetik. Eine Einführung* (Springer, Berlin)

Hunt, G. M. K. (1987): Determinism, Predictability and Chaos. Analysis **47**, 129-132

Martin-Löf, P. (1966): The Definition of a Random Sequence. Information and Control **9**, 602-619

Peitgen, H.-O. et al. (1992): *Bausteine des Chaos - Fraktale* (Springer und Klett-Cotta, Berlin and Stuttgart)

Pesin, Ya. B. (1977): Characteristic Lyapunov Exponents and Smooth Ergodic Theory. Russian Mathematical Surveys **32**, 55-114

Russell, B. (1953): On the Notion of Cause In: H. Feigl and M. Brodbeck (Eds.), *Readings in the Philosophy of Science* (Appleton-Century-Crofts, New York)

Schurz, G. (1989): Different Relations between Explanation and Prediction in Stable, Unstable and Indifferent Systems. In: P.Weingartner/G. Schurz (eds.), *Philosophy of the Natural Sciences (Proceedings of the 13th International Wittgenstein Symposium)*, Hölder-Pichler-Tempsky, Vienna, pp. 250-258

Stone, M. (1989): Chaos, Prediction and Laplacean Determinism. American Philosophical Quarterly **26** (2), 123-131

Suppes, P. (1985): Explaining the Unpredictable. Erkenntnis **22**, 187-195

Weingartner, P. (these proceedings): Under what Transformations are Laws invariant?

White, H. (1993): Algorithmic Complexity of Points in Dynamical Systems. Ergodic Theory and Dynamical Systems **13**, 807-830

Zachmann, H. G. (1973): *Mathematik für Chemiker* (Verlag Chemie, Weinheim, 3rd edition)

Discussion of Gerhard Schurz' Paper

Batterman, Chirikov, Miller, Noyes, Schurz, Suppes, Weingartner

Chirikov: According to authorized schedule, we begin with the discussion of the talk of Gerhard Schurz about various kinds of unpredictability in deterministic systems including his own new conception of pragmatic unpredictability.

Suppes: It is not that I really disagree with what you have to say but I want to emphasize that for a good pragmatic sense of unpredictability - as for example in gambling - we must have a more detailed notion than a general notion of randomness. For instance, if we have a roulette wheel as in your example, we would have randomness under the definitions being given here for example of complexity when the distribution of the frequency of numbers was very uneven. Under the terminology we are using, we got that it is random. As I understand your terminology you would want to say pragmatically that it is unpredictable. But there are sharper notions. There is a sharpening of that notion that is very important in gambling, so it is very pragmatic. For example, to know what the actual distribution is, as for roulette wheels at Las Vegas. The actual empirical distribution is rather important, so it is not just enough to have a general notion. Because when you bet according to whether frequencies are higher than average, for example, the actual distribution is rather important. It is the same with horse-races. It is not at all important in gambling to have a general notion of randomness, but it is absolutely everything to have a detailed notion with a quantitative concept. It seems to me, that is one aspect of pragmatic unpredictability, that is not sufficiently emphasized in your discussion. Under the Kolmogorov complexity definition, consider a 0-1-sequence of the following sort: At each position of the sequence which is a power of two - 2^1, 2^2, 2^3, ... - and just at these exact powers we have a randomization by flipping a coin. All other positions are 1's. Now by this complexity definition this is an unpredictable sequence, because you cannot code any actual sequence with less than infinite length, but most of the time you could do an extremely good job in predicting. So that is why I am emphasizing that for gambling purposes a much more detailed notion is needed, it seems to me, for a serious pragmatic notion of unpredictability.

Schurz: I completely agree. My definition was pragmatic only in the sense that the accuracy level of the initial data is fixed and it depends on the background, on the background of practical possibilities - and it was pragmatic also in the sense that the significant difference level in the outcome depends on the background. And I wanted to show that if I have such a notion of unpredictability, then even those systems where we have a linear relation between initial states and outcomes might be unpredictable if the linearity coefficient is very small. So this was the point of my definition. I agree fully that to have a good concept of unpredictability one has to consider also the actual distribution and only if this distribution is really symmetric then there will be unpredictability in the statistical sense.

Noyes: On the pragmatic aspects of gambling I would like to point out that - not for coin tossing but for a rolling dice - a normal person, if you bank the dice off a corner so it hits one corner, hits another side and then comes out, cannot control the die. But I am told that people who have practised that for a period of several years developed a kinesthetic sense that allows them to throw the die and get a desired result even on two banks anywhere. So this is not a mechanical thing but it is a human thing. So if you have a human operator that has developed the skill then your statistics are not going to do you any good because he could fool you. You know this is supposed to be unpredictable, supposed to be guaranteed unpredictable because any normal person, throwing two die and banking them off two corners will not be able to control the result. But a person, who has developed the skill can and can make a lot of money out of it.

Chirikov: I have a brief remark that your ideas seem to me too pragmatic. Of course, it is a very important thing of life, I believe, but I'd like to attract the attention that there is also an opposite goal, namely, to understand the fundamental properties of predictability and unpredictability, to understand properly the phenomenon which we call dynamical chaos. So I would rather agree with your understanding of pragmaticity in the sense that you don't know how to fix this arbitrary accuracy. This, in my opinion is a disadvantage of this particular notion, because you need to fix it somehow from physical point of view or mathematical or whatever. And in classical mechanics there is no limit. In quantum mechanics you mentioned there is indeed a physical limit for accuracy but not in classical mechanics. So, we need to discuss other interpretations of this notion.

Schurz: I think that when the distances get so small that they arrive at quantum distances then we have a natural physical limit of measurement accuracy. So this is one limit. But I don't know whether this limit is practically very significant. Actually, in practical situations the limit of measurement accuracy will be determined by the context and by the particular kind of system and experiment, and so in this sense it is pragmatical. My point was that the limit definitions of predictions - positive Lyapunov exponents, diverging trajectories and so on - they are independent of any pragmatical fixation. This is their benefit. But on the other hand, if you are interested in finite predictability, they don't tell you much about the possibility of making finite predictions. This was the reason why I suggested these other notions.

Suppes [to Chirikov]: I want to comment on your comment on pragmatic. You say in very many of the physical situations you certainly cannot derive from fundamental considerations the actual probability distribution of outcomes.

Chirikov: I disagree ...

Suppes: I would claim that in many cases you certainly cannot, but you may still have very good ideas. For example, in statistical practice what is extremely important for understanding the phenomena of a certain kind of level is the use of the normal Gaussian distribution. The normal Gaussian distribution is used repeatedly in situations where we know that this is not the exact distribution, but there are many good arguments for using it. And when you estimated the mean, variance and covariance matrix in the case of multidimensional properties, then even though you haven't derived that from fundamentals it will do an extremely good job of analysis, and you are not able to derive from fundamentals what you think the theoretical distribution really is. That is very important in terms of understanding phenomena, when it is too hard to do things from fundamentals.

Chirikov: You mean you cannot derive that technically ...

Suppes: Right ...

Chirikov: ... not principally.

Suppes: Yes, exactly. Technically you cannot derive.

Noyes: My comment may have sounded somewhat superstitious, but it does have an important aspect when one is dealing with experimental data. If you are testing any of these things and you rely on an experimental physics team or group or laboratory, then those of us who deal with experimental physics data realize that some people's results are more reliable that others. And you have to deal with this in terms of your feeling about how the people do accurate experiments or not. And often you cannot put your finger on why some groups get good experimental results that are reproducable and others do not. And so the human element I was talking about in terms of a professional gambler - he actually has a practical application in comparing physics data with theory. You have to know what the track record of a laboratory is in terms of producing a reliable results. And I had this experience myself, because I do deal with data analysis in physics. And some people you can trust and some groups you can trust, some laboratories you can trust and others you cannot. One of the best experimental physicists, Emilio Segré, made the point by saying: "You cannot measure errors." And I think that is something that has to be taken into account. You don't know what is going wrong when an experiment goes sour, and you really don't have a control over the physical situation in circumstances where you are dealing with what is called systematic error in statistics. It is not a mathematical problem, it is a much more experiencial one. I am just trying to emphasize that when we are talking about a science that relies on experiment like physics, then these considerations are often ignored in the mathematical treatment. But they are still very real when it comes to whether you trust the test of a theory or not, whether it be mathematical or physical.

Batterman: I have a question about your characterization of determinism right at the beginning. Did I understand you correctly when you said that determinism entails the time translation invariance of the laws?

Schurz: Yes.

Batterman: Suppose I have a law which states that the gravitational constant, or some other constant, will double every year which is a prime number. Wouldn't that be a deterministic law which is not time translation invariant?

Schurz: If the law says that the gravitation constant doubles every year then it is still time invariant because no initial data are known, the gravitation constant is not fixed for any timepoint. But if the law says the gravitation constant is at some arbitrary chosen time this and this and from this point it will double twice a year then it will be not invariant with respect to time translation.

Batterman: But it would still be a deterministic law.

Schurz: But in this case we will never have two states of a system which are identical.

Batterman: Right, there will be a privileged point in time.

Schurz: In this case - even if we had completely accurate observations, if we had complete knowledge about the system - we could never make a decision about whether the system is deterministic. We always have a function which describes the development of the system in time completely. So what does determinism mean if you always have such a function? It means that if two systems of the same kind are in the same initial state their future will be the same. But if the states of the system cannot be described independent from time so the actual time that is always a parameter of the state of the system, then we can never have two systems of the same state. This is a problem in this concept. Am I clear enough? - I have not really tried to prove that the translation invariance with respect to time follows from this definition of determinism. So your question is interesting. My claim is: if you have the class of all states of systems of a given kind, and you don't have two which are the same, then the condition that for same initial conditions the next condition will be the same just makes no sense. It is always trivially verified. All systems will be deterministic in this sense.

Batterman: But, there are time dependent systems, right?

Schurz: Yes, but translation invariance does not imply time independence. It does only imply that the law describing the dependency on the time is invariant with respect to translation of the time coordinates of the system.

Batterman: O.K.

Chirikov: If you have a system with a Hamiltonian explicitely dependent on time I would call it a purely dynamical system as well or deterministic in a more philosophical language. It is a standard situation and the difference from indeterministic is in that the time dependence is given explicitely and not by statistical means, not for example as a noise dependent on time when you fix only statistical properties of the noise but not the exact function of time. This is the difference I see. And as to the gravitational constant I would say it may look as depending on time, but actually it may depend on the gravitational interaction of the whole universe and in the future theory you may simply come back to the closed conservative system with some new, different interaction, different Hamiltonian and with some different new universal constants. So this is a very particular case of course, nobody even knows if it depends on time or not. But if you have the explicitely given time dependence of the Hamiltonian there is of course no difference from a conservative system as far as it concerns determinism. This is my understanding.

Batterman: I was just wondering whether, conceptually, determinism entails time translation invariance. It seems that I could imagine that there is a distinguished point in time, and I could construct some law, involving for instance the primes, which would seem intuitively deterministic but nevertheless not time translation invariant.

Chirikov: I would also say that deterministic is somewhat unclear, not very well defined the notion, because it is not physical, I am sorry.

Schurz: That was why I wanted to give some definition.

Chirikov: You see the usual notion is that of a dynamical system, sometimes we call it purely dynamical - to emphasize that it is not a stochastic system, not a statistical one.

Suppes: I have a comment on this time dependence. Certainly we can give a general characterization that we expect the existence of a unique solution for the given boundary and initial conditions and smoothness conditions.

Chirikov: ... and for this given time dependence.

Suppes: We can certainly write a system of classical particle mechanics where the forces are dependent on time explicitly and they are not translation invariant. I mean that is certainly conceptually straightforward to this and it meets counterfactual criteria. That is, you can have different initial conditions and everything still is going to work and be determined with the appropriate smoothness conditions. I think that's a generalization. It certainly is one that you can find in the physical literature. But then you could not prove invariance under time translation for such systems.

Chirikov: This simply means that then the question of time invariance is not answered.

Suppes: Yes.

Schurz: I am not sure about the answer to this question. I said it was a conjecture. My consideration was about the possibility that time is causally efficient, so the laws describing the dependency of the forces on the time are themselves not invariant with respect to translation in time. If this is the case and I consider two different states of the system, then even if they are the same in all respects, they are at different time-points, and so if the future development of these states is different then the reason for this might be just that there initial time was different.

Suppes: No, no, you have the same variation of possibilities of initial and boundary conditions so you have many possible trajectories. I mean that it is very straightforward in terms of physical conceptions, not something strange here. You have smooth forces as a function of time. When you say that the energy situation is changing, you are not saying this is fundamental physics. You are simply saying you can have ...

Chirikov: ... an open system.

Suppes: An open system, yes ... for example you are expending ...

Schurz: Maybe this is a misunderstanding. I meant that the fundamental laws of the differential equation are invariant with respect to translation in time, not those boundary conditions.

Weingartner: I want to make a small remark that there is a problem with invariance of fundamental laws with respect to time. The problem is with neutral K-mesons which are responsible for a small violation of CP-invariance (charge-parity invariance).

Chirikov: Time reversibility?

Weingartner: Yes, Past-Future invariance of the fundamental laws. The CP-violation affects indirectly time reversibility because one assumes CPT-invariance, the combination of charge, parity and time which has been always confirmed so far. So if the CP, the charge-parity is violated in some special cases with neutral K-Mesons, then the time has to outbalance this violation. And that means that there could be problems with the time invariance, time reversibility of the fundamental laws. This is a big problem and I think there is no solution to this so far and many do not even permit themselves to think what would follow in a lot of consequences from that.

Chirikov: We permit ourselves everything here ... haha. I simply would like to mention that indeed there is a problem of time reversibility or irreversibility in statistical physics, and one answer - some people speak about this - one solution of this irreversibility is including the CP-violation. But in my opinion, this is a too special event to explain the whole thing.

Weingartner: Most think that way. But I am not sure whether we should be satisfied with that.

Chirikov: You don't need to involve such a very special interaction even though it is virtually present. You don't need, you can explain it in a simpler way, in a cheaper way.

Weingartner: That would be nice.

Suppes: I want just to make a comment on your reducing the problem to fundamental laws. I would express scepticism about that. I don't know how you make an argument that there are fundamental laws. It does not seem to me that that's an easy thing to establish.

Schurz: Maybe you are right. I made these conjecture because it were interesting if the notion of determinism would imply some invariance. If one has no clear distinction between fundamental laws and boundary condition, my idea would not work.

Miller: Let's change the subject. You said - I think - that the logistic function does not produce a random sequence.

Schurz: I had a talk with an expert from our Mathematics Department in Salzburg. He is working on the logistic function and much more expert than I am and he told me that for almost all initial values the logistic function gives indeed random series if you code the outcome into zero and one, into two halfs. But the book where I read this uses a coding into three intervals, one, two and three. And with this coding it does not.

Miller: It's certainly true that it does not give you numbers with equal distribution, it is not symmetric in the three parts.

Schurz: It is not symmetric, but also certain combinations of the three digits do never occur.

Miller: But is it necessary for randomness that every combination occurs? It is necessary for the von Mises-definition of randomness that every sequence should eventually occur as a part of the whole sequence. Is it necessary on the computational complexity definition?

Suppes: Yes, because Kolmogorov sequences are von Mises sequences.

Chirikov: No, this is a question, let me just mention. This implication is only true when you have ergodicity. But the typical dynamical systems are not ergodic, yet they can be algorithmically complex which is called a chaotic component of the motion. Then, it does not work, this implication.

Suppes: But the implication I am referring to does not involve any assumptions about dynamical systems. This is a phenomenological characterization where you are only looking at the sequence, not what generates it, and there is a clear theorem that Kolmogorov complexity implies von Mises' randomness but not vice versa.

Miller: Are there not more subtle definitions of randomness that take account of the finite initial segments? A sequence beginning with a trillion ones and then becoming von Mises random is von Mises random, for it has limiting frequencies.

Suppes: No, because in the infinite case it is Kolmogorov random.

Miller: But it is not Martin-Löf random, as far as I know.

Suppes: It is if you take the infinite. You have to distinguish between the infinite sequences and the finite sequences.

Miller: But only the finite sequences can matter empirically or phenomenologically.

Suppes: You know, 'phenomenological' in an ideal sense. I agree you must distinguish, but it is quite a different matter when you talk about the finite sequences. Then there's a very different arena of discussion, with much less agreement and sharpness of results in the finite case.

Batterman: In John Earman's book, "A Primer on Determinism", there is an argument to the effect that the length relativized complexity of a finite sequence resulting from tosses of a biased coin will fail the randomness test. However, if you further conditionalize on the weight, defined as the number of heads, say, in sequence, you can restore the intuition that the sequence is random, even though the relative frequency of heads isn't $1/2$.

Chirikov: I like just to mention what the situation is in my understanding. I am not a mathematician but a person interested in this Kolmogorov theorem. The situation is quite opposite: there is a theorem that almost all finite sequences are random even though you cannot calculate their complexity.

Batterman: Oh, that's true.

Chirikov: Why? No infinite sequence is random according to the prior definition. If you assume that the Kolmogorov complexity, not per unit time but the whole complexity of the sequence must be equal, like for finite sequences, just to the length of the sequence, then this never happens. It is a very interesting theorem. Always you can find some sections of the infinite sequence which would have less complexity, namely the length of this segment minus its logarithm. This was one reason why physicists and some mathematicians turn from the total complexity of the sequence to the specific complexity per step, or per unit time. For the finite sequences everything remains OK., nothing special happens besides that you cannot calculate whether a particular finite sequence is random or not.

Schurz: My trouble on this notion was whether the notion of algorithmic complexity makes good sense in the finite case, because it depends on the program, on the commands, and the translation into the universal Turing machine introduces some constants. The length differs from the length of a Universal Turing machine by a constant.

Suppes: There is a beautiful article by Kolmogorov in an Indian Journal which gives a very nice characterization for finite sequences and the whole point is, the delicate computation there by Kolmogorov is, depending on the length of the sequence, the number of algorithms that you must test. Clearly, if you test all algorithms, a finite sequence will not be random and the whole point of Kolmogorov's ...

Chirikov: No, no, no. It depends on the length of the algorithm. You always can find an algorithm. One is just the copying machine.

Suppes: But the important thing about Kolmogorov's delicate computation is exactly how much you should test a finite sequence. It is quite a subtle argument as to exactly how to characterize the testing. And I don't think it is really completely agreed upon even now, but the definite problem is far beyond the simple Turing machine with a constant to analyze ...

Miller: These ideas about finite random sequences were already in Popper's "Logik der Forschung", sections 51-64, in 1934. If you test a finite sequence too much, then it will, by von Mises's standards, turn out to be non-random. For sequences of length 2^n, if you just select by subsequences of length less than n, then you can give a definition of finite random sequence. Now Popper's own definition was shown by Ville not to be satisfactory.

Suppes: That is a rather complicated problem.

Miller: Yes, but the idea that you are mentioning from Kolmogorov is quite old.

Suppes: The virtue of Kolmogorov is a very detailed calculation. That is the virtue.

Miller: Do we agree that the logistic function produces random numbers?

Schurz: One of my considerations was also that, if I am right, then no deterministic system with discrete finite space can be algorithmically random in the infinite case. I think a consequence of this is that there can be no computer generated random numbers, because you always have a finite program which knows all the transitions between the discrete states and so we have a finite recursive procedure which may compute every sequence of every length. So if one takes this concept seriously then no computer generated random numbers exist. This troubled me because I have read a lot of books where some mathematicians are proud to be able to produce random sequences with help of computer programs and so with help of discrete and finite spaces. Maybe someone has an answer to this.

Chirikov: You are speaking about infinite sequence on a discrete lattice of finite size?

Schurz: Yes.

Chirikov: You may consider this but you also may consider a finite discrete lattice and finite sequences. Then this is a question ...

Schurz: But this is a difference.

Chirikov: In this limiting case you are right. But also you may consider discrete lattice, but infinite.

Schurz: With no boundaries...

Chirikov: Then I don't know the answer.

Schurz: Then it would also be possible, because there is no finite description of all the transitions.

Chirikov: Maybe, maybe not, I don't know.

Schurz: I had this picture because in the usual statistical definitions of randomness or in the usual statistical setting you have a discrete state space and you consider infinite sequences.

Chirikov: Discrete space is only in quantum mechanics.

Schurz: No, I mean in statistical computation - you have zeros and ones, so you have a finite number of possibilities.

Suppes: It seems to me that everyone who really talks about this in detail introduces the modifying phrase 'pseudo'. So you talk about pseudo-random generators and there have been recently some very large pseudo-random sequences generated. I was just at a conference, and they have found that one has to be extremely careful about these generators for very long sequences, because after all the whole point is, that you have a relatively short code for generating the pseudo-random numbers. So you know already that you have a short description of how to get them. And I don't think that anybody thinks ... I mean, what you say is right, they are pseudo-random numbers. Everybody recognizes that and it is not really a conceptual problem.

Noyes: You said, I believe, at the start of your talk, that in order to talk about lawfulness and determinism you have to make counterfactual statements, implying that this is a well known fact. It is not well known to me, and I need guidance.

Schurz: In the literature on philosophy of science it was discussed from the 40s to the 70s all the time. Also in the discussion of Charles Sanders Peirce, he claims that you need to define these things by using counterfactuals. There is lots of literature.

Noyes: Do you have a specific reference?

Schurz: Well, Nelson Goodman, "Fact, Fiction and Forecast" is one of the classical books. Maybe you can give him some additional.

Suppes: Somebody like Stalnaker, for example. I am personally sceptical.

Schurz: You are sceptical? Let's put it this way. Consider the theoretical description of a probability space discussed here. They are natural counterfactual devices interpreted in a way that requires no special counterfactual logic. They are described in a perfectly straightforward extensional set-theoretical way. So the reduction is quite straightforward for ordinary probability statements. The issue is, for some of the logicians like Stalnaker, as to whether you can get along with purely set-theoretical and probabilistic accounts or if you need something more. That's where one gets into an argument. But for the sort of thing we are talking about here, I think the extentional reductions so to speak are very good, relatively straightforward and quite easy. For example the probability space characterizing all possible paths.

Miller: All paths, but not all possible paths.

Suppes: I use the ordinary way of speaking. But possible here is a redundancy.

Schurz: Well, I mean, the problem comes in if you want to define what it means that a particular system is deterministic. All what you can do is to describe its actual trajectory - but its actual trajectory, is it really deterministic? What does this mean? It means that if the system would [emphasized] have the same state at two different times, then its future developments would be the same, too.

Batterman: You have a deterministic system when it is impossible for a single trajectory to branch.

Schurz: No, no.

Batterman: Otherwise there will be from some state, two possible states at some future time.

Schurz: But that makes no sense - I am sorry - in the actual description of reality because what should it mean that a single trajectory branches? It means that if you have two systems and it could be that both systems are in the same state here then one system goes along this line and the other system goes along that line. And if you look at the particular single system and you always have such a single trajectory, it has only one state at one time.

Batterman: Presumably, quantum mechanics is indeterministic because if you know the exact state at a given time - the psi-function - then at best you can predict probabilities of future states. So, there is a certain probability that at some future time the system will be in one state, and also a degree of likelihood that it will be in some other state at that time.

Schurz: So you define your determinism with respect to what is implied by the set of laws describing the system. If the set of laws differential equations imply that such trajectories are possible these laws are not deterministic. But I want to have a definition of determinism which is independent of any given set of laws - a definition which may be applied without that I know what are the real laws.

Chirikov: You want to consider it analyzable as a function.

Schurz: Yes.

Decoherence, Determinism and Chaos Revisited*

H.P. Noyes

Stanford University, USA

Abstract. We suggest that the derivation of the free space Maxwell Equations for classical electromagnetism, using a discrete ordered calculus developed by L.H.Kauffman and T.Etter, *necessarily* pushes the discussion of determinism in natural science down to the level of relativistic quantum mechanics and hence renders the *mathematical* phenomena studied in deterministic chaos research irrelevant to the question of whether the world investigated by *physics* is deterministic. We believe that this argument reinforces Suppes' contention that the issue of determinism versus indeterminism should be viewed as a Kantian antinomy incapable of investigation using currently available scientific tools.

1   Introduction

I am delighted to have had the opportunity to bring to this Symposium the question of whether recent work connecting relativistic quantum mechanics to the classical relativistic theory of fields sets the interpretation of "deterministic chaos" in a rather different — and possibly illuminating — context. My title comes from the fact that I had already raised this point at the 15th annual international meeting of the Alternative Natural Philosophy Association (Noyes (1994a)). I did not gain much enlightenment on the significant and difficult issues raised from the resulting discussion. I hope that, thanks to the passage of time, subsequent work with L.H.Kauffman (Kauffman and Noyes (in press)) and the different types of expertise present at this Symposium, I will gain a broader perspective from your comments.

The first argument I mount against the relevance of "chaos research" to the issue of determinism rests on the fact that physics is a science of measurement. If one accepts the operational methodology implied by this statement, and recognizes that the smallest space interval Δx and time interval Δt which one can measure is *always* bounded from below by the current state of technology, then there is a limit to the accuracy to which the initial conditions for prediction using a classical, deterministic system of equations can be stated. What chaos research has demonstrated is that there are many non-linear classical systems which require as much input information to obtain a "prediction" as can be obtained from the result "determined" by solving the deterministic equations. Hence the issue becomes irresolvable from the point of view of *physics* once one

---

* Work supported by the Department of Energy, contract DE–AC03–76SF00515

is asked to make a "prediction" that requires more accuracy in the input than is available from current technology. The next section tries to spell this out by invoking NO-YES events, and in particular the not-firing or firing of a recording counter, as the paradigm for measurement in physics.

So far, this states a point of view, and may not sound particularly compelling. But when one asks where the classical equations come from, the argument can be tightened. So far as I can see, the *only* classical systems of equations which do not depend in detail on the structure of matter — and hence on quantum effects — are electromagnetism and gravitation. Here an ancient piece of work by Feynman, recently resurrected by Dyson (Dyson (1990)) and extended by Tanimura (1992), comes to our aid. Dyson derives electromagnetism and Tanimura also derives gravitation from Newton's second law and the commutation relations of non-relativistic quantum mechanics! This paradoxical result is shown by our analysis to depend only on the assumption that measurement accuracy is *finite, fixed* and *bounded from below*. By an appropriate and significant extension of the calculus of finite differences to a non-commuting *discrete ordered calculus* (DOC), due to Etter and Kauffman (Etter and Kauffman (in preparation), and Kauffman and Noyes (in press)), this derivation becomes rigorous in a very general context. Accepting this derivation, the *classical* equations require finite and discrete measurement accuracy to ground them in *physics*. But then, to treat them as deterministic goes beyond the range of applicability of their foundation. This puts bite into the argument that classical, deterministic equations are *always* approximate, and hence that the context in which chaos research is usually set has no validity within the world of *physics* as I understand the term. This argument is presented in more detail in Section 3.

A second reason for taking the classical equations to be approximate is the underlying non-determinism of quantum mechanics. Strictly speaking classical equations apply only at large enough distances so that the particles which probe the fields are *decoherent* in the quantum mechanical sense. Hence, we argue that "deterministic chaos" is *always* an approximation, and that any fundamental discussion of *determinism* must be conducted at the quantum level. This pushes the discussion back to the level of Bell's Theorem, which is often interpreted as showing that demonstrable laboratory effects (e.g. Aspect's experiment) preclude the possibility of a local, deterministic description of natural science. The relationship between measurement accuracy and "decoherence" in our context is discussed in Section 4.

The conflict between quantum mechanics and Einstein locality raises a third issue about the approximate character of classical physics. This is the problem of how to construct a relativistic quantum mechanics which has classical field theory as a well defined correspondence limit. The specific *measurement* limitation involved is clearly the fact that when one attempts to measure distances shorter than $\hbar/2m_e c$, either directly or indirectly, one must take proper account of the degrees of freedom corresponding to electron-positron pair creation. We note that going below these bounds requires a *relativistic* quantum mechanical analysis. This provides a *third* reason why the deterministic interpretation of classical physics can never be more than an approximation. We explore, briefly, in Section 5 how a novel theory based on *bit-strings* might meet this problem. Our concluding section returns to the philosophical issues.

2   No-Yes Events as a Relativistic Measurement Paradigm

My approach to the questions of law and prediction in *physics* — rather than in the broader context of (Natural) Science used in the title of this Symposium — starts from the trite comment that *physics is a science of measurement*. I take this characterization of physics as a *methodological requirement*. Unfortunately, from my operational and pragmatic point of view, this dictum is much more often honored in the breach than in the observance. In my practice of physics I do not allow my fundamental paradigms for how theoretical physics should be connected to laboratory experience to rest on considerations that are not in some sense bounded by the actual experimental accuracy of current measurements.

This statement of methodological principle is unabashedly taken from Bridgman's (Bridgman (1928)) heroic attempt to rescue physics from the philosophers. It is usually assumed that his program failed to provide a proper conceptual foundation for the startling and enormously fruitful developments in relativistic cosmology and elementary particle physics which have provided contemporary scientists with such a rich picture of the physical world accessible to precise measurement. But the actual reconciliation of quantum mechanics with relativity, and in particular the creation of a theory of "quantum gravity" that commands consensus among the specialists, still eludes us as this century draws to a close. I have argued in more detail elsewhere (Noyes (in press)) why a return to Bridgman's principles might help resolve some of the thorny problems that still face us.

My approach is also informed by the S-Matrix program of Chew and Heisenberg which — according to Schweber (1994) — really started with Dirac. The basic point for me is that by going to large enough distances (and hence, necessarily, times) in the experimental setup, momentum and energy can always be measured to arbitrarily high accuracy using essentially classical physics techniques and concepts. In contrast, direct space-time measurement at short distance is always restricted by the uncertainty principle and looses direct operational meaning. Hence the *formal* symmetry between position and momentum measurement in quantum mechanics is destroyed *in practice*. As Chew used to put it, short distance space-time is an *artifact* of Fourier transformation and cannot have physical significance. Unfortunately, from my point of view, he did not take the next step and reject *continuum mathematics* as well.

This next step has, for me, a long history which is briefly explained in my contribution to *PhysComp'94* (Noyes (1994c)). The fundamental mathematical position comes from a necessary aspect of the practice of *computer science*, namely that you must name a largest integer N and the fixed, finite memory size *in advance*. If you need or wish to introduce larger numbers into the calculation, or change the size of the memory, you *must* re-examine everything you have done up to that point. This obvious fact has been particularly emphasized by David McGoveran (McGoveran and Noyes (1989)); in effect, he makes it into a methodological principle. Note that this not only rules out the continuum, but also mathematical induction. Few theoretical physicists and almost no mathematicians are willing to take such a drastic step. In elementary particle physics, whenever a theory is examined empirically, the events analysed, the model of the apparatus used in the analysis, and the theories under consideration are *necessarily* reduced to a finite number of bits on magnetic tape or some other digital form of memory. That this procedure must be used in order to test any *empirical* aspect of any theory may, perhaps, make our methodological purity seem less outrageous.

This much discussion seems necessary to justify my *measurement paradigm* based on what I call NO-YES events. The model I have in mind is a laboratory counter and associated memory storage which records whether an event *did not* take place in a time interval Δt in a volume Δx^3 with relevant linear dimension Δx (a NO event) or *did* take place (a YES event). I emphasize that, when it comes to precise measurement, the *absence* of a counter firing is often more important (eg in measuring "background") than its presence. For our paradigm we assume that the temporal resolution of the measurement Δt and the spacial resolution Δx are the *best* that can be achieved with current technology either by direct measurement, or *indirectly* as when one uses a Michelson interferometer to measure relative positions. Note that in order to relate such relative measurements to macroscopic laboratory coordinates, we would have to discuss the measurement accuracy with which we can connect the different space-time scales.

Up to this point we have treated length and time measurement as distinct. But the *System International*, employed universally by physicists in reporting the results of measurement, defines the *ratio* of space to time units by the *integer*

$$c \equiv 299\ 792\ 458\ meter/second \tag{1}$$

Thus, following current practice, we are no longer allowed to define Δx and Δt separately when specifying our lowest bound on measurement accuracy. In fact, we must make the *scale invariant* statement that

$$\frac{\Delta x}{c\Delta t} = 1 \tag{2}$$

in *any* system of units which allows us to talk about NO-YES events in a precise way.

We can summarize the content of this section by the phrase:

Physics is counting

3   Classical Relativistic Fields from DOC

In 1948 Richard Feynman showed Freeman Dyson a remarkable "proof" of the Maxwell Equations starting from the non-relativistic quantum mechanical commutation relations and Newton's second law (Dyson (1989)). Dyson no longer retains contemporary records of this conversation, but was able to reconstruct and publish the proof using notes he had made at a later date (Dyson (1990)). Although Dyson finds the proof paradoxical, we have claimed (Noyes (1991)) that in fact it makes good sense in terms of the new, fundamental theory discussed in (Noyes (1994c)).

Briefly, the argument goes as follows. The Feynman postulates are that

$$[x_i, x_j] = 0; \quad [x_i, m\dot{x}_j] = \frac{\hbar}{i}\delta_{ij}; \quad F_i(x, \dot{x}; t) = m\ddot{x}_i; \quad i, j \in 1, 2, 3 \qquad (3)$$

However, the use made in the proof of the second postulate (i.e. of the commutation relation between position and velocity) in no way requires the constant on the right hand side to be imaginary, or scaled by Planck's constant. The linearity in the mass parameter m allows us to divide through by m and replace it by the postulate

$$[x_i, \dot{x}_j] = \kappa\delta_{ij} \qquad (4)$$

with κ any constant with dimensions of area per unit time. For a particle acting under any force which obeys Newton's third law with respect to a reference particle, we know that the area (measured in units of Δx^2) swept out by the line from some appropriate center to the particle in a constant time interval (measured in units of Δt) is constant. This observation fixes κ in an *scale invariant* manner. Note that this generalization of Kepler's second law is *kinematic* rather than dynamic. It leaves both the mass standard and the mass ratio between the particle of interest and the reference particle *arbitrary*. Similarly, since Newton's second law is linear in mass, we can replace it by the assumption that the acceleration ($\ddot{x}$) is a function only of position, velocity and time. Finally, for any single particle for which the charge per unit mass is a Lorentz invariant, we can *also* divide the mass out of Maxwell's Equations, and find that the whole derivation is *scale invariant* because it depends only on fixing, *arbitrarily*, the units of length and time.

As is noted in (Kauffman and Noyes (in press)):
".... this aspect of scale invariance had already been introduced into the subject by Bohr and Rosenfeld in Bohr and Rosenfeld (1933). In their classic paper, they point out that because QED depends only on the universal constants $\hbar$ and c, the discussion of the measurability of the fields can to a large extent be separated from any discussion of the atomic structure of matter (involving m_e and e^2). Consequently, they are able to derive from the *non-relativistic* uncertainty relations the same restrictions on measurability (over finite space-time volumes) of the electromagnetic fields that one obtains directly from the second-quantized commutation relations of the fields themselves. Hence, to the extent that one could "reverse engineer" their argument, one might be able to get back to the

classical field equations and provide an alternative to the Feynman derivation based on the same *physical* ideas."

This point of view is also discussed in more detail elsewhere (Noyes (in press)).

Unfortunately, this *physical* argument has not proved compelling for many people in the relevant professional communities. We have therefore been forced to invoke the aid of a first rate mathematician and to go deeper into the mathematical foundations of the calculus of finite differences (see Kauffman and Noyes (in press)) than might be expected. This suggested further developments to T. Etter, which are now being pursued (see Etter and Kauffman (in preparation)).

The basic physical point from which the discussion of the impact of finite measurement accuracy on the relation between position and velocity starts is that velocity has to be defined as the ratio of a finite space interval to a finite time interval. We also restrict the problem to the "trajectory" of a single particle, and a finite shift along that trajectory. Then measurement of velocity must involve either first the specification of position and then the finite shift to a new position from which the velocity can be calculated, or first the shift from a previous position at some velocity and then the specification of the new position consistent with that velocity. These two velocities will not, in general, coincide. Note that this operational definition of velocity *precludes* the possibility of specifying both position and velocity at the same time. Thus the possibility of non-commutativity arises, and careful investigation of the possibilities leads to the discrete ordered calculus (DOC) of Etter and Kauffman. This (non-commutative) calculus of finite differences does, indeed, provide a rigorous mathematical context for the Feynman-Dyson "proof", allowing us to drop the quotation marks.

Exploring the mathematical niceties of this generalization of the calculus of finite differences would distract us from the thrust of this paper. Fortunately, when I recently showed (Kauffman and Noyes (in press)) to my colleague, M.Peskin, he rapidly rewrote the proof in the context of the Heisenberg representation of quantum mechanics. I had already suspected that this might be possible, and T. Etter informs me that his starting point for the construction of the DOC was essentially the same. The difficulty with adopting this point of view is that what *operational* context the Heisenberg formalism fits into is by no means obvious. So, for mathematical and physical clarity, one needs to invoke the DOC and discuss the relationship between measurement accuracy and the DOC. To make this paper self-contained I reproduce here the simpler (but conceptually problematical) derivation carried through in the context of the Heisenberg representation. I am much indebted to Peskin (1994) for allowing me to quote the following proof.

Define

$$\dot{X} = XU - UX = [X, U] \tag{5}$$

where U is the time shift operator from X to X' in time Δt (eg $U = e^{-iH\Delta t}$).

Notice that

$$(AB)^{\cdot} = [AB, U] = [A, U]B + A[B, U] = \dot{A}B + A\dot{B} \tag{6}$$

as required.

Postulate:

$$1. \quad [X_i, X_j] = 0$$

$$2. \; [X_i, \dot{X}_j] = \kappa \delta_{ij}$$

Rewrite 2 as

$$[X_i, [X_j, U]] = -[X_j, [U, X_i]] - [U, [X_i, X_j]] \tag{7}$$

and noting that $[U, [X_i, X_j]] = [U, 0] = 0$ we find that

$$\kappa \delta_{ij} = [X_i, [X_j, U]] \; symmetric \; in \; i, j \tag{8}$$

Now *define*

$$H_l = \frac{1}{2\kappa} \epsilon_{jkl}[\dot{X}_j, \dot{X}_k] \tag{9}$$

Then

$$\nabla_l H_l = \frac{1}{2\kappa} \epsilon_{jkl}[[\dot{X}_j, \dot{X}_k], \dot{X}_l] \tag{10}$$

But this cyclic sum vanishes by the Jacobi identity. Thus

$$\nabla_l H_l = 0 \tag{11}$$

which is one of the two Maxwell equations we set out to derive.

Finally, *define*

$$E_i = \ddot{X}_i - \epsilon_{ijk} H_k \tag{12}$$

We wish to prove that

$$\frac{\partial H_i}{\partial t} + \epsilon_{ijk} \nabla_j E_k = 0 \tag{13}$$

First we need to *define* $\partial/\partial t$ by

$$\dot{H} = \frac{d}{dt} H = \frac{\partial H}{\partial t} + (\dot{X} \cdot \nabla) H \tag{14}$$

Then

$$\frac{\partial H_i}{\partial t} = \dot{H}_i - \dot{X}_j \nabla_j H_i$$

$$= \frac{1}{2\kappa} \epsilon_{ikl}([\dot{X}_k, \dot{X}_l])^{\cdot} - \dot{X}_j \frac{1}{\kappa} [\frac{\epsilon_{ikl}}{2\kappa}[\dot{X}_k, \dot{X}_l], \dot{X}_j]$$

$$= \frac{1}{\kappa} \epsilon_{ikl}[\dot{X}_k, \ddot{X}_l] - \frac{1}{2\kappa^2} \dot{X}_j \epsilon_{ikl}[[\dot{X}_k, \dot{X}_l], \dot{X}_j] \tag{15}$$

$$\epsilon_{ijk} \nabla_j E_k = \epsilon_{ijk} \frac{1}{\kappa} \left[\left(\ddot{X}_k - \epsilon_{klm} \dot{X}_l H_m \right), \dot{X}_j \right]$$

$$= \frac{1}{\kappa} \epsilon_{ijk} \left[\dot{X}_j \ddot{X}_k \right] \cdot (-1) - \epsilon_{ijk} \epsilon_{klm} \epsilon_{mab} \frac{1}{2\kappa^2} \left[\dot{X}_l \left[\dot{X}_a, \dot{X}_b \right], \dot{X}_j \right]$$

$$= -\frac{1}{\kappa} \epsilon_{ijk} \left[\dot{X}_j, \ddot{X}_k \right]$$

$$-(\delta^{il}\delta^{jm} - \delta^{im}\delta^{jl}) \epsilon_{mab} \frac{1}{2\kappa^2} \left(\left[\dot{X}_l, \dot{X}_j \right] \left[\dot{X}_a, \dot{X}_b \right] + \dot{X}_l \left[\left[\dot{X}_a, \dot{X}_b \right] \dot{X}_j \right] \right)$$

$$= -\frac{1}{\kappa} \epsilon_{ijk} \left[\dot{X}_j \ddot{X}_k \right] + \frac{1}{2\kappa^2} \epsilon_{iab} X_j \left[\left[\dot{X}_a \dot{X}_b \right], \dot{X}_j \right]$$

$$- \epsilon_{jab} \frac{1}{2\kappa^2} \left[\dot{X}_i, \dot{X}_j \right] \left[\dot{X}_a, \dot{X}_b \right] \tag{16}$$

now

$$\epsilon_{jab} \left[\dot{X}_i, \dot{X}_j \right] \left[\dot{X}_a, \dot{X}_b \right] = \left[\dot{X}_i, X_1 \right] [X_2, X_3]$$

$$+ \left[\dot{X}_i, \dot{X}_2 \right] \left[\dot{X}_3, \dot{X}_1 \right] + \left[\dot{X}_i, \dot{X}_3 \right] \left[\dot{X}_1, \dot{X}_2 \right] \tag{17}$$

for $i = 1$, eg

$$= \left[\dot{X}_1, \dot{X}_2 \right] \left[\dot{X}_3, \dot{X}_1 \right] + \left[\dot{X}_1, \dot{X}_3 \right] \left[\dot{X}_1, \dot{X}_2 \right] = 0 \tag{18}$$

so

$$\epsilon_{ijk} \nabla_j E_k = -\frac{1}{\kappa} \epsilon_{ijk} \left[\dot{X}_j, \ddot{X}_k \right] + \frac{1}{2\kappa^2} \epsilon_{iab} X_j \left[\left[\dot{X}_a, \dot{X}_b \right] \dot{X}_j \right] \tag{19}$$

$$= -\frac{\partial H}{\partial t} \qquad QED .$$

We conclude that the free field Maxwell Equations are a formal consequence of assuming finite time shifts along a single particle trajectory and showing that the changes in velocity (accelerations) have the form of the Lorentz force law (i.e. $mF = eE + ev \times H$) for electromagnetic fields acting on a particle. This formula allows us to separate the acceleration into a vector which is a function of position and time (electric field) and produces an acceleration in that direction, and a second vector — also a function of position and time — which acts at right angles to the velocity and is proportional to the magnitude of the velocity (magnetic field).

We emphasize that *given* the fields, we can calculate the motion of a single particle passing through them, or *given* the trajectory, we can calculate the fields which would produce that trajectory. Invoking Newton's third law, and treating the field as a carrier of both energy and momentum, we can treat this second calculation as either the absorption of the radiation by the particle producing its motion or as the emission of the field by the particle when its motion is known. This language then allows us to treat single particle trajectories as either the sources or sinks of the fields *but not both at once*. The (insoluble) "self energy" problem cannot be met this way. One can achieve consistency at the classical level only by separating sources and sinks, as was done by Feynman and Wheeler in their "relativistic action at a distance" theory (Feynman and Wheeler (1945); Schweber (1994); Schweber (1986)). But then, in a closed system, the source and sink are made macroscopically (and non-locally) *coherent* by the energy-momentum conservation laws. Thus, treating the field as a locally defined and causally efficacious agent is only possible in the *decoherent approximation* in which we can ignore where the radiation is coming from and where it is going.

We will discuss this intricate question of coherence and decoherence further in the next section. For the moment, we emphasize that our *derivation* of the field equations from measurement accuracy *necessarily* limits their applicability as deterministic predictors to situations in which the boundary conditions and the predictions are made to less accuracy than the $\Delta x = c\Delta t$ restriction which allows us to derive the "differential" form of the field equations in the first place. Hence, *if our understanding of the classical electromagnetic field is accepted,* "deterministic chaos" cannot enter the system, and the distinction between determinism and indeterminism eludes us.

To complete the argument of this section, we need to extend the argument to the only remaining classical field, namely gravitation. At least within the framework of the Feynman-Dyson "proof", this has already been done by Tanimura in (Tanimura (1992)). Tentatively, at least, we accept this extension, but will not be sure of our conclusion until we have a rigorous equivalent using the DOC. The novelty here is that we must consider not only non-commutativity between position and velocity but the connectivity between oriented *areas*. This gives (at least formally) the usual tensor field in free space and the resulting *non-locality* of general relativity. Again, the field as a local, causal agent appropriate to think of as "deterministic" can only be a *decoherent approximation*. Thus, independent of details, we again find the phenomena of "deterministic chaos" irrelevant to what we can know *physically*.

4   Decoherence; Periodicity from Measurement Accuracy

[Spelling out in operational terms just what we mean by "decoherence" requires some care. I have already done this in (Noyes (1994a)) The next four sub-sections repeat these considerations with a few modifications.]

4.1   The Geometrical Paradigm for Decoherence

To give form to our discussion of coherence and decoherence, we use the devices schematically illustrated in Fig. 1. We assume, initially, that the "source" labeled by a question mark emits charged particles with a unique charge-to-mass ratio and a unique velocity v. Note that these particles, taken one at a time, fit into our understanding of "particle" and "field" as established in Section 3. Devices which we will use to insure that, to some finite accuracy, these assumptions are true are included in the figure, and will be discussed in more detail subsequently. For the moment we omit the "path extender".

We start from the case when the detection screen beyond the double slit (see Noyes (1980) for a detailed discussion of the conventional experiment) exhibits a double slit interference pattern whose envelope is the single slit diffraction pattern for a slit of width Δw and a distance D from the detector array. We

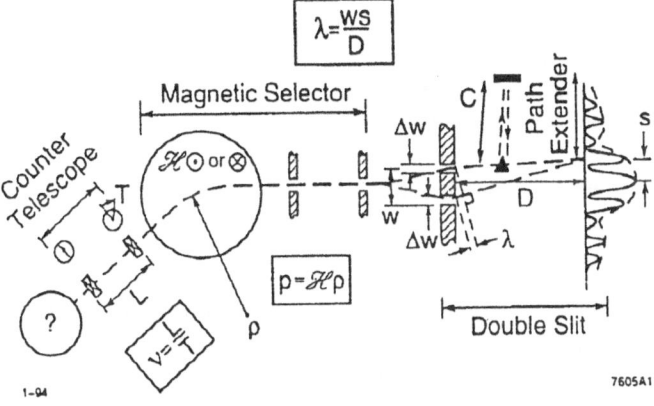

Fig. 1. *Measurement of coherence and decoherence of de Broglie waves using a counter telescope, magnetic selector, and a double slit with a path extender in one arm.*

set the parameters such that the spacing from the center of the pattern to the first interference fringe is s. Then the "wavelength" λ exhibited by this coherent interference between the beams from the two slits is measured and can be calculated from the equation

$$\lambda = \frac{ws}{D} \qquad (20)$$

We note that w, s and D are length intervals that can be measured by conventional macroscopic methods such as rods calibrated against international standards. We take this as the paradigmatic case for specifying what we mean by "coherence". We emphasize that, so far, only *length* measurements are implied and hence that our diagram is *scale invariant*.

In order to measure the "coherence length" we insert into the hypothetical "path" of the particle coming from one of the slits a "path extender", schematically represented by a wedge whose sides are mirrors. One face of the wedge reflects the beam to a second mirror which returns it to the second face of the wedge, which in turn returns it to the direction it followed in the paradigmatic case. The distance C from the wedge to the mirror is adjustable. $C = 0$ corresponds to the simplest double slit paradigm (wedge omitted). We find experimentally that for a source of a particular type the (double slit) fringe system disappears when we reach a value C_{max} or larger. We can then define the *coherence length* C_{coh} by

$$C_{coh} \equiv 2C_{max} \qquad (21)$$

Note that the definition still depends directly on geometrical measurements. Indirectly the specification depends on the *sensitivity* of the detector array, since the *intensity* of the pattern along the detector array and (if the array records individual particulate events) the *probability* of a particular region of the array being activated decreases as C increases. The disappearance of the interference pattern is our paradigm for *decoherence*.

To go further in our analysis, we must measure the velocity v, or if this velocity is close to the limiting velocity for information transfer — for which we use the conventional symbol c — the momentum. Then we can define a second critical parameter called the *coherence time* and symbolized by T_{coh} by the relationship

$$C_{coh} = vT_{coh} \tag{22}$$

Here we assume that the measurement of v using the recording counters in the first counter telescope and the time from the firing of the first counter telescope to the firing of one counter in the detector array are consistent with each other, and that all three clocks associated with the counters are synchronized using the Einstein convention.

In the situation where the interference fringes have disappeared, we can distinguish two paths emerging from the double slit by noting that all particles which follow the longer path arrive at the detector with a time delay greater by at least $T_{coh} = C_{coh}/v$ compared to the particles which traverse the shorter path. Then we *know* that the two trajectories are decoherent and (in the stated context) are *classical, decoherent* trajectories of classical particles (ignoring the single slit interference pattern which takes higher precision to see).

Various checks on the confidence with which we can make the above statements can depend on the measurement accuracy to which we can establish all the relevant parameters. Several such checks will occur to any experimental particle physicist. Since these checks are irrelevant to our main theme, we stop our articulation of the basic paradigm at this point, and focus on the accuracy to which we can measure velocity or momentum. The main point we wish to establish is simply that in a carefully specified context, *outside* of some coherence length or coherence time, particles can be said to follow two (or more) distinct trajectories for at least part of their history between production and detection. Inside that length, two coherent beams of the same type of particle can be made to interfere with a characteristic wavelength that can be measured geometrically. But asking where *within* that pattern of two coherent trajectories the "particle" is located cannot find an answer within the experimental setup. This is an example of the "complementarity" between the wave and the particle description in our discrete context.

4.2 Space-Time Velocity Measurement

The "counter telescope" we have included in figure 1 consists of two devices which *record* the time of firing *or of* not *firing* during some time interval. This is the next step in bringing the measurement paradigm presented in Section 2 closer to laboratory practice. The distance between the two counters is L and the time delay between the two recordings is T. These two recordings are NO-YES *events* in that whether the individual counters do not fire ("NO") or do fire

("YES") is recorded by two distinguishable symbols in two correlated records. These records can be repeatedly examined without destroying this distinction or the sequential ordering. In this context the velocity of a particle v is measured by a YES_1, YES_2 pair of events and is calculated by the ratio

$$v = \frac{L}{T} \tag{23}$$

The *accuracy* to which this constitutes — or can constitute — a *measurement* of this velocity cannot be adequately discussed in an article of this length. We simply note that what are called "particles" in high energy elementary particle physics have never been demonstrated to have velocities greater than the scale parameter $c \equiv 299\ 792\ 458\ m\ sec^{-1}$. Further, there is no accepted situation in which *information* in the physical or computer science sense has been transferred at a velocity greater than this value. Demonstrable exceptions to these statements would be of extreme interest to the physics and computer science communities.

4.3   Energy-Momentum Velocity Measurement

The "magnetic selector" using a magnetic field $\mathcal{H}$ perpendicular to the plane of figure 1 can also be considered to be a device capable of measuring velocity when it is properly calibrated. Its action is compatible with the Lorentz force law we explained in Section 3. The calibration procedures are more complicated than the direct calibration of rods and clocks which suffice for space-time velocity measurement. It is here that our restriction to a particular type of particle begins to become important.

If the particle is electromagnetically neutral, or if the space-time velocity is not distinguishable from c (up to the maximum value of $\mathcal{H}$ available to us), no deflection is observed and the inverse radius of curvature ρ^{-1} is indistinguishable from zero. We exclude these cases for the moment because the measuring device invoked gives no information not already provided by the counter telescope. However, when a deflection (finite, non-null ρ) is observed, we find that for fixed $\mathcal{H}$ the radius of curvature ρ changes with velocity. To cut a long story short, we find that if we measure velocity in units of c by defining

$$v \equiv \beta(v)c \tag{24}$$

and keep the magnetic field fixed,

$$\rho^2(v) \propto \frac{\beta^2}{1 - \beta^2}; \quad \rho^{-2}(v) \propto \frac{1 - \beta^2}{\beta^2} \tag{25}$$

This clearly allows us to calibrate our magnetic field to space-time measurements and, for a particular class of particles, to specify higher and lower magnetic fields over some range by the velocity-independent (over that range) definition

$$\mathcal{H} = \frac{\rho(v)}{\rho_0(v)} \mathcal{H}_0 \tag{26}$$

leaving open the units in which we ultimately decide to measure magnetic fields.

If, as is often the case in high energy physics, it is more convenient to measure radius of curvature rather than space-time velocity, we can relate this approach to the space-component of the "four velocity" $(u_0, \mathbf{u}) = (\gamma, \gamma\boldsymbol{\beta})$ with $\gamma^2\beta^2 = \gamma^2 - 1$ and

$$\beta^2(u) = \frac{u^2}{1 + u^2}; \quad \gamma^2(u) = 1 + u^2; \quad u = \pm|\mathbf{u}| \tag{27}$$

For a particular type of particle, this tells us that u^2 is proportional to ρ^2, and in a more articulated theory allows us to measure momentum by radius of curvature in a calibrated magnetic field. In this context we can ignore the (fixed) rest-mass of our "test particles" and keep our "momentum" measurements restricted to the "space-component of four velocity" or "momentum per unit mass".

Similarly, if we measure energy by the temperature rise in a calorimeter calibrated to the ideal gas law for particles of the same mass, i.e. measure pressure per unit mass rather than pressure, we can verify that this is consistent with the usual relativistic single particle kinematics

$$\frac{E^2}{m^2} = 1 + u^2; \quad \frac{E^2}{m^2} - \frac{p^2}{m^2} = 1 \tag{28}$$

and so on.

4.4 Scale Invariance

We have been at some pains to remove the mass scale from our basic paradigm for "coherence" and "decoherence" because the basic argument by which we gave meaning to classical electromagnetic fields (Section 3). used only measurement of space and time with accuracy bounded from below. To *break* scale invariance requires us to model some *physical* phenomenon involving Planck's constant and the reconstruction of relativistic quantum mechanics consistent with our operational methodology. Quantum mechanics can be arrived at in a number of ways, eg historically by the analysis of black body radiation, photo-effect, line spectra of atoms, finite size and stability of atoms measured using deviations from the ideal gas law, and so on. This is possible because the whole idea of a "test-particle" is basic to the classical definition of "fields", and is consistent with the understanding of electromagnetic fields we developed in Section 3. But *why* the same constant $\hbar$ should appear in these diverse empirical contexts remains unanswered.

The cleanest breakpoint for the *relativistic* quantum mechanics which concerns us is the creation of electron-positron pairs or the less direct but predicted and confirmed effects (eg Lamb shift, vacuum polarization in p-p scattering,...) of these degrees of freedom (Schweber (1994)). Once the degrees of freedom due to the possibility of particle-antiparticle pair creation have to be included in the theory, even the concept of a "test particle" generates nonsense. This is obvious in the case of pair creation in a system containing electrons because, thanks to the *indistinguishability* of electrons, in any system which contains one or more electrons initially whether the electron in the created pair and some initial electron are on the same or different trajectories becomes ambiguous and empirically irresolvable at distances less than $\hbar/2m_ec$. That this parameter occurs and can be measured even when there are *no* electrons in the system under examination is evidenced by the "vacuum polarization" contribution to both the energy and the angular distributions measured in proton-proton scattering below 3 Mev.

4.5   Velocity Resolution, Periodicity and "Wavelength" in a Discrete Theory

As already noted, we assume that *information* cannot be transmitted from one distinct location to another at a velocity greater than $c = 299\ 792\ 458\ m/sec$. By information we mean anything which reduces the number of possibilities at the second location relative to a previously accepted, understood, finite and countable number of possibilities. This allows us to specify velocities v in units of c by rational fractions $\beta(N,n) = v/c = n/N$ with N a fixed, finite positive definite *integer* which can be context sensitive. We distinguish *massive particles* from other modes of communication by the requirement that n be an integer in the range $-N + 1 \leq n \leq N - 1$.

We can now define *velocity resolution* by $\Delta v = c/N$. This is, clearly, a context sensitive definition, which requires a careful investigation of the experimental tools at our disposal in that context, and can have unexpected consequences such as the connection between fixed measurement accuracy and the formal structure of the classical, relativistic field equations we discussed in Section 3.

The context which we wish to explore first is when velocity is measured by the distance between two counters at positions $x_1\Delta x$ and $x_2\Delta x$ which fire sequentially at times $t_1\Delta t$ and $t_2\Delta t$. We assume finite and fixed *measurement accuracy* to mean that x_1, x_2, t_1, t_2 are *integers*, as discussed in Section 2. Then these four integers can be related to our previous definition of velocity by

$$\beta(N,n) = \frac{n}{N} = \frac{x_2 - x_1}{t_2 - t_1} \tag{29}$$

Because we took $N > 0$ in our earlier definition, we will use the definitions

$$If\ t_2 - t_1 > 0\ then\ N = t_2 - t_1,\ n = x_2 - x_1;\ \ else\ N = t_1 - t_2,\ n = x_1 - x_2 \tag{30}$$

This convention specifies positive spacial directions to be $x_2 > x_1$ and positive time evolution to be $t_2 > t_1$ in a finite and discrete 1+1 "space-time" with origin $(x_0, t_0) = (0, 0)$.

It is important to realize that, provided $\Delta v(N) \equiv c/N$ is not the best velocity resolution we can achieve in the context of interest, and a resolution $\Delta v(N_X) \equiv c/2N_x$ is at least conceivably within our grasp, $\beta(N,n)$ defines a *periodic* function with up to $2N_T$ periods, provided $NN_T < N_x$. To see this, we need only note that

$$\beta(N,n) = \beta(n_tN, n_tn) = \beta(n_tN, n + (n_t - 1)n) = \frac{n + (n_t - 1)n}{n_tN} \tag{31}$$

But this "periodicity" can have some unexpected restrictions, if we take our physical restriction on Δv seriously. In particular, for the two counter firings specified in the last paragraph, and the $\Delta v(N_x) = c/2N_x$ just assumed, we are restricted to space and time intervals between the two firings which satisfy the constraint

$$\left|\frac{x_2 - x_1}{t_2 - t_1}\right| > \frac{c\Delta t}{2N_x\Delta x} \tag{32}$$

Otherwise the two counter firings would measure a velocity to a resolution better than $c/2N_x$, contrary to hypothesis. We also have the further restriction $\frac{\Delta x}{c\Delta t} = 1$ from the general argument given in Section 2 justifying Eq. 2.2. Then we can define an *event horizon* $R_x = N_x\Delta x$, and a *time boundary* $T_x = N_x\Delta t$ which restrict the 1+1 integer coordinate space-time points we consider to the integer square in 1+1 space-time

$$-N_x \leq t \leq +N_x; \quad -N_x \leq x \leq +N_x \tag{33}$$

Thus any velocity measurement we consider restricts the "integer coordinate intervals" we consider by the equations

$$|x_2 - x_1| = n_t N < |t_2 - t_1| = n_t N \tag{34}$$

Our next concern is to understand in more detail the "state" of a particle with "constant velocity" implied by the concept of fixed, finite velocity resolution we are developing. In a continuum theory the two "point events" (x_1, t_1), (x_2, t_2) determine a line in 1+1 space-time which, according to Newton's first law, can be extrapolated to include all points between $-\infty$ and $+\infty$ outside the interval so defined, and interpolated to include all points within this interval, so long as no "force" acts on the particle. In contrast, our assumption of fixed velocity resolution restricts the positions where a constant velocity particle can appear, once the two counter firings are measured, to a very small set of integers. Assume first that n and N have no common integer factor other than 1. Then *no* interpolated positions between the two counter firings are allowed for the velocity state $\beta(N, n)$. The only coordinate pairs we are allowed (by the construction developed so far) are the extrapolated event positions

$$(x(N, n; n_t), t(N, n; n_t)); \tag{35}$$

where

$$x(N, n; n_t) = x_1 + n(n_t - 1); \quad t(N, n; n_t) = t_1 + n(n_t - 1) \tag{36}$$

Here we allow n_t as well as n to be negative, so long as the event horizon constraints

$$-N_x < x(N, n; n_t) < +N_x; \quad -N_x < t(N, n; n_t)) < +N_x \tag{37}$$

are met.

Note that this periodic sequence of space-time positions where a third counter might (but need not) fire *also need not* include the "origin" $(0,0)$. A naive interpretation of our formalism would allow us to include this origin as "physical"; in a more detailed discussion we would show why we must use caution in making this part of our construction. When we have shown how to measure quantum interference phenomena using only counter firings as our paradigm, we will see that the inability to locate the origin "absolutely", but only with an uncertainty $|x_2 - x_1|n_t \Delta x$, and positions "relative" to some unique reference event only with an uncertainty $|x_2 - x_1|\Delta x$ is the analog in our theory of the inability to measure "absolute phase" in conventional quantum mechanics.

Here we can take only the preliminary step of relating this finite and discrete model of positions where counters can fire sequentially to the paradigmatic case of the measurement of coherence length illustrated in Figure 1. Suppose (x_1, t_1) and (x_2, t_2) are the space-time coordinates for the firing of the entrance and exit counter before the magnetic selector, and that the counters are thin enough and the clocks accurate enough so that all four numbers are integers in units of Δx or $\Delta t = c\Delta x$, making $N_{12} \equiv L/\Delta x$ and $D_{12} = T/\Delta t$ integers and $\beta(D_{12}, N_{12})$ a rational fraction. If N_{12} and D_{12} have a common factor N_T, so that $N_{12} = N_T n_{12}$, $D_{12} = N_T d_{12}$ and $\beta_{12} = n_{12}/d_{12}$, we could obviously postulate that the signal emerging from the counter telescope is a periodic phenomenon with N_T periods, spacial periodicity $\lambda = n_{12}\Delta x$ and temporal periodicity $\tau = d_{12}\Delta t$ and start articulating this model in such a way that the phenomena described in our paradigm defining coherence-decoherence can be reproduced.

We cannot flesh out this model in detail here, and stick to a few elementary points, confined to modeling the positions of the peaks in the double slit interference pattern. Two cases need to be distinguished. If the source contains a pseudo-random distribution of particle velocities which happens to include cases with v_{12}, the coherence time is $T_{coh} = N_T d_{12}\Delta t$. On the other hand, the source may be independently specified using some other part of the theory (eg. the decay of an excited atom). We must insure that our model properly includes both possibilities. Another complication is that we must distinguish in our modeling the fact that there are *two* kinds of space and time periodicities corresponding to the group velocity (v_{12}) of the "wave packet" and the "phase velocity" defined by $v_{12}v_{ph} = c^2$. A third is that the interference pattern wave length is given by $h/p(\beta_{12})$ and must be computed using the proper relativistic formula given above relating β to 4-velocity u. Spelling all this out will take a textbook — which is being written (Noyes (in preparation)). We take a few steps in the next sub-section toward specifying what we mean by *finite and discrete Lorentz invariance* in a theory which takes measurement accuracy seriously.

4.6   Initial Steps Toward Constructing Finite and Discrete 1+1 Lorentz Boosts in "Space-Time"

Keeping in mind the fact that we must eventually return to an examination of the experimental context in which our "origin" of coordinates is specified, we now develop "Lorentz boosts" between two velocity states $\beta_i(N_i, n_i)$, $\beta_f(N_f, n_f)$ under the assumption that the corresponding event coordinates are

$$t_i = N_i, \ x_i = \beta_i t_i = n_i; \quad t_f = N_f, \ x_i = \beta_f t_f = n_f \tag{38}$$

for a boost velocity $\beta(N, n) = n/N$. The obvious constraint we must satisfy is that

$$\beta_f = \frac{\beta + \beta_i}{1 + \beta \beta_i} \tag{39}$$

The less obvious constraint is, that is the minimum number of periods of each of the three velocities must allow us to insure that all three events are "physical" when referred (as we implicitly have) to a fourth "reference event" at $(0,0)$ *and* that the counter firings which allow the velocity to be measured lie within the event horizon. This we start to work out here and will complete elsewhere.

By assigning (integer) coordinates (x_1, t_1) and (x_2, t_2) in a theory with a limiting velocity c, we have implicitly assumed that the clocks which record t_1 and t_2 at these two distinct locations have been synchronized using the Einstein convention. We now include the possibility of this synchronization explicitly in our construction. We let $x_2 - x_1 = 2X > 0$, $t_2 - t_1 = 2T > 2X > 0$ and (formally) fix the space time coordinates of firings 1 and 2 at $(-X, -T)$ and (X, T) respectively. Then the two counter firing bracket our (formal) "origin" $(0,0)$. To synchronize the clocks, we place a "mirror" at some position $(-T_X)$ with $T_X > T > 0$, and require that a light signal sent from $(-X, -T)$ to this mirror and reflected back along the same line will arrive at $(+X, +T)$. Then the time it takes for the light signal launched at the time of the first firing to reach the mirror is $T_X - T$, while the time interval from the reflection to the arrival in coincidence with the second counter firing is $T_X + T$. This insures that the time interval between the two firings is, in fact $2T$, consistent with our formal assignment of coordinates, *independent* of where along the line we place our reference "mirror" $(-T_X)$ and consistent with the Einstein synchronization convention. Note that the velocity measured by the two sequential counter firings is $\beta(T, X) = X/T$.

In a continuum, classical theory of space-time measurement, it is possible to specify both position and velocity simultaneously at any instant of time t. In our context, which we have constructed by paying careful attention to the constraints imposed by finite velocity resolution, this is no longer possible. If we use the times of the two counter firings and their previously measured positions (and clock calibrations) to measure the velocity, all we can say from the point of view of measurement is that the particle position and time (x, t) *during* the measurement of velocity is subject to the constraints $-X \leq x \leq +X$, $-T, \leq T \leq +T$ (with both x and t integer). If we are willing to assume that a third counter placed on the line *between* the first two does not interfere with the velocity measurement — an assumption that can only be checked "statistically" by repeated measurement — we can reduce this uncertainty considerably and check the assumption of "constant velocity" between the counters to limited accuracy. Place this counter at a position x which satisfies the position constraint, and assume it fires at time t, and hence with the time intervals $T^- = T + t$, $T^+ = T - t$. We have now made two, rather than one, velocity measurements which give the values

$$\beta^- = \frac{X + x}{T + t}; \quad \beta^+ - = \frac{X - x}{T - t} \tag{40}$$

Of course, if we can place our counter precisely at $x = 0$ and it always fires at $t = 0$, we will confirm the classical, continuum model. *But this assumption would violate our initial hypothesis of finite velocity resolution.* Clearly the detailed exploration of what we mean by *finite and discrete Lorentz invariance* would take us too far afield. We intend to develop it elsewhere (see Noyes (in preparation)).

4.7 Conclusions About Decoherence in a Discrete Theory

We hope that the discussion in this section at least gives the flavor of how we intend to develop a complete relativistic quantum mechanics of single particle phenomena which will give precision to question of where the limitations on the Feynman-Dyson-Tanimura- Kauffman derivation of the classical relativistic fields will arise due to quantum effects. In a more conventional vein, we could say that we can only apply classical considerations to systems where the "collapse of the wave function" has changed quantum states from a coherent superposition to a mixture. Lacking our own theory for this, and noting that there is considerable controversy in the literature both about the "correspondence limit" of relativistic quantum mechanics and whether there is such a thing as "quantum chaos", we again conclude that whatever the outcome of research pursued on current lines, it is bound to remove the question of the *physical* meaning of "deterministic chaos" still farther away from the practice of physics when examined operationally at the level of fundamental theory.

In the next section we describe a promising theory which *does* have correspondence limits in non-relativistic quantum mechanics, relativistic (classical) particle physics, and (if the derivation of classical relativistic field theory given in Section 3 is accepted) in the classical relativistic field theories of electromagnetism and gravitation.

5   Bit-String Physics: A Novel Relativistic Quantum Theory

[Since we have recently completed (in Noyes (1994c) a fairly complete and systematic presentation of our "theory of everything" we content ourselves here with quoting the introduction to that paper and the essential results, and refer the reader to the longer publication, and references therein, for details.]

"Although currently accepted relativistic quantum mechanical theories incorporate many discrete phenomena, they are embedded in an underlying space-time continuum in a way which guarantees the creation of infinities. Despite many phenomenological successes, they have as yet failed to achieve a consensus theory of 'quantum gravity'. We believe that these two difficulties are connected, and that both can be circumvented by basing fundamental physical theory directly on the computer tools of bit-strings and information theory based on bit-strings. This has the further advantage that we can base our model for space and time on finite intervals between events (eg. counter firings) measured to finite (and *fixed* in any particular context) accuracy. This operational methodology then allows us to avoid such metaphysical questions as whether the 'real world' is discrete or continuous (see Noyes (in press)), or whether the 'act of observation' does or does not require 'consciousness' " (Noyes (1994b)).

"By a 'theory of everything' (ToE), we mean a systematic representation of the numerical results obtained in high energy particle physics *experiments* and by *observational* cosmology. The representation we use employs a growing but always finite assemblage of bit-strings of finite length constructed by a simple algorithm called *program universe* explained" (Noyes (1994c)).

"More conventional ToE's are based on the mathematical continuum and the structures of second quantized relativistic field theories (QFT). They ignore the flaws of QFT (infinite answers to physically sensible questions, unobservable 'gauge potentials', and no well defined correspondence limit in either classical relativistic field theory, non-relativistic quantum mechanics or nuclear physics). The most ambitious of these theories assume that non-Abelian gauge theories in the form of 'string theory' succeed in explaining "quantum gravity". Comparison with practical metrology is made by identifying $\hbar$, c and G_{Newton} in their theoretical structures. It is then an act of faith that everything else is calculable. Less ambitious ToE's (eg. GUT's = grand unified theories) fix the third parameter as a universal coupling constant at an energy of about a thousandth of the Planck mass-energy and then 'run' it down in three different ways to energies a factor of 10^{15} smaller where these three distinct values are identified as the measurable fine structure constant ($\alpha = e^2/\hbar c$), weak interaction constant (G_{Fermi}) and strong coupling constant α_s; because the strong (QCD) coupling 'constant' is supposed to diverge at zero energy, models must include its energy dependence over a finite energy range. In practice, such theories contain a fairly large number of phenomenological parameters."

"In contrast, we employ a structure in which we need only identify $\hbar$, c and m_p (the proton mass) in order to make contact with standard MLT metrology, using the kilogram, meter and second as arbitrary but fixed dimensional units. α, G_{Fermi}, G_{Newton} and a number of other well measured parameters can be computed and the quality of the fit to experiment evaluated in a less problematic way. While these comparisons are very encouraging, with accuracies ranging from four to seven significant figures, they are not perfect. So far as we can see the discrepancies could arise from the concatenation of effects we know we have so far not included in the calculations, but we are prepared to encounter 'failure' as we extend the calculations. However, the quality of the results achieved to date lead us to expect that such 'failure' would point to where to look for 'new physics' in *our* sense. Since we leave no place for 'adjustable parameters', such a crisis should be more clear cut for us than in a conventional ToE. We do not believe that it is possible to make a 'final theory', and might even welcome a failure serious enough to allow us to abandon this whole approach and turn to more conventional activities."

"We start from a universe of bit-strings of the same length which grow in length by a random bit, randomly chosen for each string whenever XOR between two strings gives the null string; else the resulting non-null string is adjoined to the universe. Then recurse. Because of closure under XOR (Amson (1979)), and a mapping we present (Noyes (1994c)) of the quantum numbers of the 3-generation standard model of quarks and leptons onto the first 16 bits in these strings, we can model discrete quantum number conservation (lepton number, baryon number,charge, weak isospin and color) using a bit-string equivalent of 4-leg Feynman diagrams. Quarks and color are necessarily confined. All known elementary fermions and bosons are generated, and no unknown particles are predicted. The scheme implies reasonably accurate coupling constants and mass ratios, calculated assuming equal prior probabilities in the absence of further information. The combinatorics and the standard statistical method of assigning equal weights to each possibility provide an alternative interpretation of results previously obtained from the *combinatorial hierarchy* (Bastin (1966); Noyes and McGoveran (1989); McGoveran and Noyes (1991)), including the closure of these bit-string labels at length 256, and the prediction of the Newtonian gravitational constant. Baryon and lepton number conservation then gravitationally stabilizes the lightest charged (free) baryon (the proton) and lepton (the electron) as rotating black holes of spin 1/2 and unit charge."

"The growing portion of the bit-strings beyond the quantum number conserving labels can be interpreted as describing an expanding 3-space universe with a universal (cosmological) time parameter. Within this universe pairwise collisions produce products conserving relativistic 3-momentum (and, when on mass shell, energy) in terms of quantized Mandelstam parameters and masses. The baryon and lepton number, ratio of baryons to photons, fireball time, and ratio of dark to baryonic matter predicted by this cosmological model are in rough accord with observation. The model contains the free space Maxwell equations for electromagnetism and the free space Einstein equations for gravitation as appropriate macroscopic approximations for computing the motion of a single test particle."

COUPLING CONSTANTS

Coupling Constant	Calculated	Observed
$G^{-1}\frac{\hbar c}{m_p^2}$	$[2^{127} + 136] \times [1 - \frac{1}{3 \cdot 7 \cdot 10}] = 1.693\,31 \ldots \times 10^{38}$	$[1.69358(21) \times 10^{38}]$
$G_F m_p^2/\hbar c$	$[256^2\sqrt{2}]^{-1} \times [1 - \frac{1}{3 \cdot 7}] = 1.02\,758 \ldots \times 10^{-5}$	$[1.02\,682(2) \times 10^{-5}]$
$sin^2\theta_{Weak}$	$0.25[1 - \frac{1}{3 \cdot 7}]^2 = 0.2267 \ldots$	$[0.2259(46)]$
$\alpha^{-1}(m_e)$	$137 \times [1 - \frac{1}{30 \times 127}]^{-1} = 137.0359\,674 \ldots$	$[137.0359\,895(61)]$
$G^2_{\pi N\bar{N}}$	$[(\frac{2M_N}{m_\pi})^2 - 1]^{\frac{1}{2}} = [195]^{\frac{1}{2}} = 13.96 ..$	$[13,3(3),> 13.9?]$

MASS RATIOS

Mass ratio	Calculated	Observed
m_p/m_e	$\frac{137\pi}{\frac{3}{14}(1+\frac{2}{7}+\frac{4}{49})\frac{4}{5}} = 1836.15\,1497 \ldots$	$[1836.15\,2701(37)]$
$m_\pi^\pm/m_e$	$275[1 - \frac{2}{2 \cdot 3 \cdot 7 \cdot 7}] = 273.12\,92 \ldots$	$[273.12\,67(4)]$
m_{π^0}/m_e	$274[1 - \frac{3}{2 \cdot 3 \cdot 7 \cdot 2}] = 264.2\,143 \ldots$	$[264.1\,373(6)]$
m_μ/m_e	$3 \cdot 7 \cdot 10[1 - \frac{3}{3 \cdot 7 \cdot 10}] = 207$	$[206.768\,26(13)]$

COSMOLOGICAL PARAMETERS

Parameter	Calculated	Observed
N_B/N_γ	$\frac{1}{256^4} = 2.328.... \times 10^{-10}$	$\approx 2 \times 10^{-10}$
M_{dark}/M_{vis}	≈ 12.7	$M_{dark} > 10 M_{vis}$
$N_B - N_{\bar{B}}$	$(2^{127} + 136)^2 = 2.89 \ldots \times 10^{78}$	$compatible$
ρ/ρ_{crit}	$\approx \frac{4 \times 10^{79} m_p}{M_{crit}}$	$.05 < \rho/\rho_{crit} < 4$

Table I. **Coupling constants and mass ratios** predicted by the finite and discrete unification of quantum mechanics and relativity. Empirical Input: $c, \hbar$ and m_p as understood in the "Review of Particle Properties", Particle Data Group, *Physics Letters*, **B 239**, 12 April 1990.

[This paper ends with the following caveat]

"We warn the reader that detailed and rigorous mathematical proof of some of the statements made above is still missing. We wish to thank David McGoveran for pointing out to us that this caveat is particularly relevant for the use we make of the corrections he derived in the context of the combinatorial hierarchy construction. For him, constructing our bit-strings using program universe and bringing in the identification of the labels from 'outside' — i.e. from known facts about quantum number conservation in particle physics — amounts to creating a *different* theory. While we have confidence that mixing up the two approaches in this way can, eventually, be justified in a compelling way, it may well turn out that our confidence in this outcome is overly optimistic.

"To summarize, by using a simple algorithm and detailed physical interpretation, we believe we have constructed a self-organizing universe which bears a close resemblance to the one in which physicists think we live. It is not 'self-generating' — unless one grants that the two postulates with which Parker-Rhodes begins his unpublished book on the 'inevitable universe', namely: *'Something exists!'* and *'This statement conveys no information'* suffice to explain why our universe started up."

6   Philosophical Implications

We now return to the question of how this work in foundations of particle physics and physical cosmology relates to the question determinism versus indeterminism in *physics*. Since the theory we present is, to put it mildly, controversial, it is obvious is that any conclusions must be tentative. Nevertheless, we believe that the fact that a "theory of everything" (i.e. of particle physics and physical cosmology) using only finite and discrete observations and sticking to this methodology is at least possible is relevant to the issue of determinism versus indeterminism. Clearly the theory is computational, and in that sense "deterministic". Yet, because it rigorously excludes both the continuum and mathematical induction, it provides a *physical* theory in which "deterministic chaos" simply cannot arise.

Of course the operational methodology on which our approach to physics is based cannot be argued for to the exclusion of more conventional approaches. But even those approaches provide three reasons why "deterministic chaos" should not be considered a *fundamental* theory and hence relevant to metaphysical conclusions. The first is that the classical theories of electromagnetism and gravitation can be *derived* by accepting a lowest, finite and fixed bound on the accuracy to which space and time intervals can be measured. Hence assuming boundary conditions known to an accuracy needed to reach "deterministic chaos" is logically *inconsistent* with using these equations to establish it. The second is that the non-relativistic uncertainty principle removes "deterministic chaos" from consideration as a physical theory in any case. The third is that once we take into account the observed phenomenon of electron-positron pair creation, *any* theory which tries to specify distances to better than $\hbar/2m_ec$ is *ipso facto* operationally meaningless.

Patrick Suppes has argued on quite general grounds in his paper entitled "The Transcendental Character of Determinism" Suppes (1993) that modern work on what are called "deterministic systems" has shown that "Deterministic metaphysicians can comfortably hold to their view knowing that they cannot be empirically refuted, but so can indeterministic ones as well." He then proposes a fundamental reinterpretation of Kant's Third Antinomy, claiming that "Both Thesis and Antithesis can be supported empirically, not just the Antithesis." We offer this paper as support to his claim, using a very different body of physical theory and experimentation as our context.

References

Amson, J. (1979): Appendix in T.Bastin, H.P.Noyes, C,W.Kilmister and J.Amson. Int.J.Theor.Phys. **18**, 455

Bastin, T. (1966): On the Scale Constants of Physics. Studia Philosophica Gandensia **4**(77)

Bohr, N., Rosenfeld, L. (1933): Det. Kgl. Danske Videnskabernes Selskab. Mat.-fys. Med. **XII**(8)

Bridgman, P. W. (1928): *The Logic of Modern Physics* (MacMillan, New York)

Dyson, F. J. (1989): Physics Today, No. 2, 32

Dyson, F.J. (1990): Amer J. Phys. **58**, 209

Etter, T., Kauffman, L.H. (in preparation): Discrete Ordered Calculus

Feynman, R. P., Wheeler, J. A. (1945): Rev. Mod. Phys. **17**, 157

Kauffman, L. H., Noyes, H. P. (in press): Discrete Physics and the Derivation of Electromagnetism from the Formalism of Quantum Mechanics, *Proceedings of the Royal Society* **A**

McGoveran, D. O., Noyes, H. P. (1989): Foundations for a Discrete Physics. SLAC-PUB-4526

McGoveran, D. O., Noyes, H. P.(1991): Physics Essays **4**, 115

Noyes, H. P. (1994a): Decoherence, Determinism and Chaos. In *Alternatives: Proc. of ANPA 15,* C.W. Kilmister, ed.; published by ANPA, c/o Prof. C.W. Kilmister, Ted Tiles Cottage, High Street, Barcombe, Lewes BNH 5D8, UK, 89-108

Noyes, H. P. (1994b): STAPP's QUANTUM DUALISM: The James/Heisenberg Model of Consciousness. In *Mind-Body Problem and the Quantum; Proc. ANPA WEST 10,* F. Young, ed.; published by ANPA WEST, 112 Blackburn Ave., Menlo Park, CA 94025, 1-17

Noyes, H. P. (1980): An Operational Analysis of the Double Slit Experiment. In *Studies in the Foundations of Quantum Mechanics,* P.Suppes, ed., manufactured by Edward Brothers for the Philosophy of Science Association, East Lansing, Michigan, 77-108

Noyes, H. P. (1991): On Feynman's Proof of the Maxwell Equations. Presented at the *XXX Internationale Universitätswochen für Kernphysik,* Schladming, Austria, and SLAC-PUB-5411

Noyes, H. P. (1994c): BIT-STRING PHYSICS: A Novel 'Theory of Everything'. In *Proc. Workshop on Physics and Computation,* Dallas TX, Nov. 17-20, 1994, D.Matzke, ed., IEEE Comp. Soc. Press, Los Amitos, CA, 88-94

Noyes, H. P. (in press): OPERATIONALISM REVISITED: Measurement accuracy, Scale Invariance, and the Combinatorial Hierarchy. Physics Philosophy Interface

Noyes, H. P. (1994d): On the Measurability of Electromagnetic Fields: A New Approach. In *A Gift of Prophecy*, E. G. Sudarshan, ed., World Scientific, incomplete; for the unmutilated text, see SLAC-PUB-6445

Noyes, H. P. (in preparation): DISCRETE PHYSICS: A New Fundamental Theory, J.C. van den Berg, ed.

Noyes, H. P., McGoveran, D. O. (1989): Physics Essays **2**, 76

Peskin, M. (1994): private communication, Nov. 4

Schweber, S. S. (1986): Rev. Mod. Phys. **58**, 449

Schweber, S. S. (1994): QED and the men who made it (Princeton)

Suppes, P. (1993): The Transcendental Character of Determinism. Midwest Studies in Philosophy **XVIII**, 242-257

Tanimura, S. (1992): Annals of Physics **220**, 229

Discussion of H. Pierre Noyes' Paper

Batterman, Chirikov, Miller, Noyes, Suppes, Weingartner, Wunderlin

Noyes: There seems to be some confusion between the fact that in my theory I have finite space-time intervals between space-time events, and the fact that in the conventional continuum theory of strong interactions (i.e. QCD or quantum chromodynamics), that theory is approximated by introducing a rigid space-time lattice with a fixed interval between the lattice points. This arbitrary cutoff procedure is needed in QCD because the strong coupling constant in that theory is infinite at zero energy, producing confined quarks and what is called infrared slavery. So far as I know, even using the most sophisticated super-computers the lattice calculations have yet to produce quantitative results of sufficient quality to compare with experiment. In contrast, my theory gives a good non-perturbative result for the pion-nucleon coupling constant — one of the basic strong interaction coupling constants that can be measured experimentally — in a very simple calculation. Eventually I should be able to calculate low energy quark phenomena in a similar way by developing a relativistic Faddeev-Yakubovsky type of quantum scattering theory. But that technical development lies in the future. It is also superior to the conventional lattice calculations in that quantized angular momentum does not have to be approximated. Eventually, this should lead to technical advantages over the conventional continuum theories. But that is only a hope so far.

Weingartner: You had no lattice on the basis you said. But do you have such an integer geometry?

Noyes: I have an integer geometry, but it does not refer to the lattice used for QCD calculations, or to any rigid lattice.

Weingartner: Yes, right. But does this give you enough possibilities? I mean these selected proportions which you can get out of this integer geometry - does it give you enough basis to do all this physics?

Noyes: I obtain standard angular momentum quantization. So my treatment is equivalent to the standard quantum mechanical treatment of rotations. The basic difference is that my masses are quantized in units Δm which are in the same ratio to the proton mass as the proton mass is to the Planck mass ($M_{Planck} = [\hbar c/G]^{\frac{1}{2}} = 1.3 \times 10^{19} m_{proton}$). Until one can predict and measure dimensionless ratios to better than a part in 10^{19}, there will be no *direct* way to distinguish my theory from a continuum relativistic quantum mechanics. At that level of accuracy there would be genuine differences and one could ask whether there was a lattice at that scale which has absolute significance in contrast to the relativistic scheme I have presented here.

Wunderlin: How many free parameters are in your theory and of what kind are these?

Noyes: There are *no* free parameters in my theory. *Any* physical theory must make contact with the meter, kilogram and second in order to make experimental predictions which can be tested. For this purpose I identify the parameters called $\hbar$, c, and m_p in my theory with the values given in meters, kilograms and seconds in the *Particle Data Book*. But beyond those three empirical numbers I have no parameters in any sense.

Wunderlin: Perhaps it lies in the standard theory. If I remember correctly, there were 19 such parameters in the standard theory.

Noyes: I agree that in the standard theory, there are something like 19 adjustable parameters. In principle, *all* of these parameters are calculable in my theory, but so far I have only calculated those given in the table in my paper. I have not calculated the QCD coupling constant (which is not directly observable), but I have calculated the pion-nucleon coupling constant which is related to it. I have not calculated the Kobayashi-Maskawa mixing parameters. That will be a real test of my theory. In order to do so, as already noted, I will have to develop a low energy QCD scattering theory. And that hasn't been done. But in principle every observable parameter is computable. The calculation of the pion-nucleon coupling constant as 13.96 in good agreement with experiment shows that in some sense I am doing low energy QCD right. What I have to do next, now that I have a good value for the muon mass (207 electron masses) is to calculate the kaon and hyperon masses, and then go on to the charmed, bottom and top systems. None of these values are adjustable parameters for me. And that is why this theory really puts me in a box. If I come up with the wrong values I have no place to hide. This theory either stands or falls as a piece.

Weingartner: Is there a smallest distance between elementary particle collision for instance?

Noyes: Well, there certainly will be. If you go down to the Planck length ($\hbar/M_{Planck}c \approx 10^{-33}$ cm), then you would see the quantization of mass discussed above. In that sense the shortest distance is the Planck length. But long before you hit the Planck length, you are producing enormous numbers of particle-antiparticle pairs. So to talk about length in a classical sense is very misleading. What you are really looking at is the structure in energy-momentum space, which is only theoretically connected to structure in space-time. What you are exploring is the probability of producing all the different kinds of particles of standard high energy physics. To call that a measurement of short distance is, I think, the wrong way to talk about it. I know high energy physicists in public pronouncements often talk about testing quantum field theory at distances of 10^{-16} cm. As far as I am concerned they are testing a model which has that dimensional parameter in it, not physical space.

Weingartner: And what about lowest energy, is there such a border?

Noyes: The lowest energy would be, again, a proton's mass-energy divided by 1.3×10^{19}, that is Δmc^2 with the same Δm defined above as the unit in which mass is quantized. This cannot be divided further, just as the Planck constant $\hbar$ is the smallest change in angular momentum which can be measured. Of course, if new *experimental* phenomena show up on our way to the Planck energy, the theory might have to change. One thing which would be relevant is the neutrino masses indicated in the solar neutrino experiments, and in a recent Los Alamos experiment. If my theory cannot predict these masses correctly, this would kill it right there. Mine is *not* an *a priori* theory. It definitely could be refuted by experiment.

Chirikov: One question: You have two parameters or whatever in your theory?

Noyes: I have *no* parameters. *Any* theory must make contact with the meter, kilogram and second by some means. For that purpose I use $c, \hbar$ and m_p. In other words, my unit of length is $\hbar/m_pc$, my unit of time is $\hbar/m_pc^2$ and my unit of mass is m_p.

Chirikov: No, the units do not matter, of course. There are three fundamental constants in physics. Does it mean you can calculate them?

Noyes: Yes, I calculate them.

Chirikov: All of them?

Noyes: Yes, in principle (other than $c, \hbar$ and m_p, as already explained). I have, for instance, calculated Newton's constant from $G^{-1}\hbar c/m_p^2 = 1.693\,31 \times 10^{38}$ as compared to the empirical value $1.693\,58(21) \times 10^{38}$.

Chirikov: And what about Planck's constant?

Noyes: No, I am taking $\hbar, c$ and m_p from experiments consistent with the way these constants occur in my theory. But then I calculate Newton's constant G.

Chirikov: O.K.

Noyes: For example (see table in paper), I can calculate the Fermi constant, the weak angle, the fine structure constant, and the pion-nucleon coupling constant.

Chirikov: O.K. You begin with three parameters, $c, \hbar$ and m_p. I would say that c and $\hbar$ are really fundamental. Proton mass is certainly not, because the proton has a very complicated structure.

Noyes: Because I can calculate the proton-electron mass ratio (see table in paper) it does not matter whether I use the proton mass or the electron mass.

Chirikov: Yes, O.K.

Noyes: So, if you prefer, I could quote my results in terms of the electron mass rather than the proton mass. But then I would have to justify my calculation (originally due to Parker-Rhodes) of the mass ratio. Actually, the mass unit which occurs in the calculation of both the Fermi constant and Newton's constant is the proton mass.

Chirikov: This is just what seems strange to me.

Noyes: I know that a lot of people think of the proton as a complicated object, but for me it is simply the lightest, charged baryon. Since my theory conserves baryon number, perhaps this makes it less strange.

Chirikov: So, you exclude the violation of baryon-number conservation?

Noyes: I exclude the violation of baryon number.

Chirikov: In your theory it does not exist?

Noyes: As is explained in more detail elsewhere [H.P.Noyes, "Comment on 'Statistical Mechanical Origin of the Entropy of a Rotating Charged Black Hole"', SLAC-PUB-5693 (Nov. 1991)], if one starts from $2^{127} + 136 \approx 1.7 \times 10^{38}$ proton-antiproton pairs plus a proton inside a sphere of radius $\hbar/m_pc$, the energy will rapidly be dissipated by proton-antiproton annihilations (our version of Hawking radiation) leaving behind a charged rotating black hole (see W.H. Zurek and K.S. Thorne, *Phys. Rev. Letters* **54**, 2171 (1985), for the classical calculation) which is indistinguishable from a proton, and *stable* provided baryon number is conserved. So from my point of view, the proton is an elementary particle. If you wish to observe the quark degrees of freedom, you must supply the energy and momentum from outside the system, in the same way that you must supply the energy to a hydrogen atom in the ground state in order to see the levels above the ground state and below ionization threshold (Franck-Hertz experiment). The basic difference is that the quarks remain confined, corresponding to an infinite ionization threshold.

Chirikov: Yes, but your theory is inconsistent with the grand unification theory.

Noyes: It is inconsistent with unstable protons. Since grand unification theories predict that protons are unstable, and proton decay is not observed, it is the grand unification theories that are in disagreement with experience. My theory is not.

Chirikov: Not yet observed, perhaps...

Noyes: The original grand unification theories predicted that the proton lifetime should be about 10^{24} years. When the experimental lower limit was pushed above that value, they found ingenious ways to raise their predictions. By Herculean efforts, the experimentalists have now pushed the experimental lower limit up to 10^{35} years. After considerable work, the grand unification theorists have again caught up, but the last I heard was that they could not raise the limit any further than a decade or at most two. In the light of this history, I would claim that the "prediction" of proton instability is not worth much. In contrast, my theory in its current form would be disproved by the observation of proton decay.

Miller: I would not pretend that I understood everything that you said, but I should like to ask you a nasty question nonetheless. Suppose that the next calculations that you do conflict with experiments - that is, your theory is refuted. Which of the postulates of your theory would you give up?

Noyes: Then I would have to think pretty hard about whether it is possible to save the theory by some basic modification, or whether I should abandon the whole approach. Depending on where such a failure occurs, it could be an indication of some basic structure that has been omitted. Of course this theory, like any theory, will eventually be disproved. That is how we make progress, by finding out why the theory fails to explain phenomena. I would have to take my clues from where the failures occur.

Miller: May I force my question. I did not see, apart from your general remarks about units, what the detailed postulates are.

Noyes: Suppose, for instance, that Tony Leggett is right in his conjecture that the transition from quantum to classical mechanics occurs for systems that have approximately 10^{15} atoms. This would be a new phenomenon, and would define a parameter that does not occur either in my theory or in conventional theories. This is what Leggett is looking for in the quantum mechanical barrier penetration of one flux quantum between two SQUIDs. If he succeeds both conventional theories and mine would have to understand that parameter.

Miller: Yes, but that would be a refutation, wouldn't it?

Noyes: Rather, that would mean that the theory has to be extended.

Miller: That's most theories' problem.

Noyes: True, but that would not be the same as a refutation. What I would count as a serious difficulty with my theory would be if I were forced to predict the wrong mass for the K-mesons. If that does not come out approximately correct, I would know that there is something fundamentally wrong with the scheme and I would have to think very hard. So that's one of the next calculations I want to make. I have already take the first step into the second generation of quarks and leptons by calculating the muon mass as 207 electron masses, which agrees with experiment to within one electron mass (see table). Since the charged kaon is partly a bound state of a muon and a muon neutrino, this means I should be able to calculate this second generation mass. But I also have generation number in my scheme, so I should be able get the J/ψ ("gipsy") particle and the D-mesons as well. I can already see why the first generation — up and down — quark masses should be only a few electron volts, but those cannot be measured directly. That is why I prefer to talk about the meson masses which are measurable. The trouble is that I have yet to find a high energy particle physicist to collaborate with on what most of them consider a unpromising theory. And I wanted to get the foundational stuff right first. The kaon masses are not a problem in principle. But I have to work out my relativistic particle scattering theory in more detail before I will trust the answer. After I have articulated the scattering theory and can use it to calculate the kaon masses, an answer that differs from experiment by more than a few electron masses would pose a very serious problem for me.

Weingartner: Does your theory say anything about the interpretation of the Bell inequalities and to these problems of locality?

Noyes: As to the problem of locality, I know how to use a bit-string model to send information that not only synchronizes distant clocks but suffices to set up local directions at distant locations which have a global and not just a local meaning. I also know how to model the detection statistics of the polarimeters used in the Aspect experiment. If I use the same key word, sometimes called a "seed", in the local pseudo-random number generators which generate the detector statistics at space-like separated locations, I will produce not only the correct local statistics, but an absolute correlation between the two distant "detections". It seems to me that I have in hand all the ingredients needed to construct a computer model of the Aspect experiment that could actually be built and tested. Pat Suppes, and my colleague Tom Etter, are very sceptical. It is true that each attempt of mine to complete the coding for the model has turned out to be flawed. But I am still optimistic. If I succeed, Henry Stapp agrees that I will have found out something new about quantum mechanics. But so far, I cannot make that claim.

Weingartner: There is a new interpretation of Aspect's experiments by Kwiat, Steinberg and Chiao (Physical Review 47, R 2472) who did also new experiments. One type of experiments were proposed by J.D. Franson: Two photons, simultaneously emitted by a parametric converter (an optical non-linear cristall) pass (each of them) two separated interferometers where they might choose a short or a long route, the short way leading to a detector. The experiment shows that the photons when leaving the instrument "seem to know" immediately which way (the longer or the shorter) the other has taken. Both results of measurement depend non-locally upon each other. Another experiment was that of tunnelling photons compared to photons without a barrier. It led to the surprising result that those which went through the tunnel were quicker such that one had to assume superluminarity in the tunnel. This appearing non-locality (super luminarity) was interpreted not as really increasing the velocity (there is no superluminar signal) but as a shift (towards the front) of the maximum of the amplitude of the wave packet by running through the "tunnel".

Noyes: O.K. I will study that in detail. Thank you for giving me the reference. That will be worth thinking about.

Batterman: What kind of equation of evolution do you get for describing, you know, macroscopic context?

Noyes: There are two types of time evolution in my model. The first corresponds, roughly, to what would be called cosmological time in a conventional model. For me this is the program which generates an evolving universe of bit-strings. At any stage there are a finite number of strings of the same, finite length. Two are picked at random and compared using XOR. If they are the same a random bit, randomly chosen for each string, is adjoined to each string, increasing the bit-length of all strings by one. If they are different the string produced by XOR will differ from both. It is adjoined to the bit-string universe, increasing the number of strings by one.

Miller: You say that you choose two at random, and that if you get something that is different from those two then you get a new one. That does not follow.

Noyes: Why not?

Miller: Perhaps it was already there.

Noyes: If it is the same as a string that was produced earlier, it will be adjoined to the memory in a new location. In that sense it will be a new string. The action is only locally deterministic. I have no guarantee that the strings produced are globally unique. In fact, for the length of the strings to increase, I must have two strings in different locations which are otherwise identical, and pick them for the XOR comparison. For a more familiar type of time evolution, consider first two strings which combine by XOR to produce a third which could also be produced by two different strings. This can be interpreted as a four leg Feynman diagram, with a single intermediate particle. It can be shown that the standard energy-momentum-angular momentum conservation laws hold in the finite and discrete (mass-quantized) form appropriate to my theory. Further, the coupling constants (like the fine structure constant and the Fermi constant) are given by the statistics with which the strings are produced. Then, comparison with experiment can be made in the usual way by computing cross sections from S-matrix amplitudes. To go beyond this single particle exchange approximation, these amplitudes can be used as driving terms in relativistic Faddeev-Yakubovsky integral equations. Then the solution of these equations, which is guaranteed to be unitary, gives the appropriate time evolution for strongly interacting three and four body systems.

Batterman: Suppose I want to describe the behavior of a fluid. What kind of limiting relations presumably do you want for the equations?

Noyes: To go to the Navier-Stokes equations from relativistic quantum particle theory would be a Herculean task, which would require classical or semi-classical approximations that might not be under control. What I do claim to have done in my paper is to show how to go from measurement accuracy to the Maxwell equations for electrodynamics. Perhaps the optimism this engenders about more complicated classical systems is not justified.

Batterman: But if you want to use it you said something like chaos that appears in the Stokes-equation.

Noyes: Then I would have to ask for more detailed information as to what accuracy I need in specifying the boundary conditions for chaotic results to appear in the solution. If that accuracy is not obtainable, my position would be that we cannot say that the equations *predict* chaos.

Miller: Suppose that you just have a random mechanism in a black box that calculates the logistic function and then produces a sequence of 0's and 1's that we all think of as deterministic chaos.

Noyes: O.K. What is the physical system you are modeling?

Miller: I'm not an engineer. I do not have to design anything that calculates this. It is just a piece of metal inside a box that is producing the calculation.

Noyes: This looks like a computer model to me. What is it modeling?

Miller: It is not modeling a system. It is a physical system. Could you accept that coin tossing is a physical system. Spectral lines or the light going through a halfsilvered mirror or something like that.

Noyes: If it is not modeling anything, you are talking about a macroscopic system. Then I need to know the detailed specification of the system, and the accuracy to which the boundary conditions are known, before I can say whether predicting that the behavior of that the system will exhibit "deterministic" makes sense in the first place.

Miller: I didn't understand your remark that deterministic chaos is just an artifact.

Noyes: Deterministic chaos is a mathematical statement about the numerical solution of certain non-linear equations. To make this result into a prediction about a physical system which the mathematics is supposed to model, it is necessary to investigate (a) whether the calculation is vitiated by round-off error, and (b) whether the initial boundary conditions can be measured to sufficient accuracy for the chaotic situation to develop. If this accuracy requires you to measure positions to better than half an electron compton wavelength, I assert that your mathematics which predicts deterministic chaos *cannot* be used to correctly model that *physical* system in that context. Then, for me as an empiricist, deterministic chaos becomes a vacuous concept in that context.

Miller: I did not have a model. All I have is a physical process that is happening in the world.

Noyes: You are describing the world mathematically. That, from my point of view, amounts to the mathematical modeling of a physical system. I claim that it is then legitimate for me to ask whether that model is appropriate or not. If the calculation is to apply to the world, it is necessary to explain *how* the mathematical model is connected to the world you observe. As a physicist, I have to ask that. I always must ask the question what is the connection between the mathematical model and the physical system being modeled. Until you answer that question I cannot say whether your model is appropriate or inappropriate. Certainly there are situations in which models involving deterministic chaos are appropriate. All I am saying is that then they are approximations from the point of view of the underlying relativistic quantum mechanics. Consequently such models cannot be brought to bear on the question of whether the physical system being modeled is deterministic or not. I don't understand where the confusion is arising.

Suppes: I think with your system you also made this remark about deterministic chaos. Those are things that don't fit together.

Noyes: No, the connection between my particle physics and classical systems is not very tight.

Suppes: In your theory you cannot talk about solutions of the Navier-Stokes equation.

Noyes: That is true at present. But my current success in arriving at the classical Maxwell equations for electromagnetism as the correspondence limit of my theory allows me to speculate that this will ultimately be possible for the rest of classical physics. This will take a lot of work. Currently I can talk about two particle systems, and with more difficulty, three and four particle systems. To go beyond the relativistic Faddeev-Yakubovsky finite particle number equations will be tough.

Chirikov: I would put this question another way. Is the whole system you construct, your whole theory, describing everything, is it a dynamical theory or a statistical one?

Noyes: It is a statistical theory because the bit-strings are (if you like) the underlying ontological elements. They are generated by a random process which will return a zero or a one with equal probability. My model is quite definite on that score. Whether the universe is really like that is another question.

Chirikov: In this theory you must have some chaos from the beginning.

Noyes: I am assuming that all particles in the universe can be described by bit-strings.

Chirikov: Is there a finite number of particles?

Noyes: The number of protons, or more precisely the number of baryons minus the number of anti-baryons is about 2.9×10^{76} (see table). This is in approximate agreement with observation. My cosmology predicts this finite number. This is the number of protons you end up with a few hundred million years after the start of the process that generates space, time and particles. This is also approximately the time when the universe changes from being optically thick to optically thin, sometimes called "fireball time". Then the radiation breaks away from the matter and goes on expanding for fifteen billion years, providing the currently observed $2.7^\circ K$ cosmic background radiation.

Chirikov: This is the number of protons?

Noyes: Approximately, yes.

Chirikov: But what is the number of all particles?

Noyes: To a good approximation, 75% of the matter is hydrogen and 25% helium. There is one electron per hydrogen atom, two protons, two neutrons and two electrons per helium atom, and a small percentage of other stuff. So, to about a factor of two, the number of protons is the same as the baryon number, is the same as the number of particles.

Chirikov: But at the beginning there were no protons at all.

Noyes: That is true of standard inflationary models for cosmology. My model is somewhat different at the start because I have no space or time either, and have to generate a bit-string model that describes all three at the same time. But by the time my model reaches fireball time, it looks very much like the standard cosmological models.

Chirikov: Why do you need random numbers?

Noyes: Why do I? Because the basic program I use (a) picks two strings at random to generate the next string, and (b) adjoins an arbitrary bit to the growing end of each string if the result of the first process is a null string. This model implies a statistically predictable distribution of bit-strings and hence of particles. This has not been worked out in detail, and may fail to be in accord with observation in the same sense that any specific cosmological model can fail. But the first order results look promising (see table). What I have yet to do is to compute the power spectrum of the early fluctuations of the background radiation, as measured by the COBE satellite. These results have already killed the majority of pre-existing cosmological models, and mine might suffer the same fate. Fortunately, my cosmology is to a certain extent independent of my elementary particle physics, and one might survive without the other.

Chirikov: Have you quarks in your theory?

Noyes: Yes. I have the full standard model containing three generations of quarks and leptons. I have all the observed particles and predict no unobserved particles such as the Higgs bosons.

Chirikov: Can you explain quark confinement?

Noyes: Yes. That's easy if you first grant my proof that there are at most three asymptotic homogeneous and isotropic dimensions in a bit-string theory. This is a special case of McGoveran's Theorem, which applies to any finite and discrete metric theory. The basic result on which it rests is due to Feller (*Probability Theory and its Applications*, Wiley, 1950, 9. 247) and states that that for D independent Bernoulli sequences of length n the probability that they will accumulate the same number, n, of ones after n steps is proportional to $n^{\frac{-(D-1)}{2}}$. Consequently, this criterion — which is used to define D homogeneous and isotropic reference directions in bit-string physics — can continue to keep on being met for n indefinitely large and $D = 2$ (probability proportional to $n^{-\frac{1}{2}}$) or $D = 3$ (probability proportional to n^{-1}), but that it cannot be met for four or more sequences. Hence *spacial* dimensions greater than three compactify, and are not encountered asymptotically. We are therefore allowed three conserved quantum numbers to label three independent asymptotic dimensions. Guided by experience, and the way our model provides conserved quantum numbers, we can take these to be charge, baryon number, and lepton number. Then color, and consequently colored quark and gluon systems, are confined. Colorless systems such as baryons and mesons can still be observed asymptotically in my theory.

Chirikov: So, why only quarks?

Noyes: Only three asymptotically conserved particulate quantum numbers can be defined because of the way I construct space and time.

Chirikov: But why only quarks? Space- time for all particles is the same.

Noyes: That is not true in my theory. That is just the point. My space, time, and particles are constructed in very different ways from bit-strings. And those bit-strings specify both the discrete quantum numbers and the space-time character of the particle states. For example, the first sixteen bits can be used for the quantum numbers and the remainder of the string for the space-time state (until we come to multi-particle systems and/or gravitation). [For a more detailed description see my paper and references therein.]

Chirikov: Feynman's diagrams is a perturbation theory.

Noyes: I disagree. Once one can handle strong interactions, either using QCD, or in my way, all observable processes can be computed from the Feynman diagrams. This is Heisenberg's and Chew's basic reason for espousing S-Matrix theory. In fact, because of the mathematical ambiguities of quantum field theory, Tini Veltman has recently asserted that the Feynman diagrams *are* the theory.

Chirikov: There are many of them.

Noyes: I do not need many of them to get accurate and interesting results.

Chirikov: You simply truncate the perturbation series.

Noyes: It is true that an approximation is being made. but in contrast to expansions in powers of the coupling constants — which are at best asymptotic, and don't work for strong interactions in any case — my expansion is in terms of particle number, and is often rapidly convergent. You see, the masses in this theory are finite to begin with, and close to the empirical values. The corrections are small.

Chirikov: Back to quarks. Quarks have in your theory a different space, yes?

Noyes: They reside inside rotating black holes (protons, neutrons, mesons) or in the space near such sources as virtual particles.

Chirikov: But why are they confined, I missed.

Noyes: Because you cannot define color in addition to baryon-number, lepton-number and charge at large distances from interaction regions. You cannot get there.

Chirikov: And what about mesons?

Noyes: Mesons are bound states of quark-antiquark pairs, in a good first (valence quark) approximation.

Chirikov: Are they also confined in your theory?

Noyes: Mesons are a *colorless* combination of quarks and anti-quarks. Therefore they are *not* confined, in my theory, or in conventional theory, or in the laboratory. Of course they are unstable, and decay ultimately to colorless leptons and gamma rays. So long as they exist, the quarks inside them *are* confined. To a good approximation, you can often treat them as composed of a quark-antiquark pair.

Chirikov: Why?

Noyes: That is what my theory gives. The correct interpretation of the coupling constant is obtained from the consistency of the model.

Chirikov: Which coupling?

Noyes: The coupling of any particle anti-particle pair which, to a good approximation forms a stable or long lived particle. Call the particle a, the anti-particle $\bar{a}$, their individual masses $m_a = m = m_{\bar{a}}$, the mass of the meson to which they bind μ, and the coupling constant $g^2_{a\bar{a}\mu}$, or g^2 for short. Then taking s to be the Mandelstam variable (square of the invariant four momentum of the particle-antiparticle system), and calling the bound state wave function, which is a pole in the S-matrix, $\phi(s) = \frac{g^2\mu}{s-\mu^2}$, the unitarity of the S-matrix requires that

$$\int_{(4m^2-\mu^2)}^{\infty} \phi^2(s)ds = 1 = \frac{g^4\mu^2}{4m^2 - \mu^2}$$

Thus this first approximation relates the coupling constant to the masses. For more details see D.O.McGoveran and H.P.Noyes, *Physics Essays*, **4**, 115 (1991).

Chirikov: But what are the original particles? Not quarks?

Noyes: What I am doing is quite general. In fact as noted in the reference this argument gives Bohr's 1915 relativistic formula for the energy levels of hydrogen, and also a good value for the pion-nucleon coupling constant.

Chirikov: But not for quarks.

Noyes: Why not, in an S-matrix theory? Valence quark-antiquark pairs bind to form colorless mesons. But the formula can also be used for describing a positive pion as the bound state of a proton and anti-neutron as in the Fermi-Yang model. In fact this give a good value for the pion-nucleon coupling constant, as given in the table in my paper.

Chirikov: But why you cannot split this system into quarks? This is the problem.

Noyes: Because of my boundary condition from McGoveran's Theorem, quarks can only be evidenced indirectly as intermediate states in Feynman diagrams, or S-matrix elements. It is like exciting the states in a hydrogen atom using a probe which can only deliver an energy below ionization threshold. In effect, color confinement makes the equivalent threshold in the quark systems infinite. So quarks are never seen as free particles. But they have indirect consequences which are evidenced by the energy dependence of scattering cross sections at high energy.

Chirikov: Do you mean that you adjusted this interpretation in such a special way as to obtain quark confinement? Because there are many particles but only quarks are confined. How is that possible in a general system.

Noyes: In my system, as in QCD, quarks can only be excited as internal degrees of freedom. This is the meaning of color confinement in both theories. In the conventional theory there are many models for the confining force, but so far none have been derived from first principles. In my theory, I do not need to provide a model for the confining force. A more detailed dynamical theory than I have developed so far should be able to calculate effects due confinement. As I have tried to explain already, my construction of space, time and quantum numbers only allows three absolutely conserved quantum numbers to be conserved asymptotically, and the others must be confined. I take the three to be charge, lepton number and baryon number. This leaves color confined. That is my boundary condition.

Chirikov: Why not colour?

Noyes: Empirically, the first three are conserved asymptotically, and color is not observed asymptotically. This forces my interpretation of McGoveran's theorem.

Chirikov: Which theorem?

Noyes: McGoveran's theorem, as explained above, leaves me only three choices for asymptotically conserved quantum numbers. Having picked three that, so far as we know today *are* conserved, there is no room left for color. I conclude that it must be confined. Now, maybe this isn't a good argument.

Chirikov: But in this sense any of those three particles might become confined in your theory.

Noyes: No, not if I compare the structures given by my construction of conserved quantum numbers with what is known from experiment. I find a neutrino-antineutrino pair at level one, and the associated charged leptons at level 2, which gives me two of my three conserved quantum numbers. When I go to level 3 I find quantum numbers that can be interpreted as baryons, but also more structure that can be interpreted as colored quarks and gluons. The quarks couple to the same neutrinos, and the gluons allow me to form colorless combinations that can be interpreted as protons, neutrons, pions, and so on. But McGoveran's theorem allows me to choose only one of the quantum numbers I can form to be asymptotic, and I take it to be baryon number. The rest, including color, must be confined.

Chirikov: But if for example proton would be unstable, then the quarks were not confined?

Noyes: Then the whole thing does not work.

Chirikov: O.K.

Noyes: Or, depending on what the decay mode is which is observed, I suspect it would not contain enough stable quarks to conserve baryon number. I also doubt that we would see asymptotic color. So my scheme would have to be modified. Perhaps the three conserved quantum numbers would be fermion number, charge, and weak isospin. I might be able to make a consistent theory along those lines, but I would have to see. I think the observation of proton decay would be as much of an embarrassment for me as the many failures to observe it at a predicted rate should be for the grand unifiers.

Photons, Billiards and Chaos

P. Suppes, J. Acacio de Barros

Stanford University, USA

1 Introduction

In this paper we continue the foundational study of photons as particles without wave properties (Suppes and de Barros (1994a), Suppes and de Barros (1994b)). In the earlier work we assumed: (i) Photons are emitted by harmonically oscillating sources. (ii) They have definite trajectories. (iii) They have a probability of being scattered at or absorbed in the near presence of matter. (iv) Detectors, like sources , are periodic. (v) Photons have positive and negative states which locally interfere, i. e. annihilate each other, when being absorbed. In this framework we are able to derive standard diffraction and interference results. We thereby eliminate in this approach wave-particle duality for photons, and give nonparadoxical answers to standard questions about interference. For example, in the two-slit experiment each photon goes through only one slit.

In the earlier work we did not construct a stochastically complete model of the monochromatic harmonically oscillatory point source, but only assumed an expectation density for the emission of positive and negative photons, namely $n_\pm(t) = \frac{A}{2}(1 \pm cos\omega t)$ with $t > 0$. In the present paper we derive this equation (see (8) in Section 1) as the expectation density in free space-time from a probabilistic model of a two-level atom as source, which easily generalizes to N atoms. We also can go further and derive from the model the cross-correlation of two arbitrary space-time points, but we do not include that calculation in this paper.

In Section 2 we look at photons as particles which have, in certain special environments, ergodic motion. In particular, we study the way in which photons having definite trajectories can move in ergodic fashion like billiard balls on a rectangular table with a convex obstacle in the middle. Such billiards are called Sinai billiards after the Russian mathematician Ya.G.Sinai. Their ergodic motion is strongly chaotic.

Finally, in Section 3 we examine the isomorphism of deterministic and stochastic models of photon ergodic motions. Here we use important results of D.S.Ornstein and his colleagues on the indistinguishability of these two kinds of models of ergodic behavior.

The positive and negative photons we introduce can well be thought of as virtual photons, for in a detector they locally interfere with each other, and only the excess of one or the other kind is observable (for further details see Suppes and de Barros (1994b)).

2   Two-Level Atom as Source

We have several processes at the source, which we initially treat as a single atom. In this version we begin by making time discrete, with the time between the beginning of successive trials on the order of the optical period, 10^{-15} s.

Process I. Pure Periodic Process. On an odd trial a photon in the positive state may be emitted or absorbed, and on an even trial a photon in the negative state may be emitted or absorbed. This process is defined by the function

$$f_{\pm}(n) = n \bmod 2, \tag{1}$$

where n is the trial number. Intuitively, we use (1) to make the probability zero of an atom emitting or absorbing a negative photon on trial n if n is odd, and probability zero for a positive photon if n is even. This is our periodicity.

If, on an even-numbered trial, the atom is in the excited state, which we label 1, at the beginning of the trial, then there is a positive probability, but not in general probability 1, of emitting a negative photon, and similarly on odd trials for a positive photon. Correspondingly we use 0 as the label for the ground state

Process II. Exponential Waiting Times. We use a discrete Markov chain in the two states 0 and 1 to give us in the mean the geometric distribution of waiting for absorption or emission, but with different parameters. The geometric distribution is, of course, the discrete analogue of the exponential distribution in continuous time. The transition matrix is:

$$
\begin{array}{c|cc}
 & 1 & 0 \\
\hline
1 & 1 - c_1 & c_1 \\
0 & c_0 & 1 - c_0
\end{array}
\tag{2}
$$

Thus, c_0 is the probability of absorbing a positive or negative photon when in the ground state at the beginning of a trial. In our simple model meant for low-energy experiments, e.g., those dealing with the optical part of the electromagnetic spectrum, we exclude the possibility of multiple-photon absorption or emission on a given trial. The parameter c_1 is just the probability of emitting a positive or negative photon when in the excited state at the beginning of a trial.

Processes I & II together. The description just given of absorption and emission of photons is for Process II alone. Combined with the periodicity of Process I, we can write a single matrix, but one that depends on whether the trial number is odd or even:

$$
\begin{array}{c|cc}
 & 1 & 0 \\
\hline
1 & 1 - c_1 f_{\pm}(n) & c_1 f_{\pm}(n) \\
0 & c_0 f_{\pm}(n) & 1 - c_0 f_{\pm}(n)
\end{array}
\tag{3}
$$

where $f_+(n) = 1$ if n is odd and 0 if n is even, and contrariwise for $f_-(n)$.

Asymptotic Distribution of States. The Markov chain characterized by (2), for $c_0, c_1 \neq 0$ is obviously ergodic, i.e., there exists a unique asymptotic stationary distribution independent of the initial probability of being in either state.

For computing this distribution we can ignore the distinction between the even and odd numbered trials, as expanded in (3), and consider only the process characterized by (2).

The asymptotic distribution is just obtained from the recursion:

$$p_0 = c_1 p_1 + (1 - c_0) p_0. \tag{4}$$

Solving,

$$p_0 = \frac{c_1}{c_0 + c_1} \tag{5}$$

and

$$p_1 = \frac{c_0}{c_0 + c_1}. \tag{6}$$

So, asymptotically, for N atoms, the expected number in the ground state is $\frac{c_1 N}{c_0 + c_1}$.

Properties of Photons. Omitting here polarization phenomena, a photon is a 3-tuple $(\omega, c, \pm)$, where ω is the frequency of oscillation of the source, c is the velocity, and $\pm$ are the two possible states already discussed.

Process III. Direction of Emission. We assume statistical independence from trial to trial in the direction of emission of photons by a single atom. We also assume that the probability of direction of emission is spherically symmetrical around the point source. For this analysis we further restrict ourselves to two dimensions and a scalar field, as is common in the study, e.g., of optical interference in the two-slit experiment, or the "billiards" case discussed in the next section.

Periodic Properties of $f_\pm(n)$. Various properties are needed.
(i) If φ is even, $f_\pm(n + \varphi) = f_\pm(n)$.
(ii) If n is even, $f_\pm(n + \varphi) = f_\pm(\varphi)$,
(iii) $f_\pm(n + \varphi) = f_\pm(n + 1) f_\pm(\varphi) + f_\pm(n) f_\pm(\varphi + 1)$.
This is easy to prove by considering the four cases: n is odd or even, and so is φ.

For $r \neq 0$,

$$P(\text{photon being at } (r, \theta, t) | \text{ emission at } t') = \frac{1}{4\pi r} \delta(t - t' - \frac{r}{c}).$$

We now compute the unconditional probability of emission at t', where gr.st. is the event of being in the ground state.

$$
\begin{aligned}
P(\pm \text{ emission at } t') &= \sum_{t'' < t'} P(\pm \text{ emission at } t' | \text{ absorption at } t'') \\
&\quad \cdot P(\text{absorption at } t'') \\
&= \sum_{t'' < t'} \sum_{t''' < t''} P(\pm \text{ emission at } t' | \text{absorption at } t'') \\
&\quad \cdot P(\text{absorption at } t'' | \text{gr.st. at } t''') \, P(\text{gr.st. at } t''') \\
&= \sum \sum c_1 (1 - c_1)^{t' - t''} f_\pm(t') c_0 (1 - c_0)^{t'' - t'''} \frac{c_1}{c_0 + c_1}. \tag{7}
\end{aligned}
$$

The continuous analogue of (7) is obvious, where β_0 corresponds to c_0 and β_1 to c_1.

$P(\pm \text{ emission at } t') =$

$$= \int_{-\infty}^{t'} \int_{-\infty}^{t''} \beta_1 e^{-\beta_1(t'-t'')} (\frac{1}{2} \pm \frac{1}{2} \cos \omega t') \beta_0 e^{-\beta_0(t''-t''')} \frac{\beta_1}{\beta_0 + \beta_1} \cdot dt''' dt''$$

$$= \frac{\beta_1}{\beta_0 + \beta_1} (\frac{1}{2} \pm \frac{1}{2} \cos \omega t') \left[e^{-\beta_1(t'-t'')} \Big|_{-\infty}^{t'} \; e^{-\beta_0(t''-t''')} \Big|_{-\infty}^{t''} \right]$$

$$= \frac{\beta_1}{\beta_0 + \beta_1} (\frac{1}{2} \pm \frac{1}{2} \cos \omega t'). \qquad (8)$$

Let us abbreviate the space-time position as $i = (r_i, \theta_i, t_i)$. Then the random variables for positive and negative photons are defined as:

$$X_i(\pm) = \begin{cases} 1 \text{ if there is a } \pm \text{ photon (from the single atom) at } (r_i, \theta_i, t_i), \\ 0 \text{ otherwise.} \end{cases} \qquad (9)$$

Then, since $t_i - t'_i = \frac{r_i}{c}$, we have

$$E(X_i(\pm)) = \frac{1}{2\pi r_i} \frac{\beta_1}{\beta_0 + \beta_1} (\frac{1}{2} \pm \frac{1}{2} \cos \omega (t_i - \frac{r_i}{c})), \qquad (10)$$

which is the form of the expectation density, $h_\pm = E(X_i(\pm))$, for positive and negative photons derived in Suppes and de Barros (1994b), but here for a single-atom source.

3   Photons as Billiards

We now move to the study of photons as particles executing ergodic motions. Let us begin with a rectangular box that has reflecting sides. We assume the classical law of reflection, that is, the angle of reflection equals the angle of incidence. An ideal laser could execute the periodic motion shown in Figure 1 in such a box. In fact, in classical mechanics it is always in terms of such motions that billiards are used as standard examples of mechanical systems. It is only in the modern study of billiards that matters become much more complicated. Intuitively it is easy to describe how to get such additional complication. We add a convex obstacle to the rectangular box with reflecting sides as shown in Figure 2.

Now the path of the photon emitted by an ideal laser executes the motion of a photon as a Sinai billiard. Sinai and other investigators have studied very thoroughly the mechanical motion of a point particle in such a rectangular box with a convex obstacle and with the collisions of the particle observing the classical law of reflection as well as the law of perfect elasticity. In the case

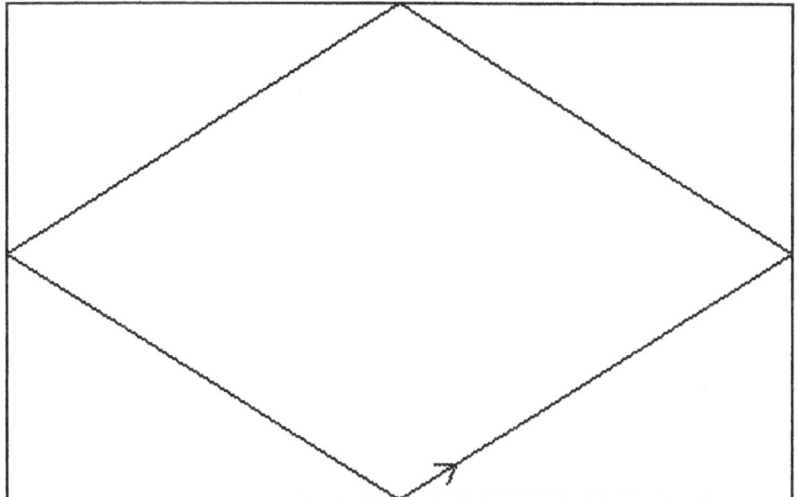

Fig. 1. Periodic photon billiard motion

of photons we drop the concept of elasticity but have the reflection take place without loss of energy.

It is worthwhile to analyze the law of reflection more carefully in our theory. We give a semiclassical derivation in that we assume the reflecting walls are continuous perfect conductors in the sense of classical electromagnetic theory. The boundary condition for a scalar field is that it be zero at the conductor surface. We can show how we obtain this result most clearly by returning to the temporally discrete model introduced at the beginning of Section 1. For a perfect conductor we change the basic assumptions as follows. We modify the transition matrix (3) to reflect the assumption that in reflection a photon changes its state from positive to negative and vice versa.

This single change for perfect conductors enables us to show that the defined scalar electric field at the reflecting surface is zero. For a single photon, we have at once from the new transition matrix below that at a point of the surface, $h_+ = h_-$ and therefore $\mathcal{E} = 0$. We use here the definition of the field $\mathcal{E}$ given in Suppes and de Barros (1994b), i.e., $\mathcal{E} = \frac{\mathcal{E}_0(h_+ - h_-)}{\sqrt{h_+ + h_-}}$. We thus replace (3) by the

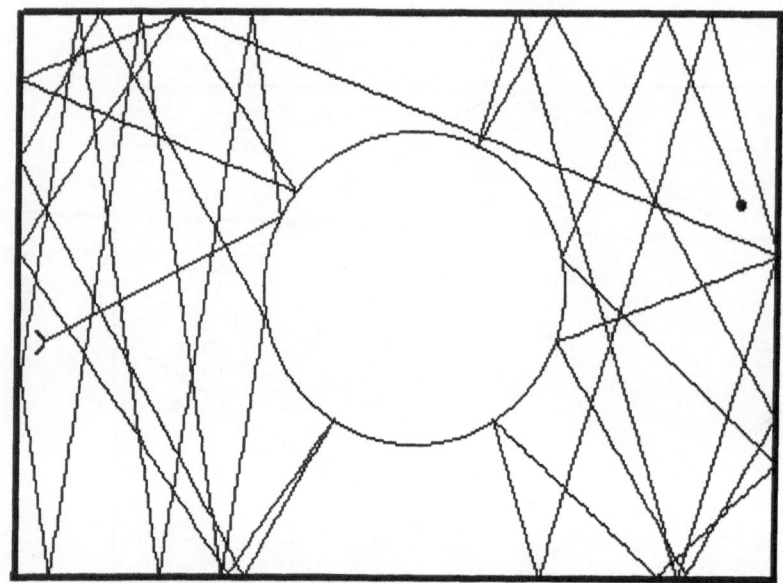

Fig. 2. Photon billiard motion with a convex obstacle

matrix

$$
\begin{array}{c|ccc}
 & 1- & 1+ & 0 \\
\hline
1- & 1 - c_1 f_+(n) & 0 & c_1 f_+(n) \\
1+ & 0 & 1 - c_1 f_-(n) & c_1 f_-(n) \\
0 & c_0 f_-(n) & c_0 f_+(n) & 1 - c_0
\end{array}
\tag{11}
$$

It follows from fundamental results of Sinai (1970) that the following theorem can be proved.

Theorem 1 *The motion of a photon as a Sinai billiard, as shown in Figure 2, is ergodic.*

We can see this even more clearly by showing the picture of a simulation of the motion of a photon as a Sinai billiard. It is intuitively clear that we get then the following corollary from the ergodic motion.

Corollary 1 *The motion of a photon as a Sinai billiard is strongly chaotic.*

We say more about this chaos later, but remember the chaos is derived here from our conception of photons as point particles executing linear trajectories between points of reflection.

Measurement of photons. The chaotic motion of photons is not a topic usually discussed in the many physical discussions of chaos. Why is this? The answer is that in terms of observation using, for example, photodetectors, we can only measure the intensity of a light source averaged over time. It takes about 10^{-9} seconds for an atom to absorb a photon. In contrast, a single period of an optical source is about 10^{-15} seconds. Thus a photon that is emitted by an optical source takes in terms the period of the source on average about 6 orders of magnitude to be absorbed. This averaging process means that there is little hope of observing directly the chaotic motion of an individual photon. These remarks about averaging apply to quantum mechanics and classical electromagnetic theory of optical phenomena, as well as to the probabilistic atom model developed in the previous section. The average intensities predicted by quantum mechanics, by semi-classical application of classical electromagnetic theory, or by the kinds of probabilistic computations developed in the previous section are all average intensities that wipe out in the measurement process any evidence of chaos in the motion of an individual photon.

This means that our straightforward "free particle" theory of photons leads directly to a theory of chaos for photons, but the chaos is not observable by standard means.

4   Deterministic and Stochastic Models

It is widespread folklore in discussions of chaos by physicists that most important physical examples of chaos are deterministic. On the other hand, there is a variety of evidence, especially mathematical arguments, that associated with chaos, particularly in the strongest chaotic examples, are phenomena that can only be regarded as genuinely random or stochastic in nature. It would be easy to argue that one has got to choose either the deterministic or stochastic view of phenomena, and at least for a given set of cases, it is not possible to move back and forth in a coherent fashion. It is this view, also perhaps part of the folklore, that we want to argue very much against in the present discussion. We shall refer occasionally to our work on photons, but we will be depending much more on general ideas from ergodic theory and in particular on the strong kind of isomorphism theorems proved by Donald Ornstein and his colleagues. Before we turn to the details, there are one or two other points we want to discuss in a very intuitive fashion. For example, if we take a billiard model of the photon, or if you want, a mechanical particle, and we consider the deterministic model in the case of an ergodic motion, that is, one, for example, where there is a convex obstacle as shown in Figure 2, then there is an empirically indistinguishable stochastic model. The response to this isomorphism might be, "Well yes, but for the case of ergodic motion where the convex object is present we should choose either the simple Newtonian model or in the case of the photon, the simple deterministic reflection model, really from geometrical optics". Because this Newtonian or geometrical optical model works so well in the nonergodic periodic case when there is no convex object, it is natural to say that it is not a real choice between the deterministic or stochastic models. Because of its generalizability the choice obviously is the deterministic model.

But this argument can run too far and into trouble when we turn to a wider set of cases. On the same line we would be pushed to argue that the only kind of complete physical model for quantum mechanics must be a deterministic one, for example the kind advocated by Bohm, but the evidence once we turn to quantum mechanical phenomena seem far from persuasive for selecting as the unique intuitively correct model the deterministic one. Here there is much to be said for choosing the stochastic model, which is much closer in spirit to the standard interpretation of classical quantum mechanics. Our point, without going into details at this juncture, is that whether we intuitively believe the model should be deterministic or stochastic will vary with the particular physical phenomena we are considering. What is fundamental is that independent of this variation of choice of examples or experiments is that when we do have chaotic phenomena, especially when we have ergodic phenomena, then we are in a position to choose either a deterministic or stochastic model. When such a choice between different models has occurred previously in physics—and it has occured repeatedly in a variety of examples, such as free choice of a frame of reference in Galilean relativity, or choice between the Heisenberg or Schroedinger representation in quantum mechanics—, the natural move is toward a more abstract concept of invariance. What is especially interesting about the empirical indistinguishability and the resulting abstract invariance in the present case, is that at the mathematical level the different kinds of models are inconsistent, that is, the assumption of both the deterministic and the stochastic model leads to a contradiction when fully spelled out. On the other hand, it leads to no contradiction at the level of observations, as we shall see in an important class of ergodic cases.

Entropy and measure-theoretic isomorphism. In order to look at the entropy of appropriate processes, we begin with some of the simplest examples. Without much thought it is clear that the simplest example is a Bernoulli process with a finite number of alternatives and discrete trials. We shall call a finite discrete Bernoulli process any stochastic process with the following features. It is a probability space with a transformation namely a quadruple $(\Omega, \Im, \mu, T)$ satisfying the following assumptions: There is a finite set S and a probability measure p_i on S such that $\sum_{i \in S} p_i = 1$ and where Z is the set of integers, $\Omega = S^Z$, $\Im$ is the product σ-algebra on Ω, μ is equal to the product measure on Ω, and T is a left shift on Ω which means that if x, y are in Ω and for every n $y_{n-1} = x_n$, then $T(x) = y$. We say more about the shift T below. Notice that in this definition x and y are doubly infinite sequences, that is, they are sequences going from $n = -\infty$ to $n = \infty$ and the product measure guarantees that we have independence from trial to trial. Continuing with our Bernoulli example, and having it in the back of our minds, but not restricted to it, if we have a stochastic process defined in terms of a doubly infinite sequence of random variables, $\ldots, X_{-1}, X_0, X_1, \ldots, X_n, \ldots$ then we define the *entropy* rate as the following limit, if it exists, for a finite sequence of random variables.

$$H(\mathcal{X}) = \lim_{n \to \infty} \frac{1}{2n+1} H(X_{-n}, \ldots, X_0, \ldots, X_n) \tag{12}$$

and for the independence that is the strong feature of the Bernoulli process we have at once

$$H(\mathcal{X}) = \lim_{n \to \infty} \frac{H(X_{-n}, \ldots, X_0, \ldots, X_n)}{2n+1} = \lim_{n \to \infty} \frac{2n+1}{2n+1} H(X_0) = H(X_0).$$
$$\tag{13}$$

By a similar line of argument for a finite-state discrete Markov chain we get for its entropy rate the following expression:

$$H(\mathcal{X}) = -\sum_i p_i \sum_j p_{ij} \log p_{ij}. \tag{14}$$

Note that of course for the Markov process always $\sum_j p_{ij} = 1$.

In the ergodic literature there has been an intense study of how the entropy rate of a process relates to the measure-theoretic isomorphism of processes. (Terminology differs in the literature; what we call entropy rate is often just called entropy, but there are several different but closely related concepts of entropy, and the differences are not just a matter of terminology.) For that purpose we need an explicit definition of isomorphism. Let us first begin with a standard probability space $(\Omega, \mathfrak{S}, P)$, where it is understood that $\mathfrak{S}$ is a σ-additive algebra of subsets of Ω and P is a σ-additive probability measure on $\mathfrak{S}$. We now consider a mapping T from Ω to Ω. We say that T is *measurable* if and only if $A \in \mathfrak{S} \to T^{-1}A = \{\omega : T\omega \in A\} \in \mathfrak{S}$, and even more important, T is measure preserving, that is, $P(T^{-1}A) = P(A)$. T is *invertible* if the following three conditions hold: (i) T is $1-1$, (ii) $T\Omega = \Omega$, and (iii) If $A \in \mathfrak{S}$ then $TA = \{T\omega : \omega \in A\} \in \mathfrak{S}$. It is the measure preserving shift T introduced above that is important. Intuitively this property corresponds to stationarity of the process—a time shift does not affect the probability laws of the process.

We now characterize isomorphism of two probability spaces on each of which there is given a measure-preserving transformation, whose domain and range need only be subsets of measure one, to avoid uninteresting complications with sets of measure zero that are subsets of Ω or Ω'. Then we say $(\Omega, \mathfrak{S}, P, T)$ is *isomorphic in the measure-theoretic sense* to $(\Omega', \mathfrak{S}', P', T')$ if and only if there exists a function $\varphi \colon \Omega_0 \to \Omega_0'$ where $\Omega_0 \in \mathfrak{S}, \Omega_0' \in \mathfrak{S}', P(\Omega_0) = P(\Omega_0') = 1$, and φ satisfies the following conditions:

(i) φ is $1-1$,

(ii) If $A \subset \Omega_0$ & $A' = \varphi A$ then $A \in \mathfrak{S}$ iff $A' \in \mathfrak{S}'$,
    and if $A \in \mathfrak{S}$
$$P(A) = P'(A'),$$

(iii) $T\Omega_0 \subseteq \Omega_0$ & $T'\Omega_0' \subseteq \Omega_0'$,

(iv) For any ω in Ω_0
$$\varphi(T\omega) = T'\varphi(\omega).$$

To show how recent fundamental results are about the relation between entropy rate and measure-theoretic isomorphism, it was an open question in the 1950s whether the two finite state discrete Bernoulli processes $B(1/2, 1/2)$ and $B(1/3, 1/3, 1/3)$ are isomorphic. (The notation here should be clear $B(1/2, 1/2)$ means that the probability for the Bernoulli process with two outcomes on each trial is that for each trial the probability of one alternative is $1/2$ and of the other $1/2$). The following theorem clarified the situation.

Theorem 2 *(Kolmogorov (1958), Kolmogorov (1959) and Sinai (1959)). If two finite-state, discrete Bernoulli or Markov processes have different entropy rates, then they are not isomorphic in the measure-theoretic sense.*

Then the question became whether or not entropy is a complete invariant for measure-theoretic isomorphism. The following theorem was proved a few years later by Ornstein.

Theorem 3 *(Ornstein (1970)). If two finite-state, discrete Bernoulli processes have the same entropy rate then they are isomorphic in the measure-theoretic sense.*

This result was then soon easily extended.

Theorem 4 *Any two irreducible, stationary, finite-state, discrete Markov processes are isomorphic in the measure-theoretic sense if and only if they have the same periodicity and the same entropy rate.*

We then obtain:

Corollary 2 *An irreducible, stationary, finite-state discrete Markov process is isomorphic in the measure-theoretic to a finite-state discrete Bernoulli process of the same entropy rate if and only if the Markov process is aperiodic.*

We can go further in terms of photons and billiards with the concept of measure-theoretic isomorphism. To keep things in the context of finite-state discrete processes, we can form a finite partition of the free surface on the billiard table, as shown in Figure 3. This constitutes a finite partition of the space of possible trajectories for the photon or billiard and we correspondingly make time discrete in terms of movement from one element of the partition to another. With these constructive approximations, the following theorem has been proved:

Theorem 5 *(Gallavotti and Ornstein (1974)). With the discrete approximation of the continuous flow just described above, the discrete deterministic model of the photon or billiard is isomorphic in the measure-theoretic sense to a finite-state discrete Bernoulli process model of the motion of the photon or billiard.*

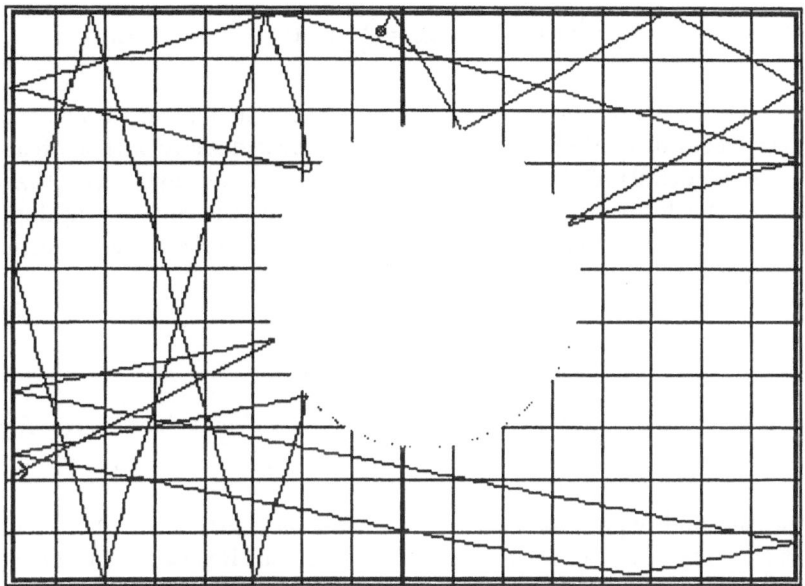

Fig. 3. Finite partition of the billiard table with convex obstacles

It should be noted that instead of this theorem we could have stated a theorem for continuous time and such results are to be found in the paper by Gallavotti and Ornstein. What the Gallavotti and Ornstein theorem shows is that the discrete mechanics of billiard balls is in the measure-theoretic sense isomorphic to a discrete Bernoulli analysis of the same phenomena. However, it is to be emphasized that in order to claim that intuitively the two kinds of analysis are indistinguishable from observation we need stricter concepts.

To show this, we need not even consider something as complicated as the billiard example but consider only a first-order Markov process and a Bernoulli process that have the same entropy rate and therefore are isomorphic in the measure-theoretic sense, but it is also easy to show by very direct statistical tests whether a given sample path of any length, which is meant to approximate an infinite sequence, comes from a Bernoulli process or a first-order Markov process. There is for example a simple chi-square test for distinguishing between the two. It is a test for first-order versus zero-order dependency in the process. The analysis is statistical and of course cannot be inferred from a single observation, but the data are usually decisive even for finite sample paths that consist of no more than 100 or 200 trials.

To spell out the details of this test, let $n_{ij}(t)$ denote the observed number of cases (for several possible runs) in state i at $t-1$ and state j at t. Further, let

$$n_i(t-1) = \sum_j n_{ij}(t), \quad n_{ij} = \sum_t n_{ij}(t), \quad n_i = \sum_j n_{ij}.$$

The Markov character of the sequence of position random variables, or of other sequences of random variables, may be tested directly without recourse to theoretical details of the process. We can test the null hypothesis that the outcomes of trials are statistically independent (zero-order process) against the alternative hypothesis that the process is a first-order Markov chain by computing the sum

$$\mathcal{X}^2 = \sum_{ij} n_i \frac{\left(\frac{n_{ij}}{n_i} - \frac{n_j}{N}\right)^2}{\frac{n_j}{N}},$$

where $n_j = \sum_i n_{ij}, N = \sum_{i,j} n_{ij}$, and n_{ij} and n_i are as defined above. Again, $\mathcal{X}^2$ has the usual limiting distribution with $(m-1)^2$ degrees of freedom. (A Bayesian modification of this test is easily given.)

A second null hypothesis is that the process is a first-order Markov chain against the hypothesis that it is a second-order chain. Rejection of the null hypothesis in this case would mean that the position probabilities can be predicted better by observing the two immediately preceding positions rather than simply the single immediately preceding one, and so on for $n+1$st-order vs nth order. Similar chi-square tests can be formulated for stationarity.

Congruence. To obtain a stricter sense of isomorphism it is natural to impose a geometric conditon, especially for a wide variety of physical examples of ergodic systems. Here we follow Ornstein and Weiss (1991). Let $\alpha > 0$ and let $\chi = (\Omega, \mathfrak{S}, P, T)$ and $\chi' = (\Omega', \mathfrak{S}', P', T')$ be two spaces isomorphic under φ in the measure-theoretic sense. Then χ and χ' are α-*congruent* if and only if there is a function g from Ω to a metric space (with d the metric) and a function g' from Ω' to the same metric space such that for any ω in Ω, $d(g(\omega), g'(\varphi(\omega))) < \alpha$ except for a set of measure $< \alpha$.

Intuitively the parameter α reflects our inability to measure physical quantities, inlcuding geometric ones, with infinite accuracy. What is significant is that α-congruence for small α, can be proved for Sinai billiards, and thus photons in a Sinai billiard box. And when α is chosen at the finite limit of our measurement accuracy, the Newtonian mechanical and a Markov process model of a Sinai billiard are observationally indistinguishable, as they are α-congruent.

Stated informally, we then have the fundamental result.

Theorem 6 *Using the discrete approximation described just before Theorem 5, the discrete deterministic model of the photon or billiard is observationally indistinguishable from a finite-state discrete Markov model of the motion of the photon or billiard.*

It is important to note that Theorem 6 is not true if the Markov model is replaced by a Bernoulli model. The observable dependencies discussed above, and for which a chi-square test was stated, rule out the Bernoulli model as a candidate for being α-congruent to the deterministic billiard model.

References

Gallavotti, G, Ornstein, D. S. (1974): Billiards and Bernoulli schemes. Comm. Math. Phys. **38**, 83–101

Kolmogorov, A. N. (1958): A new metric invariant of transient dynamical systems and automorphisms in Lebesgue spaces. Dokl. Akad. Nauk. SSSR **119**, 861–864. (Russian) MR **21** #2035a.

Kolmogorov, A. N. (1959): Entropy per unit time as a metric invariant of automorphism. Dokl. Akad. Nauk SSSR **124**, 754–755. (Russian) MR **2** # 2035b.

Ornstein, D. S. (1970): Bernoulli shifts with the same entropy are isomorphic. Adv. Math. **4**, 337–352

Ornstein, D. S., Weiss, B. (1991): Statistical properties of chaotic systems. Bulletin (New Series) of the American Mathematical Society **24** (1), 11–116

Sinai, Ya. G. (1959): On the notion of entropy of a dynamical system. Dokl. Akad. Nauk SSSR **124**, 768–771

Sinai, Ya. G. (1970): Dynamical systems with elastic reflections. Ergodic properties of displacing billiards. Usp. Mat. Nauk. **25** (2) (152) , 141–192. (Russian) Zbl. 252.58005. English transl.: Russ. Math. Surv. **25** (2), 137–189

Suppes, P., de Barros, J. A. (1994a): A random walk approach to interference. International Journal of Theoretical Physics **33** (1), 179–189

Suppes, P., de Barros, J. A. (1994b): Diffraction with well-defined photon trajectories: A foundational analysis. Foundations of Physics Letters **7**, 501–514

Discussion of Pat Suppes' Paper

Batterman, Chirikov, Miller, Noyes, Schurz, Suppes, Weingartner

Chirikov: Let's come back to the Ornstein theorem that there are some processes which cannot be distinguished as deterministic. You promised to give an example.

Suppes: Yes, the example is from this reference of Galavotti and Ornstein.

Chirikov: Sinai Billiard?

Suppes: Yes.

Chirikov: Why you cannot distinguish it?

Suppes: Well, that's exactly what the theorem is. The theorem states an isomorphism between stochastic and deterministic modelingmodel!stochastic, in the article quoted from '91.

Chirikov: But my question is whether you could explain it.

Suppes: I am going to, but I first say where the details are. The '91 article of Ornstein and Weiss gives details on the notion of α-congruence and small deviations.

Chirikov: For any time interval?

Suppes: You always have a small interval. You can observe as many times as you want but you must be observing sets of positive measure. So you are not observing an individual point.

Chirikov: Not an individual point but ...

Suppes: You're always observing a neighbourhood.

Chirikov: In a short time interval. Then why you cannot follow the exact straight line directly?

Suppes: Of course, you can. But, of course, the semi-Markov process can have that too. It's in the given states to go straight along, then it changes. It's a semi-Markov process ...

Chirikov: But can you follow any part of this trajectory?

Suppes: I think, I can follow in the same way as the semi-Markov-process ...

Chirikov: I don't think so. There must be some mathematical trick.

Suppes: No, I don't think so....

Miller: Could you remind us what a semi-Markov-process is?

Suppes: It's a semi-Markov-process, because the parameters of the holding time in a given state, vary with the state rather than being the same parameter for any state. So you have an exponential holding time for any state. And that parameter varies with the state. Remember you are not going to predict exactly. That's very important. It's different when you identify the point precisely. Then you have ideal observation points. You would not have a stochastic situation.

Chirikov: But does that mean that you cannot repeat such trajectory ...

Suppes: Yes ...

Chirikov: ... for a sufficiently long time? But then this is the general property of exponential instability, or is it something new?

Suppes: It is not the exponential instability, we all agree on that. What is new is to have a detailed isomorphism, to say you can just as well represent this by a semi-Markov-process.

Chirikov: This I understand, but what I don't understand is that you cannot distinguish for a sufficiently long time.

Suppes: But see, a semi-Markov-process can start out the same way. You can have a very good trip on the initial segment of the path. So we have that what is important ...

Miller: The conclusion you want to draw from that seems to me rather strong, if I understood it correctly. You show that there exist processes whose character is empirically undecidable: you cannot distinguish whether they are deterministic processes or semi-Markov-processes. You concluded something about determinism itself transcending experience. But what about other processes? Here we have some examples where we cannot make a distinction empirically. But you're surely not saying that in all cases no discernment is possible.

Suppes: I used a phrase before determinism, universal determinism.

Miller: But universal determinism can be falsified, can't it?

Suppes: No, the point here is what I am saying is not that it can be falsified, it is a different point. The whole point concerns deterministic hidden variables. Well, at least in the present cases you cannot falsify determinism. But what I am saying is not that it is falsified. I am showing that it has a competitor - a conceptual competitor, that is just as true as it is.

Miller: You mean, I hope, just unfalsified.

Suppes: Just as true - I mean, as they stand and fall together on these examples.

Miller: On these examples , but ...

Suppes: ... on these examples, yes. But that's the point, I mean, in other words if you believe in universal determinism then you must have metaphysical views to say this is the correct thing.

Miller: But are there other examples?

Suppes: Oh, there are a whole lot. I am giving examples here and one can generate lots of examples.

Miller: Now I'm somewhat unhappy, because we seem to disagree so radically. It's certainly not the case that all examples you can give allow either a deterministic or a stochastic interpretation.

Suppes: No ...

Miller: For instance, ...

Suppes: There may be some examples as you want it - by approximating them as finite, if you want to hold to the deterministic view. I actually think those examples are much rarer and we discuss them abstractly. I am more aggressive about that than you ...

Miller: ... and there are cases understandable only from the stochastic viewpoint.

Suppes: No, I wanted to say in the stochastic case you can always build a deterministic model.

Miller: It's always possible?

Suppes: Yes, and you would say at the end the deterministic cases, in those few cases that are certainly deterministic, are special cases of stochastic behavior. So what I wanted to say is that we can have a complete overlap of the stochastic and deterministic.

Miller: That's fine. I agree that there can be universal deterministic laws in a stochastic world. It's the other way round ...

Suppes: With universal we mean universal for some things. It's two senses. When I use *universal* of determinism I use it in a stronger sense. Everything is deterministic ...

Miller: That is how I used it too.

Suppes: That is what I am saying: that would be to transcend experience. The belief that everything is deterministic transcends experience. It is not false but you have a corresponding model that is stochastic in many cases, generally stochastic, not properly one. It does equally well. I think that is the sense in which it transcends.

Schurz: I would also like to ask a question for understanding the claim better. In my talk I had the consideration: ... isn't that effect true for all continuous deterministic processes and isn't it just a consequence of the fact that I have a limited accuracy of measurement, that I always have a positive ε. So, for any finite sets of observations of real points up to some ε, these observations will not decide whether the process is deterministic or stochastic.

Suppes: Wait a moment. Take out the convex obstacle and assume the classical billiard.

Schurz: Yes, what is the difference?

Suppes: We have the ideal process here. The ideal process would be deterministic and the semi-Markov-process would be a trivial deterministic one.

Schurz: But we could assume that the ball moves slightly within this ε-constraints.

Suppes: No, no, no. It is very important to play one game here at a time. I am saying: in the framework of Sinai billiards we now take out the convex obstacle. So, we are only looking at the mathematical model of motion. The mathematical model has its perfect periodicity in the standard ideal billiard table of a point particle. I am not saying, the real world is like that, I am saying, that's the mathematical model and with respect to that mathematical model the theorem would not hold that there is a genuine stochastic model of that. I think, David and I agreed about that. Now, you may want to claim we could now shift the framework of the discussion to behavior in the real world and you may want to claim in the real world we always have some fluctuations. I think the case could be made. I'm very sympathetic to that. But I did want to make clear what the conceptual situation is here in this special mathematical model. ... I admit ... I'll be happy with the fluctuations.

Schurz: Yes, I just wanted to know, because your general philosophical corollary was that no finite set of observations of any but positive accuracy may verify or falsify, may decide whether determinism is true ...

Suppes: For some processes - I didn't say it for all.

Schurz: I think that should be true for every process where you assume continuity. Because you may always have fluctuations within the ε-accuracy.

Suppes: But these mathematical models certainly assume continuity. What you must be assuming is a good deal more about fluctuations: there are ever and ever fluctuations in a real world. Remember something: there is no such ideal billiard table in the real world. We are assuming here completely elastic collisions, no dissipation of energy due to frictions - sort of all goes on and on and on forever. So you see, we must now move to a more detailed arena, if you want to discuss what happens in actual billiards. Because one thing that's absolutely agreed upon: the system is dissipative, the ball is not now moving. We have something different. We don't have anything like the assumptions needed for ergodicity.

Schurz: I am just saying I could make a stochastic model, predicting some small fluctuations - smaller than ε - around this trajectory and I could not discriminate by my finite set of observations between these two descriptions ...

Suppes: You mean, when we put into the model now real fluctuations, so we had a real stochastic process ...

Schurz: Unobservable ones.

Suppes: Right. But, what is interesting here - I am usually always on side of realistic models. But at the present case there is some interesting in this idealization. Because it is in the ideal case. We certainly formulate this model conceptually and intuitively as being an example par excellence of a deterministic mechanical model with extremely simple, not complicated force laws. All you've got are elastic collisions. What I find surprising is that without introducing any of these complications about the real world - like fluctuations, dissipation, regular friction - we still get a stochastic isomorphism. That's the point of the example.

Schurz: I agree.

Batterman: In the Ornstein theorem exactly what do they mean by determinism?

Suppes: They mean a system of differential equations for which you could prove the existence and uniqueness of solutions for given initial and boundary conditions.

Batterman: So do you mean a classical system. Sometimes the mathematicians seem to take a deterministic system to be one for which there is a transformation such that give a state at one time, there is a unique state at the next time.

Suppes: More general.

Batterman: Much more general.

Suppes: Yes. Because it is billiard, it is elastic collisions. They are in this case referring ...

Batterman: ... to classical systems, to Hamiltonian systems.

Suppes: They are not necessarily Hamiltonian.

Batterman: Under a broader sense of determinism, a Markov-chain would be deterministic?

Suppes: ... Semi-Markov.

Batterman: Semi-Markov. And that those are deterministic in this broader sense, right?

Suppes: Well, not ordinarily.

Batterman: Well, there is a transformation, a shift-transformation ...

Suppes: Right. The way one uses determinism in mechanical models, that's not determinism.

Batterman: O.K.

Suppes: That is an unusual definition. It's a fact that you can have a shift. Obviously you can find that it's an unusual characterization. Let me give you another characterization: I have a probabilistic process as random, if you want. For a given particle I have a sample path. That sample path is of course a unique function of time. Now you can of course regard that as deterministic - it is exhibited not in the theory but that's the sample path like, for example, if I flip a coin ten thousand times, I have a unique record of the sequence. But that is an unusual sense of determinism.

Chirikov: I don't remember exactly the definition of Markov chain. But suppose you have a stochastic system - stochastic or semi-Markov - and you observe a trajectory. Will the time when you can predict trajectory depend on the accuracy of the observation of the stochastic system.

Suppes: I am sure. Again you don't have an exact observation. In a probabilistic sense - you ask: well, what is the expected position of the particle at time t', given that it was at time t in the Borel set A. We don't know the exact position of the particle at time t. So we have the same problem for both.

Chirikov: What I do not understand: the stochastic system needs not to be exponentially unstable.

Suppes: Take the case of coin flipping - that's an example. What do you think is the analogue of exponential instability in the case of coin flipping?

Chirikov: No, I think that in the case of coin flipping the prediction time will not depend on the accuracy as in a chaotic system.

Suppes: It is true, in that example we observe independence.

Chirikov: By this criterion I can distinguish. I can distinguish dynamical chaos in deterministic system and stochastic system - by the dependence of prediction time on accuracy.

Suppes: I did not mean to introduce the coin flipping as giving us directly a satisfying isomorphism. I am not saying that - by the way that's misleading on my part. I didn't mean that.

Chirikov: What does it mean that in this stochastic system which you have formulated in your paper there is exponential instability?

Suppes: Well, we now have a continuous time process and we take two different sample paths starting in the same interval, we could certainly expect divergence - as we would expect for example in Brownian motion. Take Brownian motion for example. You certainly expect exponential divergence of the two sample paths. Take two particles. They start at the same time, same Brownian motion, with same diffusion, we expect to have exponential divergence. I think that's the same.

Chirikov: O.K., I will think about it.

Weingartner: I want to make a short remark to this observational equivalence - observational equivalence as a new kind of symmetry. I think that it shows - and one should keep this in mind - that one can look at symmetry principles from two different points of view. One point is that it is a positive thing in the sense that you have a very general principle and it's invariant against so many changes. From another point of view you can look at it in a kind of negative sense - that it makes symmetric (equivalent) too much. So that means that you get a loss of information, you loose distinctions that are important. One property is conserved, another is lost or becomes unobservable. There are some physicists who say that too much invariance - for example time symmetry in QM - makes trouble since the laws seem to be incomplete. You can always see these symmetries from these two points of views.

Suppes: Well, I agree. Every principle of invariance can be interpreted as a limitation of knowledge. So, for example, the desire to have a position of absolute rest is defeated if we accept Galilean invariance. So we have a limitation. We would like to think it were otherwise - I have some beautiful quotations of Newton's about absolute rest - maybe in some distant place of the universe, there is something at absolute rest. But in any case, we know it's not nearby. So Newton does not stop the search. But if we accept Galilean invariance, then we are dropping the search for an object that is at absolute rest. More than the contention, we drop the concept of absolute rest. But you know another example: Aristotle is very firm about orientation. You have a natural up and you have a natural down. But, surprisingly enough, Aristotle's model is not part of Euclidean geometry. So, in geometry it is not formulated which could well have been, - namely, a geometry of orientation. And so, if we just use Euclidean geometry we can't speak about orientation. We can't talk about it. So I agree with your remarks.

Chirikov: But if in nature such a symmetry exists, what to do? (Laughing) The question is not whether it would offend, but whether it is wrong or true.

Suppes: Well I mean there is nothing more central in our experience than the up and down.

Noyes: The current empirical situation with regard to absolute rest and uniform motion has changed. There is now no doubt that we are moving at something like 600 kilometers per second with respect to the cosmic background radiation. Galilean invariance is in contradiction with experience.

Suppes: I am very happy with that.

Noyes: This empirical fact does not, in principle, require one to look outside of a closed laboratory. If one measures the temperatures of the walls of an isolated laboratory, the side moving into the background radiation will be hotter, and the side moving away from it will be colder.

Suppes: Furthermore, if we look at the perceptual geometry for humans and animals, there is a general difference between horizontal and vertical. And then if you stay in the horizontal plane - in human perception and in animal perception - there is a fantastically strong effect for shortening along the depth axis as opposed to the frontal axis. This is confirmed in a variety of experiments. So, pick your geometry for whatever your problem is.

Part III

Chaos, Complexity and Order

Chaos: Algorithmic Complexity vs. Dynamical Instability*

R.W. Batterman

Ohio State University, USA

1 Introduction

The issues I want to address in this paper arise from the following question: What can one infer, if anything, about the degree of dynamical instability of a system from the unpredictability of its behavior over time? In a previous paper, "Defining Chaos" Batterman (1993) I argued that really one can infer very little. The claim there was that even the most regular of dynamical systems can yield behavior that is completely unpredictable in the well-defined sense that the observed sequence of measurement results is algorithmically complex and hence, random. Very roughly the idea is that the quantity one might choose to measure might be so unnatural to the system that the randomness in the sequence of its values and hence, the unpredictability of its future behavior, would tell us nothing about the rate of spreading of nearby trajectories in the system's phase space. The phase space of a system is the space defined by a set of variables such that values for them suffice to uniquely specify the system's state. A trajectory in such a space represents the evolution of the system's state over time. Quantities that are natural to the system must be related in an appropriate way to the system's state variables.

I argued in "Defining Chaos" that paying attention solely to the *output* of a system, ignoring the genesis of that output, can lead to the mistaken inference that the generating system is sensitively dependent on initial conditions when, in fact, it is not. This claim was supported by a particular example. I claimed that a roulette wheel or wheel of fortune can be seen to yield algorithmically random output sequences; yet, despite this the roulette wheel is, relative to its natural set of state variables, the most regular system imaginable. Its phase space trajectories are not in the least sensitively dependent on initial conditions. I still believe that the main claim of the earlier paper is correct; though for interesting technical reasons, I am no longer convinced that the example I gave supports it in the strongest possible way. I intend here to improve on my earlier discussion by first re-examining and reformulating the earlier example, and then by providing a new, though similar, example which directly supports the claim. This discussion will lead, I believe, to some further interesting results concerning the connection

* I would especially like to thank Homer White and Mark Wilson for many helpful discussions about these issues.

between unpredictability understood in terms of algorithmic complexity and chaos understood in terms of sensitive dependence on initial conditions.

The motivation for the paper "Defining Chaos" was J. Ford's claim that "in its strictest technical sense, chaos is merely a synonym for randomness as the algorithmic complexity theory of Andrei Kolmogorov, Gregory Chaitin, and Ray Solomonoff so clearly reveals." Ford (1989), p. 350 Concerning accounts of chaos that take exponentially sensitive dependence as the key definiens, Ford has said the following:

> A working definition of classical chaos which now appears in the literature with increasing frequency involves the notion of exponentially sensitive dependence of final system state upon initial state—e. g., positive Lyapunov numbers or the like. This is in fact an excellent working definition, but why is attention so sharply focused upon exponential sensitivity? ... The answer is revealing. Simple exponential sensitivity is precisely the point at which, in order to maintain constant calculational accuracy, we must input about as much information as we get out of our calculations.... But even more important, it is the point at which algorithmic complexity theory asserts that our deterministic algorithms are trying to compute variables which, in fact, are mathematically random. With this additional insight, it becomes sensible to subsume the above working definition of classical chaos under the most general possible theoretical definition: Chaos means deterministic randomness. Ford (1988), pp. 130–131

Ford's claim is, in fact, quite strong. He holds that "...exponential error growth, or the like, is completely equivalent to deterministic randomness; which definition one uses is a matter of choice." Ford et al. (1991), p. 510 While the above argument (in terms of calculational accuracy) is meant to forge "intuitively" the connection between the "conventional definition of chaos" as sensitive dependence and algorithmic randomness, Ford et al. note further that "...it is comforting to know that Alekseev, Yakobson, and Brudno have rigorously confirmed that 'sensitive dependence' and 'deterministic randomness' are but two sides of the same coin." Ford et al. (1991), p. 510

The next section will provide a explication of the theorems appealed to in the claim that sensitive dependence and deterministic randomness are equivalent notions. In the third section, I will discuss in detail the example of the roulette wheel. The example will be reformulated in a way which avoids an unclarity in the earlier presentation and which, furthermore, indicates that the nature of the connections between sensitive dependence and algorithmically complexity are best framed in terms of a concept of *conditional* algorithmic complexity. This notion differs from the concept of absolute complexity which appears in the theorems. Section 4 will offer another example; one which I believe calls directly into question the validity of inferring from algorithmic randomness to sensitive dependence on initial conditions. It is an example that is very familiar to workers on chaos, that of the delta kicked rotor. I will try to present it in a different light.

2   The Theorems of Brudno–White and Pesin

The theorem which according to Ford proves the equivalence of the two "definitions" of chaos is due to A. A. Brudno Brudno (1983) and has recently been strengthened by Homer White (1993). Let (X, T) be a compact dynamical system, where X is a compact metrizable space and $T : X \to X$ is a homeomorphism (a continuous mapping with a continuous inverse). The theorem relates the *complexity of an orbit of a point* $x \in X$ under T to the *metric entropy of the transformation*. The orbit of x under T is the bi-infinite sequence of points $\dots T^{-2}x, T^{-1}x, x, Tx, T^{2}x, T^{3}x \dots$. That is, it is the sequence $< T^{j}x : j \in \mathcal{Z} >$. Roughly construed, the theorem states that for almost all points $x \in X$, the algorithmic complexity of the orbit equals the metric entropy. I will first discuss the entropy of the transformation T and then talk about the other quantity—the algorithmic complexity of the point $x \in X$. It is perhaps easiest to understand these quantities by reference to examples. Let us first consider the so-called Baker's transformation.

This is a transformation T of the unit square $X = (0 \leq q \leq 1; \; 0 \leq p \leq 1)$ to itself defined as follows:

$$T(q, p) = \begin{cases} (2q, p/2) & \text{if } 0 \leq q < 1/2 \\ (2q - 1, p/2 + 1/2) & \text{if } 1/2 \leq q \leq 1 \end{cases}$$

The transformation can best be visualized as the composition of two successive operations. The first squeezes the square to half its height and stretches it to twice its length. The second cuts this new rectangle in half and stacks the right half upon the left half. Figure 1 illustrates this for the half square $0 \leq p \leq 1/2$.

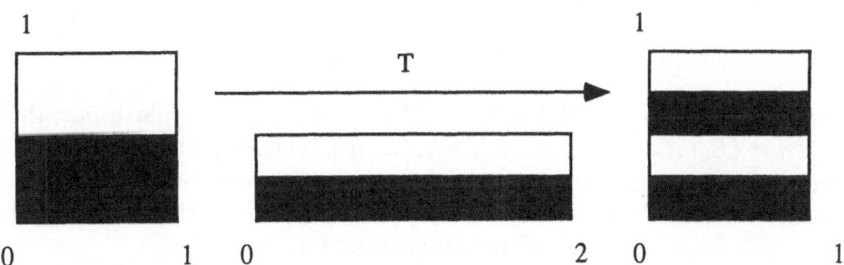

Fig. 1. The Baker's Transformation

The definition of $h_{\mu}(T)$, the entropy of the transformation T or *metric entropy*, can be understood by first defining the *entropy of a partition* α of the space X,

$h(\alpha)$, and then defining the *entropy of the transformation T with respect to the partition α*; namely, $h(\alpha, T)$.

Let $\alpha = \{A_0, A_1\}$ be a partition of the square X into its left half and its right half. (A partition of a space X is an ordered finite collection of disjoint sets of positive measure whose union is, up to sets of measure zero, all of X.) See figure 2.

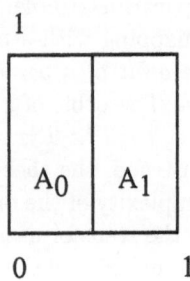

Fig. 2. The Partition α

The *entropy of the partition* $\alpha = \{A_0, A_1\}$ is simply defined as follows:

$$h(\alpha) = -\sum_{i=0}^{1} \mu(A_i) \log \mu(A_i)$$

This is a measure of our uncertainty as to where the point $x \in X$ is relative to the partition α. Clearly if $\mu(A_0) = \mu(A_1) = 1/2$, $h(\alpha)$ is maximal.

Now, $T(\alpha)$ is also a partition of X; namely, $T(\alpha) = \{T(A_0), T(A_1)\}$. Likewise, $T^2(\alpha) = T(T(\alpha)) = \{T^2(A_0), T^2(A_1)\}$ is also a partition. They are shown in figure 3.

In general, if $\beta = \{B_1, \ldots, B_n\}$ and $\gamma = \{C_1, \ldots, C_m\}$ are two partitions, their *join* $\beta \vee \gamma = \{B_i \cap C_j : i = 1, \ldots, n; j = 1, \ldots, m\}$ is also a partition. If β and γ are two partitions, γ is a refinement of β ($\gamma \geq \beta$) if each $C_j \in \gamma$ is, up to a set of measure zero, a subset of some $B_i \in \beta$. Therefore, the partition $\beta \vee \gamma$ is the least common refinement of both partitions β and γ.

Now given the partition $\alpha = \{A_0, A_1\}$, the following holds: If $T^j x \in A_i (i = 0, 1)$ for some $x \in X$, then $x \in T^{-j}(A_i)$. Suppose we have the sequence of 0's and 1's, $< j_i >$ corresponding to which set A_i the point x is in at the jth iteration of T for $j = 0, \ldots, j = n - 1$, then

$$x \in A_{0_i} \cap T^{-1} A_{1_i} \cap T^{-2} A_{2_i} \cap \ldots \cap T^{-n+1} A_{(n-1)_i}$$

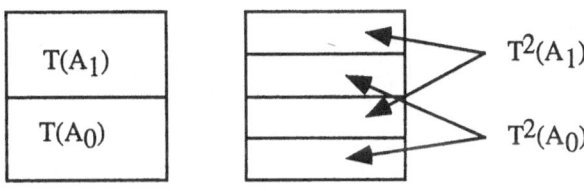

Fig. 3. The Partitions $T(\alpha)$ and $T^2(\alpha)$

In other words, the partition of which this set is a member, $\bigvee_{j=0}^{n-1} T^{-j}\alpha$, codes the first n places of the orbit of points $x \in X$. In this way, relative to a finite partition, every orbit can be coded by a bi-infinite sequence of numbers from a finite alphabet (in this case: 0, 1).

The *entropy of the transformation* T *with respect to the partition* α, $h_\mu(\alpha, T)$ can now be defined in terms of this entropy of a partition as follows:

$$h_\mu(\alpha, T) = \lim_{n\to\infty} \frac{1}{n} h(\alpha \vee T^{-1}\alpha \vee \ldots \vee T^{-n+1}\alpha)$$

This limit exists and $h_\mu(\alpha, T)$ is a measure of the average uncertainty per unit time about which element of the partition α the iterate of the point x will enter in the next instant of time given its past history relative to α. If the partition α is silly (suppose it contains a single set of measure one), then there won't be much (in fact, zero) uncertainty about which element of α, $T^j x$ will be in for each j. (It will always be in that set.) This motivates us to define, finally, the *entropy of the transformation* T, $h_\mu(T)$ as follows:

$$h_\mu(T) = \sup_{all\, \alpha} h_\mu(\alpha, T)$$

The supremum is over all partitions of X. This is known as the *metric* or KS entropy.

This number, $h_\mu(T)$, is a measure of our average uncertainty about where the transformation T will move the points of the space X. The greater $h_\mu(T)$, the more T "disorganizes" the space. (See Peterson (1983), Chapter 5 for a discussion of entropy.) *Since we care about the nature of predictability of the system, we can and should think of a partition α of X as representing possible measurement outcomes on a system.* On this way of thinking, $h(\alpha)$ is "a measure of our (expected) uncertainty about the outcome of the experiment or equivalently of the amount of information that is gained by performing the experiment." Peterson (1983), p. 234 Then $h_\mu(\alpha, T)$ measures the time average of the information content of the measurement represented by the partition α. Finally, $h_\mu(T)$ is "the maximum information per repetition that can be obtained from any experiment, so long as T is used to advance the time (i.e., to develop the system)." Peterson (1983), p. 234

Let us now discuss the other quantity appearing in the Brudno–White theorem; namely, the algorithmic complexity of a point $x \in X$. This notion depends only on T and the topology of X. In particular it does not depend on the metric or even the metrizability of the space X. Let $V = \{U_0, \ldots, U_{n-1}\}$ be any finite open cover of X. Unlike a partition, the sets in V may intersect one another. We want to use the elements of V to code the orbits of points in X. Define for $x \in X$

$$\phi_V(x) = \{\omega \in \{0, 1, \ldots, N-1\}^{\mathcal{Z}} : \text{ for all } n \in \mathcal{Z}, T^n(x) \in U_{\omega(n)}\}$$

The set $\{0, 1, \ldots, N-1\}^{\mathcal{Z}}$ is the set of all two-way infinite sequences containing numbers from the alphabet of the N "letters" $0, 1, \ldots, N-1$. A two-way infinite sequence ω, whose entries come from this alphabet is a member of $\phi_V(x)$ if and only if the nth place of ω, $\omega(n)$ is the subscript of an open set in V to which $T^n(x)$ belongs. Thus, $\phi_V(x)$ is the set of all codings of the orbit of x with respect to the open cover V. For example, suppose $\omega = \ldots 0211402 \ldots \in \phi_V(x)$ with $\omega(0) = 0$, $\omega(1) = 2$, $\omega(2) = 1$, $\omega(3) = 1$, etc. Then we have

$$x \in U_0 \cap T^{-1}(U_2) \cap T^{-2}(U_1) \cap T^{-3}(U_1) \cap T^{-4}(U_4) \cap T^{-5}(U_0) \cap T^{-6}(U_2)$$

Since the open sets in V can intersect, there may be many different codings of the same point. That is, the set $\phi_V(x)$ may contain many members.

Let ω^n denote the finite sequence $\omega(0)\omega(1)\ldots\omega(n-1)$; namely, the first n places of ω. Let $l(s)$ denote the length of any finite sequence s. (Therefore, $l(\omega^n) = n$.) Now let C be any fixed universal Turing Machine. This is a machine that can simulate any specified Turing machine given any input. The number, $K_C(s)$, where s is a finite sequence, is the complexity of s relative to the universal machine C. It is defined to be the length of the shortest sequence s^*, $l(s^*)$ such that $C(s^*) = s$. (See White (1993) and references therein for details.) We want a definition of complexity for infinite sequences. This can be done as follows: Define[1]:

$$\sup -K(x, T, V) = \limsup_{n \to \infty} \min_{\omega \in \phi_V(x)} \frac{K_C(\omega^n)}{n}$$

and

$$\inf -K(x, T, V) = \liminf_{n \to \infty} \min_{\omega \in \phi_V(x)} \frac{K_C(\omega^n)}{n}$$

These definitions need some explanation. Suppose ω^n codes the first n places of the orbit of $x \in X$ relative to the open cover V. Since the open sets U_i in V may overlap, for $\omega' \in \phi_V(x)$, one has ω'^n as a different coding of the first n places of x relative to V. We want to choose the coding sequence with the minimum algorithmic complexity. This is why the "lim sup" and "lim inf" are taken with respect to that coding of x's orbit with the least complexity. The motivation for this is the following. Consider figure 4.

---

[1] The notation here is slightly different than that used in White (1993)

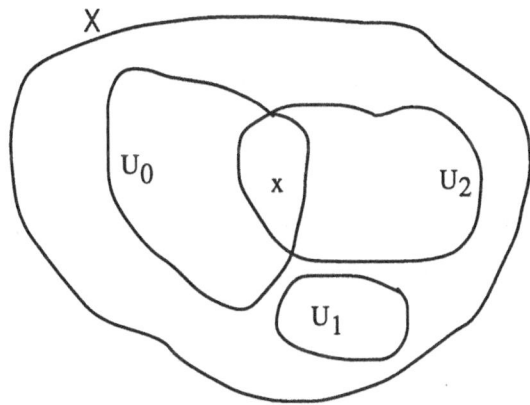

Fig. 4. Open Cover of X

Suppose x is a fixed point of the transformation T; i.e., $Tx = x$. And, suppose x lies in the intersection of two open sets U_0, U_2, both members of V. Then the following are possible codings of the orbit of x under T relative to V.

$$\dots 0000000000 \dots \tag{1}$$

$$\dots 2002022202 \dots \tag{2}$$

$$\dots 2222222222 \dots \tag{3}$$

(1) and (3) are obviously simple. They have zero algorithmic complexity. But (2) might be very complex. It is reasonable to choose a simplest coding (both (1) and (3) are equally simple), since a fixed point of a transformation has a simple orbit. Therefore, the supremum and infimum complexities of x given T relative to the open cover V chooses the coding with the smallest algorithmic complexity. Note also that in taking the limit $n \to \infty$, there is no longer any dependence on a particular universal Turing machine C.

But, neither $\sup -K(x, T, V)$ nor $\inf -K(x, T, V)$ will suffice as the definition of complexity of an orbit of a point in X under T, for basically the same reason that $h_\mu(\alpha, T)$ is not good enough to represent the metric entropy of the transformation T: It is no good to look at a single partition or a single open cover. If the open cover is sufficiently "rough", then an orbit which is intuitively very complex might remain within a single set U_i of V. Its coding sequence would then simply be $\dots iiii \dots$, and it would have $\sup -K(x, T, V) = \inf -K(x, T, V) = 0$. Therefore, *we need to look at all open covers*. We get the following definitions:

$$\sup -K(x, T) = \sup_{allV} \sup -K(x, T, V)$$

and

$$\inf -K(x, T) = \sup_{allV} \inf -K(x, T, V)$$

Finally, we can state the relevant theorems. Brudno (1983) proved the following:

Let (X, T, μ) be a dynamical system, with μ an ergodic T-invariant Borel probability measure on X. Then, for μ-almost every $x \in X$,

$$\sup -K(x, T) = h_\mu(T)$$

White strengthened the result to say the following:

For (X, T, μ) as above, $\inf -K(x, T) = \sup -K(x, T) = h_\mu(T)$ White (1993), p. 812

This theorem

is a statement about the limiting algorithmic complexity of orbits of points in measure preserving systems: the (measure-theoretically) typical orbit possesses a limiting degree of complexity with its per-element unpredictability being equal, in the long run, to the entropy of the system. (White (1993), p. 812)

Now, this result, by itself, is not sufficient to allow one to forge a connection between the algorithmic complexity of the typical orbit of a system and a claim that the system exhibits exponentially sensitive dependence on initial conditions. What is needed is a further result connecting the metric entropy of a transformation T with exponential spreading of trajectories in phase space. The relevant theorem is due to Pesin (1977). A statement of the theorem is the following:

Let (X, T, μ) be a C^2-dynamical system. That is, X is a smooth compact manifold, T is a C^2-diffeomorphism on X (a mapping such that T and T^{-1} are both twice differentiable) and μ is a measure on X compatible with the manifold structure of X and preserved by T. For each $x \in X$, let $S(x)$ be the sum of the positive Lyapunov exponents of x, then

$$\int_X S(x) d\mu(x) = h_\mu(T)$$

The Lyapunov exponents of a trajectory characterize the mean exponential rate of divergence of its nearby trajectories. Thus, Pesin's theorem connects the exponential separation of nearby trajectories (the sensitive dependence on initial conditions) with the metric entropy of the transformation T as defined above.

If we combine the Brudno-White theorem with Pesin's theorem, then Ford's account of chaos as a synonym for algorithmic complexity which subsumes the "sensitive dependence" account appears to be well-founded. But, before discussing the roulette wheel example in section 3 and how it fares in light of this fairly in-depth presentation of these two theorems, it is important to note a difference in the conditions of the two theorems.

Pesin's theorem requires that the dynamical system be a diffeomorphism on a smooth compact manifold. The Brudno-White result demands considerably less, and consequently applies to a much wider class of dynamical systems: The connection between the algorithmic complexity of a point and the metric entropy holds for any homeomorphism on a metrizable space. If we are interested in Hamiltonian systems, then of course, the stronger conditions *are* satisfied and the chain of equivalences relating with probability one, positive algorithmic complexity to positive Lyapunov exponents can be maintained. Yet, it is interesting to note that there are weaker notions of the "spreading of trajectories" in a space X than that of having positive Lyapunov exponents.

For example, there is a quantity $h_\mu(x, T)$, called the *local "entropy" of a transformation about a point*, due to Brin and Katok (1983). It provides a reasonable notion of the separation of orbits of nearby points under a transformation T in a space X. Here T need only be continuous with a continuous inverse preserving a Borel probability measure μ, and X a compact metric space. They prove a theorem relating the average exponential spreading of the orbits of nearby initial points as measured by the local entropy $h_\mu(x, T)$ to the usual metric entropy $h_\mu(T)$. Therefore, for a large class of nondifferentiable dynamical systems, one can still forge a connection between the complexity of a typical single orbit and a form of dynamical instability. For differentiable dynamical systems, the average local entropy for all points $x \in X$ is equivalent to $\int_X S(x)d\mu(x)$ and so, by Pesin's theorem it is also equivalent to the metric entropy $h_\mu(T)$.

3   The Roulette Wheel Revisited

In "Defining Chaos" (Batterman (1993)) I argued that the positive algorithmic complexity of an observed output sequence of a system does not necessarily imply that the system exhibits exponentially sensitive dependence on initial conditions. The basic idea was that looking at a system in the "wrong" way could yield the wrong answer. I tried to describe a way of looking at a rotating wheel (a roulette wheel or wheel of fortune), which should clearly count as a system not sensitively dependent on initial conditions, such that the observed output is algorithmically random. That is, I tried to characterize a "wrong" way of looking at it. However, to readers familiar with the Brudno-White results, my discussion seemed misleading since as I characterized the wheel, it is not a proper dynamical system. In this section I will first present the example as I characterized it in "Defining Chaos" and then reformulate it as a proper, albeit abstract, dynamical system. I will then discuss how the ergodic theorems discussed in the last section apply to the reformulated example.

The wheel example goes as follows: Imagine two black-boxes; the first contains an ideal hard sphere gas, the second contains a roulette wheel or wheel of fortune with a person to spin the wheel. Suppose also that in the first box a device performs measurements at unit time intervals on the ideal gas. Since, as we have seen in section 2, a partition $\alpha = \{A_i : i = 0, 1, \ldots, N-1\}$ of the phase space can be thought of as representing possible measurement outcomes, we have the box print out the number i corresponding to which cell of the partition the system's representative point is to be found at each interval of time. For the ideal gas trajectories from nearby initial states separate exponentially in time, the entropy is positive, and the typical orbit sequence will be algorithmically random (will have positive algorithmic complexity).

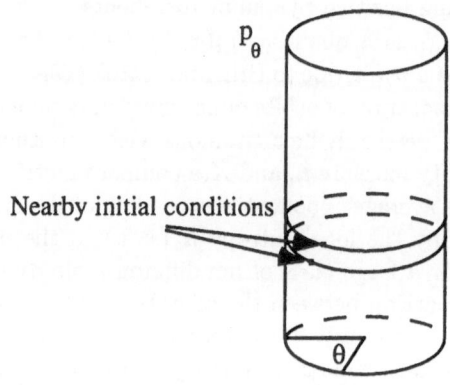

Fig. 5. Phase Space of a Conservative Wheel

The other black-box contains an N-space roulette wheel (imagine the circumference marked off in N equal segments labeled: $B_0, B_1, \ldots, B_{N-1}$). The wheel is spun once per unit time and the box prints out the number of the space opposite a pointer at the end of that time interval. Just as with the ideal gas box, we will be looking at a sequence of numbers from the alphabet $\{0, 1, \ldots, N-1\}$. Since roulette wheels are taken to be paradigm examples of the most "random" systems (they generate Bernoulli sequences with probability one), we expect that sequences arising from the pointer "measurements" will be algorithmically complex. By the Brudno-White theorem we expect that the wheel dynamical system has positive entropy and so by Pesin's theorem, nearby initial conditions will yield exponentially separating trajectories.

In "Defining Chaos" I argued that while the chain of equivalences holds for the ideal gas system, it breaks down for the roulette wheel. My reasoning depended on noting a qualitative difference in the allowed possible dynamical evolutions of the systems in the two boxes. The ideal gas is a nonintegrable ergodic system whose phase space trajectories are free to wander throughout the entire $6N - 1$ dimensional energy surface of its phase space. As noted, these trajectories typically exhibit exponentially sensitive dependence on initial conditions. On the other hand, the phase space of an "ideal" roulette wheel is a cylinder with coordinates θ, the angular position of the wheel, and p_θ, the wheel's angular momentum. See figure 5. If the wheel experiences no friction then for given p_θ

the trajectory is just a circle around the cylinder. Two initial points that are nearby in this space can never separate exponentially in time. In fact, the motion is clearly integrable. If we allow for dissipation (so that the wheel is no longer conservative, say because it is clicking against a pointer), the trajectories will be spirals from some initial point (θ, p_θ) to some final rest state $(\theta', 0)$. See figure 6. Clearly this situation also will not allow for any exponentially sensitive dependence on initial conditions. (There are no strange attractors in this dissipative system.)

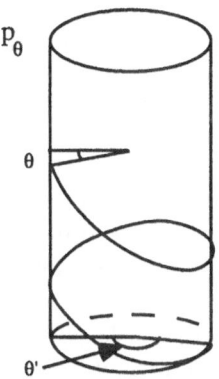

Fig. 6. Phase Space of a Dissipative Wheel

Prima facie, then it appears as if the chain of probability one equivalences— (positive algorithmic complexity) $\leftrightarrow$ (positive metric entropy) $\leftrightarrow$ (positive Lyapunov exponents) fails for this system. If the Brudno-White theorem holds, then because the output of the box is algorithmically random, the system must have positive metric entropy. But, since no trajectory has a positive Lyapunov exponent, the link between complexity and sensitive dependence fails.

This conclusion, however, is too hasty. A deeper analysis reveals that the roulette wheel as I have characterized it is really not even a dynamical system. Actually, I noted this fact as a potential objection to the example in "Defining Chaos".

[The system] does not really evolve continuously from some one initial state as, for example, the hard sphere gas does. Here in order to get the random output sequence, we need the person who keeps restarting the wheel over and over again in slightly different initial states.... (T)he output is random because of the (dynamically speaking) artificiality of the way the sequence is constructed. Batterman (1993), pp. 63–64

But I dismissed this as really no objection at all. My point was that the Hamiltonian motion of the wheel is stable so the randomness in output must be due to something other than the dynamics—namely, to our ignorance of the exact initial conditions for each spin. In other words, the randomness in the output of the wheel is due to the fact that the wheel is linked to an external source of randomness; in particular, the vagaries of the person spinning the wheel. Indeed, this is just the way real wheels of fortune and roulette wheels do yield apparently random outputs.

In fact, I now believe that the objection cannot be dismissed quite so easily. The main issue, of course, is whether for a deterministic dynamical system it is possible to infer from randomness in the output sequence to a sensitive dependence on initial conditions in the dynamics. But the roulette wheel as I have characterized it is not a deterministic system. It is not a deterministic system because its state at the end of each interval does not uniquely determine its state for the next or any other interval. The problem is that the person inside spins the wheel according to her whim; that is, according to no deterministic rule. The "system" as I have described it, is indeterministic or stochastic, and no one should be terribly surprised that an independent stochastic process can yield sequences that are algorithmically random. But for such a nondeterministic system, the Brudno-White theorem does not apply because no relevant transformation is defined for it.

Now the question is what we can learn from the wheel, if we formulate it as a proper, that is, as a deterministic dynamical system. I will concentrate on this question for the rest of this section. Then in the next section, I will consider a different, though similar, example which will directly support the claim that ignoring the genesis of the sequential output of a dynamical system which has positive complexity can lead one to mistakenly infer that the system is sensitively dependent on initial conditions.

One can formulate the roulette wheel as an abstract (non-Hamiltonian) dynamical system as follows [2]: The basic idea is to consider the combination "wheel-plus-spinner" as the system of interest, and not just the wheel as I did earlier. To do this precisely, let us identify the wheel with the unit circle S^1 in $\Re^2$. As in figures 5 and 6, we use the angle θ to represent positions on the wheel. Let $\lambda = < \lambda_0, \lambda_1, \lambda_2, \lambda_3, \ldots >$ be a sequence in the space of sequences Λ^∞—the set of all one-way infinite sequences of 0's and 1's. That is, for each i, $\lambda_i = 0$ or 1. The 0's and 1's will tell us how much angular velocity/momentum the wheel gets at the beginning of each time interval.[3] Thus, the sequence λ is our representation of the spinner. The dynamical system (X, T) is defined as follows: Let $X = \{(\theta, \lambda) : \theta \in [0, 2\pi) \text{ and } \lambda \in \Lambda^\infty\}$. X is the Cartesian product of the interval $\Omega = [0, 2\pi) \subset \Re$ with the space of sequences Λ^∞, $X = \Omega \times \Lambda^\infty$. Then X represents all possible positions of the wheel together with all possible

---

[2] Here, as in numerous other places, I am deeply indebted to Homer White

[3] We consider only one-way infinite sequences since we are, for simpicity, only concerned with predicting values of the wheel.

future spins of the wheel in the following sense: We let the wheel receive angular momentum p_0, at the beginning of the nth time interval if the nth number in the sequence λ, λ_{n-1}, equals 0 and let it receive angular momentum p_1 in that interval if $\lambda_{n-1} = 1$. Define $T : X \to X$ as follows:

$$T(\theta_0, < \lambda_0, \lambda_1, \lambda_2, \lambda_3, \ldots >) = (\theta_1, < \lambda_1, \lambda_2, \lambda_3, \ldots >)$$

where θ_0 is the wheel's initial position, and θ_1 is the wheel's position at the end of the first time interval when it receives angular momentum p_i at the beginning of the interval ($p_i = p_0$ if $\lambda_0 = 0$ and $p_i = p_1$ if $\lambda_0 = 1$). Note that the new value for λ is just the next value in the sequence. (Note also that the new θ value is determined by the completely integrable Hamiltonian motion of the wheel between spins.) It is easy to construct an ergodic T–invariant measure μ on X from the uniform Lebesque measure on the interval $\Omega = [0, 2\pi)$ and the "fair coin," Bernoulli measure on the space Λ^∞. The picture is essentially that of figure 5 again. (Actually, it is really just the two disconnected circles in figure 7.)

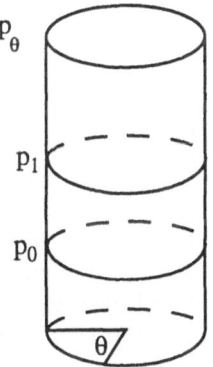

Fig. 7. The "Wheel-plus-Spinner"

Since μ-almost all sequences λ have positive algorithmic complexity (they are almost all *random* sequences of 0's and 1's), the sequence of position values (the numbers that the black-box will output) will also be algorithmically random. Thus, we have that for μ-almost all $x \in X$, $K(x, T) > 0$, and so by the Brudno-White theorem, $h_\mu(T) > 0$. This, of course, agrees with the result for the black box containing the ideal gas.

Yet there still seems to be an intuitive difference in the chaotic nature of the gas and the wheel-plus-spinner, in that the latter has a component of its evolution (the wheel) that is not sensitively dependent on initial conditions.

The sequence of θ-values, inherits its randomness through the coupling with the other variable λ, and not as a result of any instability of motion. In the gas, of course, the algorithmic complexity of the measurement sequence is the result of instability. We can, I believe, use the theory of algorithmic complexity to formalize this intuitive difference.

The first thing to notice is that we are really only interested in inferring from the unpredictability/complexity of the sequence of positions, the θ-values, to the nature of the instability of the dynamics. In other word, our "measurement data" is just the sequence $< \theta_0, \theta_1, \ldots >$; we have no information whatsoever about the conjugate "variable" λ. So, the complexity we are interested in is not really the full complexity of a point $x \in X$ relative to the transformation T, but instead, what might be called the "restricted algorithmic complexity of x relative to T"—restricted to a sequence of numbers from the space Ω. It is, in fact, possible to define the Ω-restricted complexity of a point $x \in X$ using "Ω-only" open covers of X.[4] These are open covers that can be used to code points x of X, but by looking only at their Ω-components, not their Λ^∞-components. In a similar vein, one can define the Ω-restricted metric entropy and a restricted notion of local entropy to measure the spreading in time of Ω-only orbits given two points in X with Ω-components nearby according to the metric on Ω.

Let us now see how our two black boxes compare with one another. I think that there still remains an interesting difference between the wheel as characterized above and the ideal gas. Consider first the ideal gas of N molecules. The state of each gas molecule is completely specified by providing six numbers: three position coordinates x, y, z and three momentum coordinates p_x, p_y, p_z. Suppose that for some reason we are only interested in the x-values for the ith molecule. We can think of the full phase space as the Cartesian product $\Re \times \Re^{6n-1}$ of the space of the variable of interest (call it x_i) and the space of the remaining state variables. Suppose (and admittedly this requires some stretching of our imaginations) we have a device that will measure the values of the x-coordinate for the ith molecule of the gas. Suppose further (and admittedly, this demands even more of a stretch) that we have exact knowledge of the other five coordinates for the ith molecule as well as exact knowledge of the position and momentum coordinates for every other molecule of the gas. In other words, we know all there is to know about every molecule of the gas except for the x-coordinate of the ith molecule, the x_i-coordinate. Equivalently, we know the point in $\Re^{6N-1}$. Now, at unit intervals of time our measuring device gives us a value for the x_i-coordinate. Because of the instability of the gas, and because of the fact that the typical orbit of a point on the energy surface is algorithmically complex, it is reasonable to expect that the restricted x_i-orbit sequence is algorithmically complex—it has positive $\Re$-restricted algorithmic complexity.

Now, consider two gases (or two states of the same gas) that agree on all the other values but differ in their x_i-coordinates. Let their x_i coordinates be close to one another in the standard Euclidean metric on $\Re$. The trajectories from these

---

[4] These definitions are due to Homer White; private communication.

nearby initial conditions will diverge exponentially on the full energy surface. As a result of this spreading, it seems reasonable to expect that there will be positive x_i-restricted local entropy: The x_i-only orbits will diverge exponentially with time. We can conclude the following for the ideal gas:

> If the sequence of measured values of a variable of interest is algorithmically complex, then it is likely (with probability one) that the dynamics exhibits exponentially sensitive dependence on initial conditions, *even if we are given absolute knowledge of all other relevant state variables.*

Now let us ask a similar question about the roulette wheel. For this system, as we have seen, the sequence of θ-values is algorithmically complex. Suppose we are given absolute knowledge of the variable "conjugate" to the variable of interest. That is, suppose we have absolute knowledge of the sequence λ. (This is analogous to the knowledge of exact values for all variables except x_i in the gas example.) Consider two initial conditions (θ_1, λ) and (θ_2, λ) such that θ_1 and θ_2 are close in a standard metric $d(\cdot, \cdot)$ on Ω; say $d(\theta_1, \theta_2) = \epsilon$. It is clear that the Ω-only orbits from these initial conditions will have zero restricted local entropy. In fact, for any j, $d(T^j\theta_1, T^j\theta_2) = \epsilon$. Thus, for the wheel we can conclude the following:

> There is *no* separation in Ω and *no* sensitive dependence on initial conditions *if we are given absolute knowledge of the other relevant state variable* λ.

It seems to me that the difference between these two cases is best understood in terms of the *conditional algorithmic complexity* of the respective sequences of variables of interest. The absolute complexity of any sequence was defined above in section 1. Intuitively, it is a measure of the length of the shortest program which when input to a universal Turing machine will output the sequence in question. As we saw, there is a natural way to associate a point $x \in X$ with a coding of its orbit under T. Determining the orbit complexity then involves looking at all possible codings of the orbit.

The conditional complexity of a finite sequence s given a sequence t can be understood as follows: As before let s and t be finite sequences from some alphabet, e.g. $\{0, 1\}$. Consider a universal Turing machine C which in addition to its two-way infinite work tape also has a read only tape on which a finite sequence can be recorded. Define the conditional complexity of s given t to be the length of the shortest string s^* from the set of finite sequences of 0's and 1's such that given input s^* (on its work tape) the machine C will output the sequence s when it has the sequence t on its read only tape. We can write this as follows: $K_C(s|t) = l(s^*)$. If $\alpha = \{A_0, \ldots, A_{n-1}\}$ and $\beta = \{B_0, \ldots, B_{m-1}\}$ are two partitions of X (say α is a partition corresponding to the variable we want to measure and β is a partition of the remaining variables), define $\phi_\alpha(x)$ and $\phi_\beta(x)$ as follows:

> $\phi_\alpha(x)$ is that element ω of the set of infinite sequences from the alphabet $\{0, 1, \ldots, n-1\}$ such that for all $j \geq 0, T^j(x) \in A_{\omega(j)}$. Recall that $\omega(j)$ is

that number from $\{0, 1, \ldots, n-1\}$ found at the jth place in the sequence ω.

$\phi_\beta(x)$ is defined in a like manner:

$\phi_\beta(x)$ is that element λ of the set of infinite sequences from the alphabet $\{0, 1, \ldots, m-1\}$ such that for all $j \geq 0, T^j(x) \in B_{\lambda(j)}$.

Thus, $\omega = \phi_\alpha(x)$ is the coding of the orbit of x with respect to the partition α of X, and λ codes the orbit of x with respect to the partition β of X. (In the following, "$(\phi_\alpha(x))^n$" denotes the first n places of the sequence $\phi_\alpha(x)$, similarly for "$(\phi_\beta(x))^n$".) Then,

$$\sup -K(x, T, \alpha|\beta) = \limsup_{n \to \infty} \frac{K_C((\phi_\alpha(x))^n|(\phi_\beta(x))^n)}{n}$$

and

$$\inf -K(x, T, \alpha|\beta) = \liminf_{n \to \infty} \frac{K_C((\phi_\alpha(x))^n|(\phi_\beta(x))^n)}{n}$$

For example, the $\sup -K(x, T, \alpha|\beta)$ provides an upper limit on how complex the orbit of x looks, relative to the partition α, *given* that one knows the orbit of x relative to β. Once again in taking the limit, the dependence on any particular universal Turing machine is eliminated.

Now let, $< \alpha_n >$ and $< \beta_n >$ be increasing sequences of partitions. This means that the partition $\alpha_{i+1} \geq \alpha_i$ (α_{i+1} is a refinement of α_i) and similarly for the β_n's. Let $\mathcal{F}$ be the sigma-algebra of sets generated by the sequence $< \alpha_n >$, and let $\mathcal{G}$ be the sigma-algebra generated by the sequence $< \beta_n >$. For any partition γ define:

$$\sup -K(x, T, \gamma|\mathcal{G}) = \limsup_{n \to \infty} -K(x, T, \gamma|\beta_n)$$

and finally, define:

$$\sup -K(x, T, \mathcal{F}|\mathcal{G}) = \limsup_{n \to \infty} -K(x, T, \alpha|\mathcal{G})$$

Similarly, one gets definitions for $\inf -K(x, T, \gamma|\mathcal{G})$ and $\inf -K(x, T, \mathcal{F}|\mathcal{G})$. The idea behind all of this is the following: Let the sequences of partitions $< \alpha_n >$ and $< \beta_n >$ generate the sigma-algebras $\mathcal{F}$ and $\mathcal{G}$ respectively. Suppose furthermore that the sequence of partitions $< \alpha_n \vee \beta_n >$ generates the full sigma-algebra of sets of the space X. Consider the ideal gas example. If α is a partition of the space $\Re$ of the variable of interest, x_i, and β partitions the space $\Re^{6N-1}$ of the remaining variables, then $\sup -K(x, T, \mathcal{F}|\mathcal{G})$ measures the complexity of the sequence of x_i values if we are given the sequence of future values of the remaining variables to as great a degree of accuracy as we would like. Similarly, in the case of the roulette wheel we would have α be a partition of the space Ω and β be a partition of the space of sequences Λ^∞. Then $\sup -K(x, T, \mathcal{F}|\mathcal{G})$ measures the complexity of the sequence of θ-values given full knowledge of the sequence λ defining the future spin history of the wheel. *It turns out that the conditional complexity of the sequence of x_i-values for the ideal gas is positive, but the conditional complexity of the θ-values for the wheel is zero.* Therefore, it appears that the notion of *conditional complexity* is better suited for describing the difference between the two systems, than is the notion of the absolute complexity of the sequence. Homer White has proved a theorem to the effect that if a dynamical system exhibits positive conditional complexity in a variable of interest (say x_i or θ) given future values of the "conjugate" quantities as accurately as one would like, then that system will exhibit a very strong form of restricted instability in the variable of interest.[5] In fact, the instability implies positive restricted local entropy, so *positive conditional complexity entails exponentially sensitive dependence on initial conditions in the variable of interest.*

Conditional algorithmic complexity is an appropriate concept for exhibiting the connections between the unpredictability of a system as evidenced by the high complexity of measured values, and chaos in the sense of sensitive dependence on initial conditions. It is more appropriate than absolute complexity, since if we had only considered the absolute complexity of the measured sequence of θ-values on the wheel we would be unable to explain the difference in local spreading of orbits given absolute knowledge of the other variables noted above between the roulette wheel and the ideal gas.

Now, it seems to me that the proper explanation for the difference noted between these two systems, the evidence for which is manifest in terms of conditional complexity, is the following. The wheel-plus-spinner has a component which is a regular, integrable Hamiltonian system; namely, the wheel. The absolute positive complexity of the θ-sequence, in effect ignores the perfectly regular evolution of the wheel between spins. There is, however, no analog to this "regular" evolution in the variable of interest, x_i, of the ideal gas. Despite all of this effort above to support the connection between absolute complexity and sensitive dependence on initial conditions, in the next section I would like to discuss another example which I believe clearly demonstrates that one must be very careful in applying the theorems discussed above. The system I want to consider next has received much attention in the literature on chaos in classical mechanics. It is not an unphysical abstract system as is the reformulated wheel we have just been considering. The system is called a kicked rotor and it gives rise to what is known as the *standard mapping*.

4   The Kicked and Unkicked Rotors

The object of this section is to illustrate explicitly how a system which is completely integrable, with dynamics exhibiting nothing like exponential sensitivity to initial conditions, can nevertheless yield a sequence of measurement results

---

[5] Homer White; private communication.

which is (absolutely) algorithmically random.[6] Really, what I am doing is reformulating, once again, my original roulette wheel example. But to make easy contact with the chaos literature, I shall switch to speaking of a rotor rather than a wheel. The idea is that we are going to measure the location of the rotor at strange intervals of time, so that the positions for the integrable rotor will exactly imitate those of a well known, genuinely chaotic system. The main issue here will be to demonstrate that if we want to apply the Brudno-White and Pesin theorems, we must be very careful to insure that our "observables" relate properly to the underlying structure of the system—the very structure that allows us to *define* "exponential divergence."

Let me first describe the time dependent Hamiltonian system known as the kicked rotor. This system consists of a bar of moment of inertia I which is fastened to a frictionless pivot at one end. (There is no gravity.) The other end of the bar is subject to periodic delta-function kicks of period T. See figure 8.

The amplitude of these instantaneous impulses is dependent upon the angular position of the rotor and is proportional to the value of a parameter k. Between kicks the angular momentum p_θ is constant and so throughout that interval, θ evolves linearly in time.

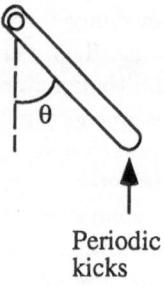

Periodic
kicks

Fig. 8. The Delta-Kicked Rotor

The system has the following Hamiltonian:

$$H(\theta, p_\theta, t) = \frac{p_\theta^2}{2I} + k\cos\theta \sum_n \delta(t - nT) \tag{4}$$

Hamilton's equations for this system are:

$$\frac{dp_\theta}{dt} = -\frac{\partial H}{\partial \theta} = k\sin\theta \sum_n \delta(t - nT)$$

---

[6] I want to especially thank Mark Wilson for a number of helpful discussions concerning this example.

$$\frac{d\theta}{dt} = \frac{\partial H}{\partial p_\theta} = \frac{p_\theta}{I} \tag{5}$$

Integrating these last two equations from a time just before the kick, $t = nT - \epsilon$, to a time just before the next kick, $t = (n+1)T - \epsilon$ and setting $\mathbf{I} = 1$ and $T = 1$, yields the so-called *standard map*:

$$\theta_{n+1} = \theta_n + p_{n+1} \quad \mathrm{mod}\ 2\pi$$
$$p_{n+1} = p_n + k \sin \theta_n \tag{6}$$

Equations (6) describe a Poincare return map at unit time intervals for the kicked rotor with the Hamiltonian (4) above. Figure (9) shows the time dependence of the angular position and of the angular momentum for this delta kicked rotor.

When $k = 0$ the map reduces to

$$\theta_{n+1} = \theta_n + p_n \quad \mathrm{mod}\ 2\pi$$
$$p_{n+1} = p_n = p_c \tag{7}$$

Thus, $p_n = p_c$ is a constant of the motion and this corresponds to an integrable rotor spinning with constant angular momentum. This is the most regular and periodic of dynamical systems.

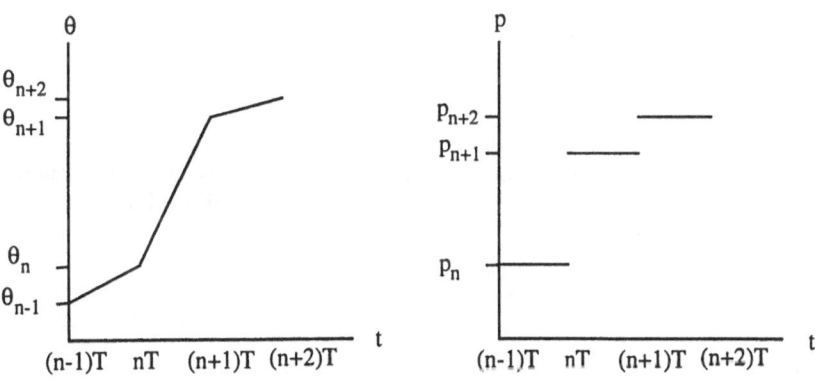

Fig. 9. θ vs. t and p vs. t Plots

For this *unkicked* rotor, if one has two nearby initial conditions, (θ_1, p_c) and (θ_2, p_c) such that $d(\theta_1, \theta_2) = \epsilon$, then for any time n, $d(T^n\theta_1, T^n\theta_2) = \epsilon$; so, there is clearly no spreading of values in angular position. On the other hand, as figure 9 suggests and as has been verified by numerical experiment, for the *kicked* rotor two points initially close to one another in phase space will yield exponentially separating trajectories for sufficiently high values of the kick parameter k. In fact, when $k > 8.888$, there are no discernible KAM tori and it appears that the entire surface of section is covered by a single ergodic orbit. Chirikov (1979) It is likely, therefore, that the sequence of θ-values will be algorithmically random.

Now let us suppose that inside a black box we have an unkicked rotor. Then its θ vs. t plot will look like that in figure 10.

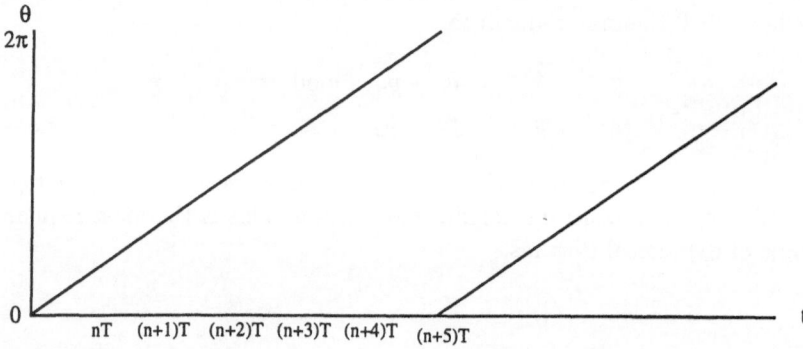

Fig. 10. θ vs. t for the Unkicked Rotor

The slope of the (θ, t) path, $(\theta_{n+1} - \theta_n)/\Delta T$, is just equal to p_c–the constant momentum. Furthermore, plotting p_θ vs. t yields the graph in figure 11.

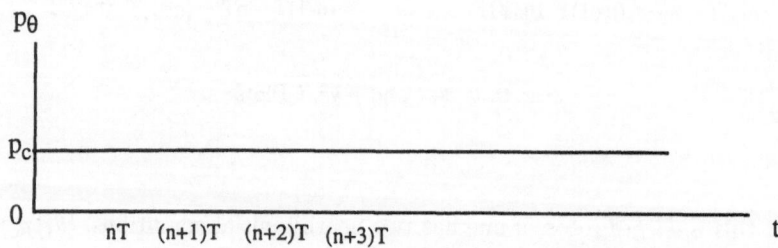

Fig. 11. p_θ vs. t for the Unkicked Rotor

Now suppose we superimpose the (θ, t) plot for the unkicked rotor (figure 10) on the (θ, t) plot for the kicked rotor with a high value for k that renders the kicked system fully chaotic. See figure 12. Although I will picture only a few iterations, one must think of the entire sequence of values $< \theta_i >$ as having positive algorithmic complexity. For simplicity we set $T = 1$.

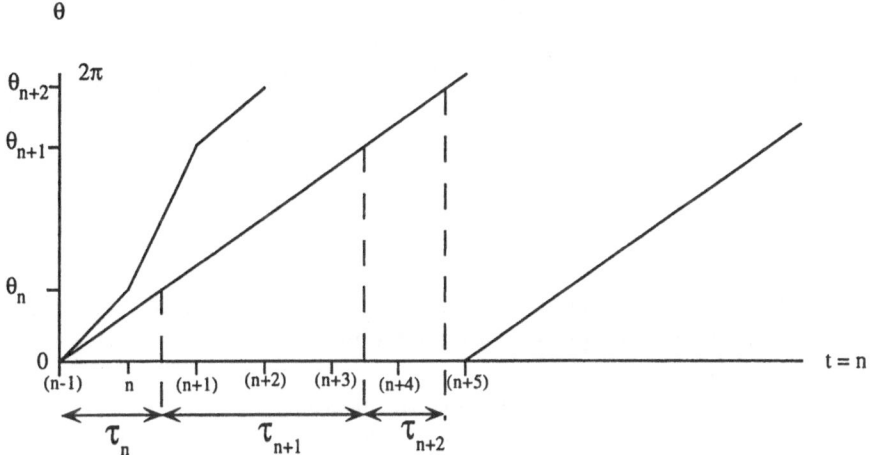

Fig. 12. θ vs. t for both Rotors

For the kicked rotor at time n, $\theta = \theta_n$; at time $n+1$, $\theta = \theta_{n+1}$ etc. Clearly, if the slope $\theta_{n+1} - \theta_n = p_{n+1}$ is greater than the slope p_c for the unkicked rotor then the kicked rotor gets to the position θ_{n+1} faster than does the unkicked rotor. Similarly, if the slope $\theta_{n+1} - \theta_n < p_c$ the unkicked rotor will arrive at θ_{n+1} from θ_n in less time than the "kicked" rotor. Let τ_n be the time taken by the unkicked rotor to get from θ_{n-1} to θ_n, and τ_{n+1} the time for it to get from θ_n to θ_{n+1}. Then, since the momentum for the unkicked rotor is a constant p_c, we must have the following relation between these times:

$$\frac{\theta_{n+1} - \theta_n}{\tau_{n+1}} = \frac{\theta_n - \theta_{n-1}}{\tau_n} = p_c$$

Hence,

$$\tau_{n+1} = \left(\frac{\theta_{n+1} - \theta_n}{\theta_n - \theta_{n-1}} \right) \tau_n$$

and using the equations of the standard map (6), we get

$$\tau_{n+1} = \left(\frac{p_n + k \sin \theta_n}{p_n} \right) \tau_n \tag{8}$$

or equivalently:

$$\tau_{n+1} = \left(\frac{(\theta_n - \theta_{n-1}) + k \sin \theta_n}{\theta_n - \theta_{n-1}} \right) \tau_n \qquad (9)$$

Now, suppose inside the black box containing the *unkicked* rotor there is a clock which opens a shutter at times $\tau_1, \tau_2, \tau_3, \ldots$ thereby allowing an angular position measurement to be performed. The box then outputs the θ-values determined by these position measurements. But, *these θ-values are exactly the same as those that could be determined by a shutter operating at unit intervals of time on a* kicked *rotor.*

The point of all of this should now be fairly clear. We have been supposing that the sequence $< \theta_i >$ from a kicked rotor is algorithmically random–the value for k is sufficiently large. Using the Brudno-White theorem and Pesin's theorem we can infer that with probability one the kicked rotor exhibits exponentially sensitive dependence on initial conditions. And, we would be correct in making this inference, as many numerical experiments have shown. But since the same algorithmically random sequence can be output by the unkicked rotor, the inference fails for that system, since as is quite clear from the data of figures 10 and 11, that system does not admit any separation of trajectories in phase space. In fact, the phase space for the unkicked rotor is, once again, the cylinder of figure 5.

In this example, we are measuring the position of the unkicked rotor at strange times. *But*, the time of measurement is something that is *completely determined* by past values of the sequence of measured values. So, unlike my original roulette wheel, this is in fact, *a genuinely deterministic dynamical system* with integrable motion which nevertheless explicitly yields a sequence of measurement outcomes that is algorithmically complex.

The reason it is possible to construct this example, in the face of the theorems of Brudno-White and Pesin is that the transformation T which supplies the θ-values does not advance the system in unit intervals of "real" time. Instead, it advances the system in unit intervals of τ-time. There is a structure imposed upon the system by the independent variable—time. In fact, this structure is in part what allows for the definition of the Lyapunov exponents in the first place. In shifting from real t-time to τ-time, the Lyapunov exponents of the trajectories evolving under the transformation T change. In effect, this mapping from t-time to τ-time installs a new differential structure on the phase space manifold. The definition of chaos advocated by Ford, does not pay sufficient attention to this underlying structure provided by the independent variable.

5   Conclusion

At the beginning of the paper I asked what one can infer about the degree of instability of a dynamical system, if one knows that its behavior over time is unpredictable. If one formalizes the notion of unpredictability in terms of the algorithmic complexity of a sequence of measurement results performed on the system, then there are theorems due to Brudno and White and Pesin that provide a partial answer to this question. If the conditions of these theorems are met, then it does look like the following chain of probability one equivalences holds: (positive algorithmic complexity) $\leftrightarrow$ (positive metric entropy) $\leftrightarrow$ (positive Lyapunov exponents).

The discussion in section 3 shows that the notion of absolute complexity which appears in the Brudno-White theorem may perhaps not be the best way to formalize our concept of unpredictability. Typically we have at least some information about other quantities of interest on our system besides the one we are measuring. This means that we really should look at the complexity of our sequence of measured values conditional upon this further information. The difference in the conditional algorithmic complexity of the deterministic "roulette" wheel compared to that of the ideal hard sphere gas illustrates why this notion is useful. If we pay attention only to the absolute complexity of the observed sequences from the two systems, the systems appear to be equally unpredictable. The difference in conditional complexity is the result of a difference in the nature of that part of the dynamics of the two systems that can be considered Hamiltonian: Once we are given the value for the momentum for a given spin of the wheel, the system undergoes a completely integrable evolution until the next spin. It is ultimately because of this that the conditional complexity of the wheel is zero, while the completely nonintegrable unstable motion of the ideal gas guarantees that its conditional complexity relative to the quantity we considered is positive.

The wheel discussed here, however, is an abstract construction. Its full phase space X is the Cartesian product of the space Ω and the space Λ^∞ of infinite sequences. My main concern both in this and in the previous paper is chaos in Hamiltonian systems. Therefore, in section 4 I presented evidence that an integrable *Hamiltonian* dynamical system, the unkicked rotor, can yield measurement results that are algorithmically random and hence, completely unpredictable. The reason this is possible is that the measurements are performed at odd times. Nevertheless, *when* the measurements are performed is completely determined by manipulating the independent variable, time, given the equations for the chaotic kicked rotor. There is nothing in classical mechanics that would rule out this performance of position measurements on the unkicked rotor at these times. There is, therefore, an important sense in which the degree of instability of a system is more fundamental than the algorithmic complexity of a sequence of measurement results for defining "chaos" in classical dynamical systems. This does not contradict the results of the ergodic theorists discussed in section 2: If we had previously determined that the transformation T advances the system in real time and not in τ-time, then the inference from algorithmic complexity to sensitive dependence on initial conditions will go through. But determining this requires that we look *first* to the dynamics of the system, to its phase space structure. For the purposes of making an inference solely from *the* unpredictability of a system's output to some conclusion about the degree of instability present in the dynamics, Ford's claim that "sensitive dependence" and "deterministic randomness" are synonymous, is simply wrong.

Finally, it is worth mentioning that Ford has, perhaps, an ulterior motive for preferring the algorithmic complexity definition of chaos. He wants an account that can easily be brought to bear on the question of the existence of chaos in quantum mechanics. In effect, defining chaos in terms of algorithmic complexity provides a "theory neutral" definition, since one need not look to the specific dynamics of the theory. One need only pay attention to system output. Ford has repeatedly argued that on his view, quantum mechanics exhibits no chaos whatsoever. (See Ford (1988), Ford (1989), and Ford et al. (1991).) Furthermore, for him this is regarded as a serious flaw of quantum mechanics.

In Ford et al. (1991), it is claimed that the failure to find "deterministic randomness" in quantum mechanics, and the failure to show how it emerges in the "classical" limit, is evidence for the failure of the correspondence principle. Though it is beyond the scope of this paper to debate the issue here, my view is that making algorithmic complexity the key definiens of chaos, has led to a rather naive and conservative view of the correspondence principle. (See Batterman (1991), Batterman (1992), and Batterman (1995).) There is much work currently being done in semiclassical mechanics which, may plausibly be construed as a detailed investigation into the correspondence relations between classical and quantum mechanics in light of questions about chaos. Ford's dynamically independent account of chaos, apparently invites us to downplay the significance of this work.

References

Batterman, R. W. (1993): Defining Chaos. Philosophy of Science **60**, 43–66

Batterman, R. W. (1991): Chaos, Quantization, and the Correspondence Principle. Synthese **89**, 189–227

Batterman, R. W. (1992): Quantum Chaos and Semiclassical Mechanics. Philosophy of Science Association **2**, 50–65

Batterman, R. W. (forthcoming): Theories between Theories: Asymptotic Limiting Intertheoretic Relations. Synthese **103**, 171–201

Brin, M., Katok, A. (1983): On Local Entropy. In *Geometric Dynamics, Lecture Notes in Mathematics 1007* (Springer, Berlin), 30–38

Brudno, A. A. (1983): Entropy and the complexity of the trajectories of a dynamical system. Transactions of the Moscow Mathematical Society **2**, 127–151

Chirikov, B. V. (1979): A universal instability of many dimensional oscillator systems. Physical Reports **52**, 263–379

Ford, J. (1988): Quantum chaos, Is There Any? *World Scientific, Directions in Chaos*, 128–147

Ford, J. (1989): What is chaos, that we should be mindful of it? *The New Physics* (Cambridge University Press, Cambridge, England), 348–371

Ford, J., Mantica, G., Ristow, G. H. (1991): The Arnold'd Cat: Failure of the Correspondence Principle. Physica D **50**, 493–520

Pesin, Ya. B. (1977): Characteristic Lyapunov Exponents and Smooth Ergodic Theory. Russian Mathematical Surveys **32** (4), 55–114

Peterson, K. (1983): *Erogdic Theory* (Cambridge University Press, Cambridge, England)

White, H. (1993): Algorithmic Complexity of Points in Dynamical Systems. Ergodic Theory and Dynamical Systems **13**, 807–830

Discussion of Robert Batterman's Paper

Batterman, Chirikov, Miller, Noyes, Suppes

Suppes: I've just a couple of remarks about the roulette wheel. As you know there is a long history starting with Poincaré and the method of arbitrary functions. Now an even better and clearer example of this method of arbitrary functions is an example of coin flipping by Keller in which you consider you have an extremely simple integrable system with a simple differential equation where you have an initial velocity vertically and then a rotation with a constant angular rotation velocity. And of course given the initial conditions and assuming that the coin lands on a soft flat surface, so it won't flip back over some of the time, then everything is integrable. So you have a deterministic system. What is interesting about this seems to me - which was very much realized already by Hopf in the 1930s - is that real systems are really unmanagable. So for example, if you flip a coin those initial conditions are simple enough but if you let it bounce it is hopeless to give a detailed analysis of the bouncing of the coin on a hard surface. Or in the case of the roulette wheel, which you mentioned, you have a standard dissipation from slowing the wheel down for otherwise it will go on forever, so there must be some method of stopping it. But the standard way of stopping it leads to a really unanalyzable dissipative system. So what I would claim is that we actually don't know how to analyze in detail the mechanical systems corresponding to real roulette wheels and real coin flopping. We cannot actually give an integrable system. So we end up really because of the inability to analyze what happens when the coin lands or when you apply friction to stopping the wheel in detail. By treating it abstractly as random, but we are not able to actually analyze in complete detail the dynamical system. Do you agree with that analysis or do you see ... I mean, I raise it because of your emphasis on the roulette wheel.

Batterman: There has actually been a fair amount of work done on the roulette wheel by chaos theorist J.D. Farmer - on real roulette wheels like they have in Las Vegas. This work is described in a book called "The Eudaimonic Pie". He was able to beat the system and predict roughly where the ball is going to fall using a computer he designed and built. So with regard to real roulette wheels, I think he had a pretty good model for what was happening.

Suppes: Some marvellous details from the mechanical standpoint: after all you can buy on a market a coin flipper that give you heads all the time, if you have a landing surface that is soft and smooth. You can buy such a device. They are available from magician's shops. So if you don't have any hard surface for it's going to pop up and turn over etc. where you have to try to analyze, then you can build a system that will guarantee heads and tails whatever you wish, with very high probability. So I'm not saying that I did mean to suggest a system was a Bernouilli shift in any serious way. It's just that we can analyze the dynamical aspects. So in this guy's study he does not what Hopf found you could not analyze - in detail the mechanical system.

Batterman: Actually, in his study - I think - he does analyze it in detail.

Suppes: Analyze with the friction, or not? You may answer, "very strict", you know. Do you claim he gives an integrable analysis mirroring exactly the friction? I am sceptical about that.

Chirikov: The remark which most of you must know of course: a real roulette wheel is not just a wheel but a ball which is very important.

Suppes: A bouncing ball in fact is rather like flipping the coin on a hard surface.

Chirikov: No, in a roulette wheel a ball may change completely the statistical properties. But I have a question [to Suppes]. Why did you say a dissipative system is not analyzable?

Suppes: I mean as a dynamical system this kind of thing I am referring I don't think there is any real improvement on Hopf's analysis in the 1930s.

Chirikov: You mean if you consider fluctuations?

Suppes: Yes, for example, when you flip a coin. You are absolutely right with the ball. That is why I am extremely sceptical in the reference you give there is any serious analysis of the details of mechanics, because the ball is exactly like flipping and landing on a hard surface. I mean in broad terms it is the same unmanageable physical phenomenon in detail.

Batterman: I just have mentioned the book where you can read about it. It is an interesting story: they actually went to Las Vegas with computers in their shoes. They observed the wheel and how the wheel spins, they timed the ball's motion to get initial conditions for the system and they were able to predict - that is, beat the odds.

Suppes: That's different from having an analysis of dynamical systems.

Batterman: I'm not sure. In order to design the program, he had to solve the problem - solve the equations of motion. But the point I am trying to make is the following: Consider flipping a coin in space and spinning a wheel forever without friction. These are rigid disks rotating, respectively, out of their plane and in their plane. They are integrable systems in the strictest sense. Nevertheless it seems to me that, since the rotor is essentially equivalent to the spinning wheel, you can have a rotor spinning without friction and look at it at funny, but deterministically determined times. One gets a sequence of measurement results which is algorithmically random. That's what I was trying to show. As a consequence, one cannot infer, by merely looking at the sequence of measurement results, the positions of the rotor, that the system generating that output is sensitively dependent on initial conditions.

Suppes: You can get the same result by simply making measurements of a real line - just taking measurements of the real line - nothing moving at all. I take, according to some intervals, measurements of the real line, and I will get the same result. I get a random sequence. Though it's numbers I get, it satisfies the same property.

Batterman: The numbers may be algorithmically random but my point is: If all you are given is a system within a black box which outputs a random sequence of numbers, can you infer that the system which generated that sequence is sensitively dependent on initial conditions? I am trying to argue that the answer is: "No!".

Suppes: But there is a fundamental distinction in the standard statistical analysis of randomness. On the one hand, there is a characterization of randomness that is very much in the spirit of complexity - namely: randomness is a phenomenological property. It only depends on properties of the numbers you are handed. And those numbers can be generated and nobody says: "Those numbers can only be generated by a dynamical system." Now the second meaning is randomness in the sense of a mechanism. Where now you have - in your case - a dynamical system with certain properties. The theorems you state do not at all state that a sequence of numbers is algorithmically complex if and only if it is generated by a dynamical system of a certain property. You have many ways you can think about generating phenomenological sequences. And I think that is really what your argument is pointing to. Is that a reasonable interpretation?

Batterman: I think so. I am concerned, in the paper, with chaos in Hamiltonian systems. I was concerned with Ford's claim that positive algorithmic complexity or randomness of a sequence of symbols (numbers) is a synonym for chaos in the sensitive dependence sense which appears in the literature. I was just trying to argue that it is not a synonym, but that the dynamical features are basic when it comes to finding a definition of chaos for dynamical systems. I do not deny that one can have abstract systems generating Bernouilli sequences or whatever. Originally, when I wrote the first paper "Defining Chaos", I was unaware of Pesin's theorem and so I believed that there was even less agreement between the complexity definition and the definition in terms of sensitive dependence. The problem was this: Suppose I had an abstract system for which one could not possibly define a notion of spreading of trajectories (since the system doesn't live on the appropriate sort of manifold). How could one possibly infer from a random output, a random sequence of numbers, that there is any kind of system inside the box which was even capable of exhibiting sensitive dependence on initial conditions. But Pesin's theorem makes the connection between entropy and the Lyapunov exponents, and so the connection between algorithmic randomness and sensitive dependence is actually stronger than I believed. The focus here, however is this: If you want to apply the Brudno-White and the Pesin theorems to infer from algorithmic randomness of a sequence of numbers to a claim to the effect that the system generating that sequence is exponentially sensitive to initial conditions, then you need to know something about the dynamics first. One cannot just take a look at the sequence alone. Unpredictability in that sense, cannot guarantee chaos in the sense of sensitive dependence.

Chirikov: You have to use that theorem in the opposite way: to infer randomness in some sense of complexity from dynamical instability.

Batterman: That is the way I think it should be used.

Chirikov: Here you should have no contradiction, yes?

Batterman: Right! I agree. It is just that Ford tends to use it the other way. What I did not get to mention (although it is at the end of my paper) is that there is another motivation for Ford to come up with the algorithmic complexity definition. It is that he wants a sort of theory neutral definition of chaos. This is because he wants to be able to talk about quantum chaos. If chaos is just a synonym for algorithmic randomness, then it is a dynamically independent definition. If quantum mechanics is to be chaotic, then all one would need to do is look for random sequences of values in, say, the energy spectrum. If random sequence is discovered then quantum mechanics would be chaotic. Ford argues that we won't be able to find such algorithmically complex outputs in quantum mechanics. In my view this is a mistaken approach. If quantum mechanics is not chaotic, it is because of dynamical features of quantum systems. There is much work being done in semiclassical mechanics (Prof. Chirikov calls this "quasi-classical" mechanics) about the connections between quantum mechanics and classical mechanics in the light of chaos. But, if you take this approach-defining chaos in terms of algorithmic randomness - then it seems that you are down-playing the significance of all the current work in semiclassical mechanics. This work, in my view, an extended investigation of the Correspondence Principle - the nature of correspondences between the two theories. If one takes chaos to be synonymous with algorithmic randomness, one ignores all of that.

Miller: I am puzzled by some of the things you say. I am certainly puzzled by why deterministic laws should be expected to lead to algorithmically complex outcomes. The outcome of the deterministic law, I take it, shows that the law compresses the information involved. If we think of a case such as the logistic function, we can work out its entire evolution from a simple formula. So I take the evolution of that function not to be in any sense complex - algorithmically. Now why is it that it nonetheless produces algorithmically complex sequences of some sort? I suppose that the simplest case is one in which - instead of taking the actual value of the function at each point - we consider only a projection of it on two points, 0 and 1 - say, is it greater than a half or is it less than a half.

Batterman: You are looking at a partition and the partition codes that order.

Miller: That's right. Now is that a characteristic feature? Is it true also for the roulette wheel and for coin tossing that what we are doing is throwing away information, deflating what is known by the system and partitioning it.

Batterman: Certain partitions - relative to their transformations - are gener-ating partitions in the sense that, roughly, if one has the entire orbit from the arbitrarily distant past to the arbitrarily distant future, then that orbit - that sequence of numbers relative to the partition - is sufficient to uniquely specify the trajectory, modulo a set of measure zero. So, in a measure-theoretic sense, you can have all the information there is to know. In other words, you can uniquely pick the point, modulo a set of measure zero of points, by looking at this orbit sequence. The idea is that if this orbit sequence is algorithmically complex, then you've got a dynamical system generating random sequences.

Miller: And in what sense is it deterministic?

Batterman: Where the point is, which cell of the partition - whether it is zero or one - is completely determined by where it's point was in the past.

Miller: Are we talking physics here? I mean, every particular sequence can be thought of as deterministic.

Batterman: If you are dealing with ordinary differential equations anyway determinism means existence and uniqueness.

Miller: Given the entire evolution from the past of the sequence, or just the previous value?

Batterman: From a point: there is one and only one possible trajectory in the past or the future.

Miller: What about indeterministic randomness?

Batterman: You cannot - as far as I know - define a notion of sensitive dependence or trajectory instability unless you have a deterministic equation.

Suppes: Oh, that's not the case. For example, a Chain is ergodic if and only if it has a unique limiting distribution independent of initial starting point. A stochastic process that is dependent on initial conditions is not ergodic. That's a standard concept of sensitivity to initial conditions. For example you have zero-one processes - I mean, they depend on the initial conditions. So I mean in other words there is a standard concept of sensitivity to the initial conditions for indeterministic random processes, and that's standard in stochastic processes.

Chirikov [to Suppes]: But dependence of what - trajectory or function?

Suppes: Dependence over trajectory in the limiting outcome.

Chirikov: The function also does not depend on initial conditions, a distribution function.

Suppes: Yes, that's different. This is an example. I mean ergodic in the sense of the Markov chains - which is very close to ergodic in dynamical systems - is exactly what is not sensitive to initial conditions. I just mentioned that because those are familiar cases I have extensively studied.

Batterman [to Suppes]: Do you mean by an "indeterministic system" a system which can be characterized, say, by having the property of being a K-system, a Bernoulli system, or some kind of Markov process.

Suppes: A Markov chain of infinite order.

Batterman: What interests me is the fact that deterministic systems, in sense of classical, Hamiltonian systems, can be shown - in certain instances, with lots of idealization - to be K-systems. This is true for a hard sphere gas in a box. My view of abstract ergodic theory is that it provides a classification, a hierarchy of statistical properties. The interesting thing from the physical point of view is that certain classical systems can actually be shown to possess statistical properties at the high end of the hierarchy. Being ergodic or mixing isn't good enough to get one the behavior one needs for, say statistical mechanics ...

Noyes: Discussion of classical systems, such as we have been having here, often makes an idealization about measurement which is unrealistic when we discuss actual laboratory practice. Actual measurement always has fixed measurement accuracy set by the current technology, and this limitation when included in the analysis has important and unexpected consequences, which are discussed in more detail in my paper. This does not have to do with quantum mechanics directly, but what I call scale invariance bounded from below ends up restricting what we can say about classical systems in a way that looks surprisingly like quantum mechanics ...

Batterman: I agree that there is limited accuracy of measurements. If we think of a partition of phase space in the way of orthodox statistical mechanics, then it may represent a possible measurement with limited accuracy: you can determine whether the system's phase point is in a cell of such and such size. So a partition represents, given certain idealizations, a measurement allowing for finite measurement accuracy.

Noyes: Actually, it is more complicated than dealing with partitions in statistical mechanics. Fixed measurement accuracy interpreted as requiring that measured results be *integers* times some fixed, smallest dimensional unit produces a lack of determinism that, from my point of view, cannot be avoided by taking a limit. I do not allow that idealization.

Chirikov: But not absolutely fixed?

Noyes: Yes, absolutely fixed.

Chirikov: Why, what is this fixed accuracy?

Noyes: Physically there is such a limitation.

Chirikov: Do you mean a global limitation?

Noyes: There is always a physical limitation on the measurement of position. If you try to measure the position of a particle to better than half an electron Compton wavelength, you will always produce electron-positron pairs with some finite probability. I am talking about the actual physical situation in a laboratory. So there is a global physical effect. But you can simulate this limitation without actually putting the electron-positron pair degrees of freedom into your model just by assuming that you have a finite, fixed accuracy to which you can measure position and time. Then you get results that look very much like quantum mechanics. It also allows a new type of discussion of how to take the "correspondence limit". That is, one can start from a finite particle number relativistic particle dynamics (not a quantum field theory) and see under what circumstances the results can be approximated by classical physics.

Chirikov: But quantum limitations are weaker because only the product of the two conjugated variables is restricted but not each variable.

Noyes: What you say may be true of the theory called non-relativistic quantum mechanics. But that theory does not accurately describe the world we live in. The world we live in is one in which pair creation actually occurs at short distance. To ignore that fact and to talk about either classical mechanics or non-relativistic quantum mechanics as if that is the real world explored by physicists in their laboratories is, I think, a mistake. I am making a fairly strong methodological claim ...

Chirikov: No, no. The limitation depends on whether you study the high-energy physics or low-energy physics.

Noyes: But this limitation applies at any energy if your precision is high enough. After all, the Lamb shift in the hydrogen spectrum requires a very low energy measurement. But it shows these effects. Or for that matter, proton-proton scattering at low energy, if you look with sufficient precision, shows a modification of the Coulomb scattering law at distances of half an electron Compton wavelength, which was predicted by L.L. Foldy. This characteristic modification of the angular distribution of the scattering cannot be due to either electromagnetic or to nuclear effects but in fact is due to vacuum polarization, i.e. electron-positron pairs. For details see H.P. Noyes, Phys. Rev. Letters 12, 171 (1965) and references therein.

On the Foundations of Synergetics

H. Haken, A. Wunderlin, S. Yigitbasi

Universität Stuttgart, Germany

Abstract. We introduce basic notions, concepts, and principles of synergetics and work out the interdisciplinary aims of this new scientific discipline. To some extend the laser as a system of physics as well as the application of synergetics to a system taken from economics will be presented. A generalization of the methods to an important class of functional differential equations will be outlined.

1   General aspects of synergetics

Synergetics can be characterized as the science of cooperation and self-organisation. As an interdisciplinary scientific field it has been founded by H. Haken in 1969 Haken (1983), Haken (1987). The aim of synergetics is to describe the macroscopic behavior of complex open systems from a unified point of view. This results, for instance, in the spontaneous emergence of spatial, temporal, and spatio-temporal structures and/or their special functioning. The complex systems which are considered can be characterized schematically by the following general properties:

(i) The systems are composed of many subsystems. It turns out that this fact allows a description of the systems under consideration on two different hierarchical levels: a microscopic point of view which is attributed to the scales of the subsystems on the one hand and, on the other hand, a macroscopic level of analysis which is attributed to the scales of the composed system.

(ii) The subsystems and their interactions form a complicated *nonlinear system*. In correspondence they follow a high-dimensional set of nonlinear evolution equations. As an important result, for example, there exists no superposition principle in the structure of the solutions.

(iii) Eventually the systems must be open. It is this fact which guarantees that the systems can be driven far away from, e.g. thermal equilibrium and appears naturally as a necessary condition that processes of self-organization can arise.

Taking into account these given conditions we still realize that systems of this kind can be observed in various places of natural as well as of human sciences. Surprisingly it turns out that close to critical regions, where new patterns of behavior are created via an instability by the system itself, all these systems show common *universal behavior on macroscopic scales*. In the following we shall present important examples of such systems from the animate as well as the inanimate world.

As an example from physics we consider a solid state laser Haken (1985), Haken (1970). Here the subsystems are identified as the laser active atoms. The laser is an open system because the device is pumped from outside. The spontaneous self-organization is manifested in the coherent action of the laser active atoms which results in the emission of coherent laser light. We emphasize that the laser has become a system of paradigmatic value for synergetics. One of the reasons is that all steps of the theory can be verified by starting from first physical principles of quantum theory and quantum field theory.

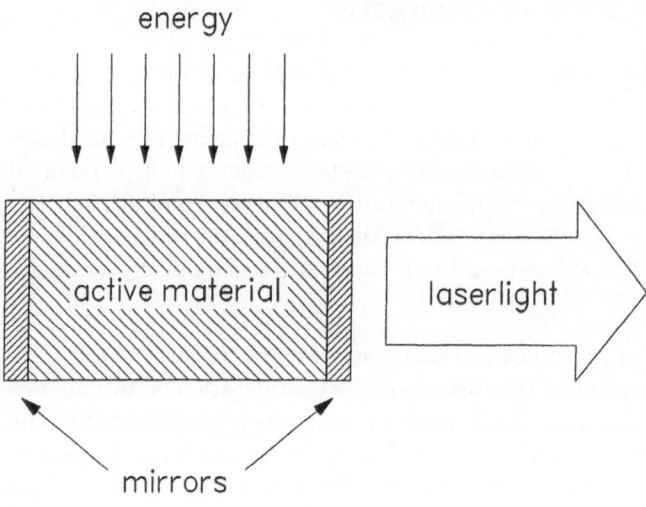

Fig. 1. Schematic diagram of a solid state laser device.

As a second example we mention a system from hydrodynamics Chanderasekhar (1981), Friedrich et al. (1990), that is the convection instability. In the Bénard problem (compare Fig. 2) a fluid layer is heated from below in the gravitational field of the earth. The heat supplied from below shows that we are dealing with an open system.

The subsystems are the molecules of the fluid, the spontaneous emergence of macroscopic ordering is reflected in the observed highly ordered convection patterns, e.g. the role pattern of Fig. 3.

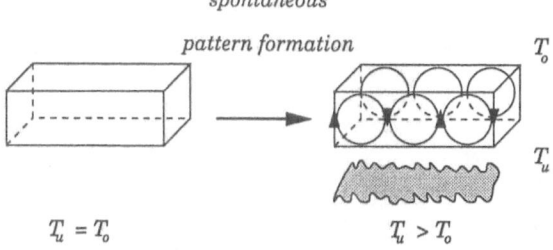

spontaneous

pattern formation

$T_u = T_o$ $T_u > T_o$

Fig. 2. Schematic diagram of a Bènard System $(T_2 < T_1)$.

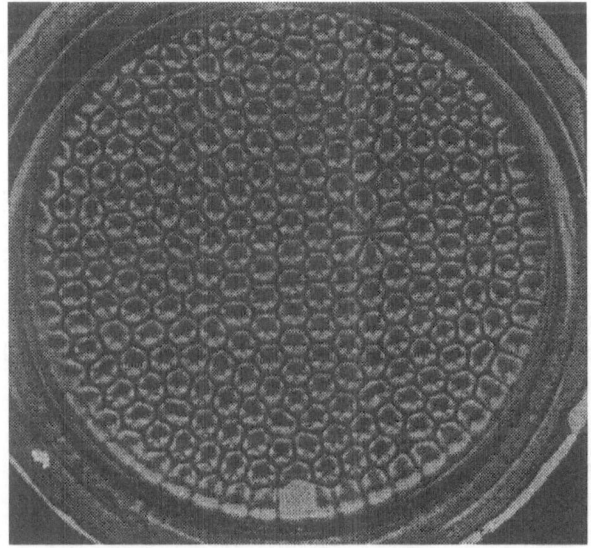

Fig. 3. Hexagonal cells of a liquid layer heated from below (Bénard-Instability).

In chemistry the autocatalytic reactions form an important class of examples. The most prominent experiment is the Belusov-Zhabutinski reaction (Haken (1983)). Again we consider an open system when matter in form of the reactands is continuously supplied and the products of the reaction are removed accordingly. The reaction can settle down in two different states which are macroscopically observed as the colours blue and red, respectively. The macroscopically ordered states are observed as coherent spatio-temporal patterns (compare Fig. 4).

The fields of biology and medicine are especially rich of examples for synergetic systems. Here we mention the coordination of movements of animals and

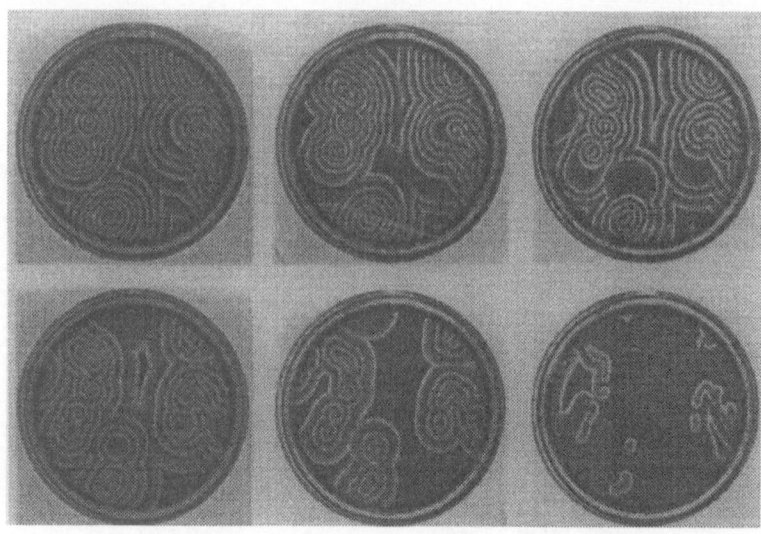

Fig. 4. Chemical spiral waves of the Belusov–Zhabutinsky–Reaction.

humans Haken et al. (1985). For instance, the different gaites of horses can be interpreted as the coherent action of muscle and nerve cells, the tissue etc. They are considered as the subsystems. The processes of metabolism classify animals as an open system.

2   The laser

Here it is our concern to introduce basic notions and principles of synergetics by discussing the laser action more closely. To that end we first start from a *'Gedankenexperiment'* which is concerned with the generation of coherent light from different points of view. In a second stage we shall discuss the laser action in more detail by giving a mathematical point of view. The results will be used to present the general synergetic theory how to cope with complex systems Haken et al. (1983).

2.1   Remarks on the notion of self-organization

During the development of synergetics, laser theory has become an important pace–maker. Indeed, the detailed theoretical understanding of laser action has become an important paradigm for the development of a general theory of the systems under consideration. As it is well–known the emergence of laser action can be derived from first physical principles. That is to say, at the microscopic level the dynamic equations of the system are perfectly known, and the resulting macroscopic behavior can be derived by using well–posed mathematical methods. Hence, the predictions of synergetics can be tested to a remarkable accuracy by using laser theory and experiments on laser devices.

2.2   The 'Gedankenexperiment'

By way of 'Gedankenexperiment', we present a simplified model of laser action in order to illustrate the concept of self–organization. When we consider light as an electromagnetic wave, we can describe it by its electric field strength $E(t)$ being a function of time t. Coherent light, then, corresponds to a single, infinitely long wave track with a definite frequency and highly stabilized amplitude (compare Fig. 5).

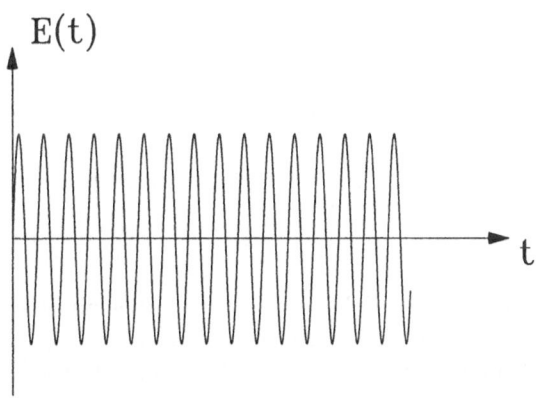

Fig. 5. Coherent laser light.
$E(t)$ represents the electric field strength as a function of time t.

In order to characterize the difference between normal and coherent light, we first consider a single atom with one outer electron, the so–called 'Leuchtelectron' (Fig. 6), the remainder of the atom being summarized as one effective nucleus.

In our drastically simplified model, it is assumed that the electron circles around the nucleus on a fixed circle with radius r, its position therefore being completely determined by angle ϕ. Since this circular motion is an accelerated uniform motion of a charge, it leads to the emission of an electromagnetic wave.

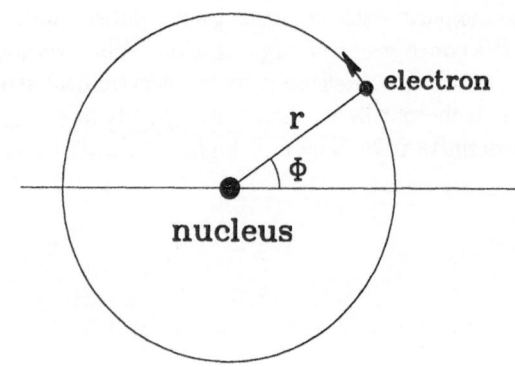

Fig. 6. 'Leuchtelectron' circling around an effective nucleus.
The electron is assumed to move with fixed radius r, its position being completely determined by angle ϕ

The amplitude of the light wave emitted by a single atom (Fig. 7), however, is much lower than the high amplitude of laser waves (compare Fig. 5). Indeed, in conventional lasers in the order of 10^{18} atoms are needed in order to produce the coherent light wave.

Let us now consider the more complicated situation where we have two atoms (Fig. 8).

The position of the two 'Leuchtelectrons' is now characterized by two independent phases, ϕ_1 and ϕ_2. In general, we will have to expect that these phases are different because the electrons are independently circling around their own effective nucleus. The resulting light field, therefore, is given by the superposition of small wave tracks, each with a different phase. We may directly extrapolate from these considerations that the light wave which is emitted by many independent atoms will never lead to coherent laser light. In order to produce laser light, the electrons have to be all in the same phase (compare Fig. 9).

There appear to exist different frameworks as to how this synchronization can be produced.

– In the *cybernetic approach* each individual atom should be steered, e.g., by a central computer. One would have to steer in the order of 10^{18} atoms

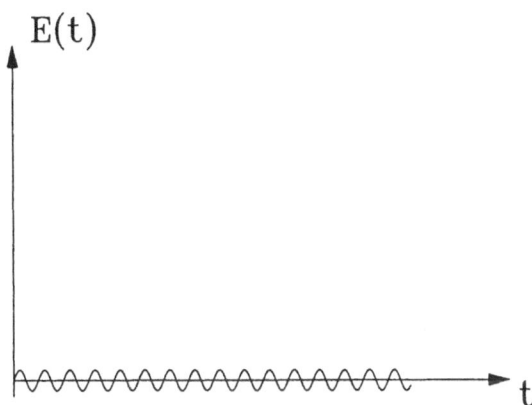

Fig. 7. The light wave emitted by a single atom.

individually. Hence, the cybernetic approach appears to be inadequate to solve our problem.

– A *pace–maker* could be conceived to produce a coherent light wave. The pace–maker, then, has to interact with all the atoms, enforcing them to take the prescribed phase. Again, this idea appears to be unrealistic. Coherent action can only be reached when the response of the atoms to the pace–maker is linear. This can only be the case when the amplitude of the pace–maker wave is low. In such a regime, however, no stabilization of the amplitude is possible.

– From a *synergetic* point of view, the idea of self-organization can be introduced. Then, the system forces itself to behave in synchrony. In other words, the nonlinear behavior of the system generates a cyclic causality between the behavior on the microscopic ('Leuchtelectrons') and the macroscopic (coherent light wave) level of description, and thereby a macroscopic ordering of the system. In the following subsection we shall discuss this phenomenon more thoroughly.

2.3   The synergetic discussion of laser action

In order to further elucidate the synergetic notion of self–organization, and to elaborate the relevant aspects of our 'Gedankenexperiment', we consider a typical

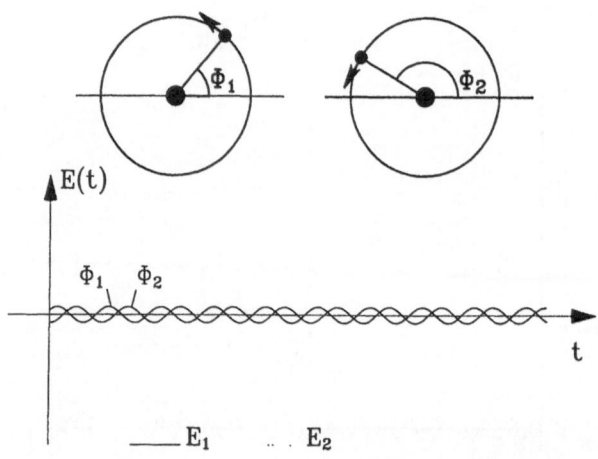

Fig. 8. Incoherent electric light field of 2 atoms.
The light field is given by the superposition of the two wave tracks.

solid state laser device. In between two mirrors which act as a resonator for the light field, the laser active atoms are embedded in a host crystal. The laser is an open system where energy is pumped from the outside in a controlled rate. As long as the energy supply is low, we observe that the laser is acting as a usual lamp. As already noted this implies in our model that the single electrons are circling independently and no phase correlation between different atoms occurs. A dramatic change is observed as soon as the energy supply exceeds a certain threshold value. Then, the laser suddenly starts to emit the coherent light wave which is depicted in Fig. 5.

In synergetics, this phenomenon is understood as the spontaneous synchronization of the 'Leuchtelectrons' which is produced by the system itself, i.e., in an act of self–organization. In other words, when we change a single unspecific control parameter, the energy supply from the outside, the system reacts with a spontaneous macroscopic ordering which is determined by the internal dynamics of the system.

The enhanced energy supply allows more and more atoms to generate their individual light waves. The nonlinear internal dynamics of the system, as well as the action of the resonator, allows the system to select a certain wave. The selected wave now acts back onto the atoms as a nonlinear pace–maker and synchronizes the movements of the electrons around their nuclei. The nonlinear

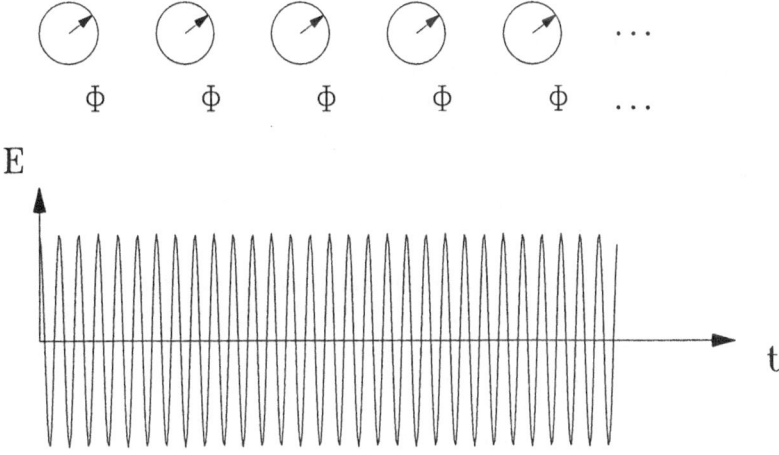

Fig. 9. Coherent light field as produced by many synchronized atoms.

behavior of the system allows for *cyclic causality* to arise: The light wave is built up by the action of the independent subsystems, the laser active atoms, and then reacts back on the subsystems to enforce their coherent motion. In this way, the coherent wave can be generated.

We summarize our observations as follows: The microscopic level of description in our model is provided by the single laser active atoms. The coherent action of the atoms beyond control parameter threshold can be understood as a spontaneous macroscopic ordering of the atoms over the whole crystal. This ordering is realized by the coherent light wave which has emerged. In this sense, we may consider the emerging light wave as an *order parameter* which forces (enslaves) the subsystems to act in concert. This exemplifies a fundamental principle of synergetics which has been called the *slaving principle* (Haken (1983), Haken (1987)).

2.4   A mathematic modelling of a laser

A solid state laser consists of a host crystal, where the laser active atoms are embedded. The crystal is placed into an optical cavity. For simplicity these atoms are treated here as two level atoms. They are assumed to be located far enough from each other in a way that their wave functions do not overlap. Obviously we have a system which fulfills all criteria of a synergetic system mentioned in the beginning. It is an open system (the laser is pumped from outside), furthermore it is composed of many (identical) subsystems, the laser active atoms, and the system has the ability of spontaneous self-organization on macroscopic scales. Self–organization will be manifested in the coherent action of the laser active atoms over the volume of the whole crystal. The light field of the laser within the cavity can be described by its electric field strength $\mathbf{E}(\mathbf{x}, t)$, where x denotes the space coordinate and t represents time. The electric field is decomposed into the eigenmodes $\mathbf{u}_\lambda(\mathbf{x})$ of the cavity according to

$$\mathbf{E}(\mathbf{x}, t) = \sum_\lambda N_\lambda [b_\lambda(t) - b_\lambda^\star(t)] \mathbf{u}_\lambda(\mathbf{x}) \,. \tag{1}$$

Here N_λ is a normalization factor and the set $b_\lambda(t)$ denotes the complex amplitudes which measure the strength of excitation and the phase of the different modes λ. The state of the two-level atoms is characterized through their polarization $\mathbf{P}(\mathbf{x}, t)$ and inversion density $D(\mathbf{x}, t)$. Both of these mesoscopic densities are divisable into the contributions from the single atoms. When we enumerate the atoms by the index μ and describe their location by the vector $\mathbf{x}_\mu$ we obtain for the polarization density

$$\mathbf{P}(\mathbf{x}, t) = \sum_\mu \mathbf{P}_\mu(t) \, \delta(\mathbf{x} - \mathbf{x}_\mu) \,, \tag{2}$$

where $\mathbf{P}_\mu(t)$ is the polarization of the atom μ. Expressing as usual $\mathbf{P}_\mu(t)$ through the atomic dipole matrix element of the two-level atoms Θ_{12}

$$\mathbf{P}_\mu = -\alpha_\mu(t) \, \Theta_{12} - c.c., \tag{3}$$

$\alpha_\mu(t)$ becomes a complex measure of the polarization of the single atom μ. Correspondingly $d_\mu(t)$, the inversion of the single atoms, constitutes the inversion density

$$D(\mathbf{x}, t) = \sum_\mu d_\mu \, \delta(\mathbf{x} - \mathbf{x}_\mu). \tag{4}$$

In a semiclassical theory, where the light field is assumed to behave classically and the atoms are treated quantum mechanically, one derives by the application of well–established approximations the following closed set of equations for the interaction of light and matter in a given optical resonator

$$\dot{b}_\lambda = -(i\omega_\lambda + \kappa_\lambda) b_\lambda - i \sum_\mu g_{\mu\lambda}^\star \alpha_\mu + F_\lambda \,, \tag{5}$$

$$\dot{\alpha}_\mu = -(i\nu + \gamma) \alpha_\mu + i \sum_\lambda g_{\mu\lambda} b_\lambda d_\mu + \Gamma_\mu \,, \tag{6}$$

and

$$\dot{d}_\mu = \gamma_\parallel (d_0 - d_\mu) + 2i \sum_\lambda (g_{\mu\lambda}^\star \alpha_\mu b_\lambda^\star - c.c.) + \Gamma_{d\mu} \,. \tag{7}$$

These equations have been written in their spatially homogeneous version, where all the atoms are treated equally. ω_λ denotes the frequencies which are associated with the cavity modes, ν the transition frequency of the single atom. Damping terms have been added: κ_λ describes the losses in the cavity, γ is attributed to the finite atomic life time, and $\gamma_\parallel$ to the damping of the atomic inversion. The quantity $g_{\mu\lambda}$ characterizes the interaction of the field and the atoms. For the sake of completeness we have added fluctuating forces F_λ, Γ_μ and $\Gamma_{d\mu}$. Their detailed properties can be derived from a complete microscopic quantum field theoretical treatment of the system, when it is additionally coupled to different heat baths which provide damping and fluctuations in a way consistent with quantum mechanics Haken (1970). The so-called unsaturated inversion d_0 takes the role of the control parameter which measures the external pumping of the system. We observe that we have a set of basically nonlinear equations which defines the mesoscopic level of consideration and can be seen as a special realization of the general form (31).

The number of equations (5-7) is enormous: about 10^{18}. Clearly, the following remark applies: It is impossible to present a complete solution for this set of equations. On the other hand it appears useless even to know the complete solutions because it is impossible to manage the huge amount of information needed to specify initial conditions and their transformation in time along the solution curves. Accepting this point of view the situation turns out to be rather similar to problems which are related to the foundations of statistical mechanics (e.g. Landau and Lifshitz (1952)). As we shall see, however, the solution here will be quite different.

2.5   Introduction of Collective Variables and Order Parameters, and the Slaving Principle

In order to demonstrate how to cope with these equations successfully, we shall choose a more or less heuristic way which will be exploited in the next section when we discuss the general formalism. We shall assume that the inversion of the atoms is homogeneous over the whole crystal

$$d_\mu = d\,. \tag{8}$$

Under this circumstance it becomes possible to prove that only one mode λ of the light field can be macroscopically excited at the laser threshold. The system prefers the mode which is in resonance with the atomic frequency ν. Taking this for granted we can suppress the index λ and introduce *collective variables* P and D in a straightforward way. We use

$$P = \sum_\mu g_\mu^\star \alpha_\mu \tag{9}$$

and

$$D = \sum_\mu d_\mu = Nd\ , \qquad D_0 = \sum_\mu d_0\,, \tag{10}$$

where N denotes the total number of laser active atoms. Transforming the variables into a rotating frame by setting

$$P = \tilde{P}\exp\{-i\nu t\} \quad \text{and} \quad b = \tilde{b}\exp\{-i\nu t\} \tag{11}$$

we drop the tilde and choose the coupling constant g independent of the atom μ to obtain

$$\dot{b} = -\kappa b - iP + F, \tag{12}$$
$$\dot{P} = -\gamma P + i|g|^2 Db + \Gamma, \tag{13}$$
$$\dot{D} = \gamma_{\parallel}(D_0 - D) + 2i(Pb^* - c.c.) + \Gamma_D. \tag{14}$$

Here the fluctuations are defined in correspondence to the transformations (9-11). For the time being we shall neglect these fluctuations and concentrate on the deterministic part of motion inherent to these equations. The following observation will become important for the further simplification of the still complicated equations (12-14): For a good cavity laser there exists a pronounced hierarchy in the damping constants

$$\kappa \ll \gamma_{\parallel} \ll \gamma. \tag{15}$$

It is of great value to note that these damping constants can be attributed to characteristic time scales on which the different variables change.

At this stage the most fundamental principle of synergetics, the slaving principle, comes into play. On the level of our heuristic considerations we are led to the conclusion that on macroscopic scales the slowest variable will rule the behavior of the complex many body system. Indeed we can use the observation expressed through (15) to approximately solve the equations for the fast variables. In a first step we formally integrate (13) to obtain

$$P(t) = i|g|^2 \int_{-\infty}^{t} \exp[-\gamma(t-\tau)]\, D(\tau)b(\tau)\, d\tau. \tag{16}$$

Here we have confined ourselves to the long term behavior to avoid transient solutions resulting from the initial conditions. Now we make use of the hierarchy in time scales (15). The characteristic time scales corresponding to the changes of the inversion and the field mode are much larger than that of the polarization. As a consequence we expect that the polarization has already reached an equilibrium value which, however, because of the nonlinear interactions is prescribed by the slow variation of the inversion and the field mode. Mathematically this fact can be expressed through the following approximation

$$P(t) \approx i|g|^2 D(t)b(t) \int_{-\infty}^{t} \exp[-\gamma(t-\tau)]\, d\tau. \tag{17}$$

Equation (17) reflects that the polarization follows the inversion and the field instantaneously (adiabatic principle). In other words, the long lived variables enslave the behavior of the short lived ones. Performing the elementary integration remaining in (17) we obtain

$$P(t) \approx i|g|^2 D(t) b(t)/\gamma. \tag{18}$$

Relation (18) yields a considerable simplification of the set of differential equations (12-14):

$$\dot{b} = -\kappa b + (|g|^2/\gamma) D(t) b(t) \tag{19}$$

and

$$\dot{D} = \gamma_\|(D_0 - D) - 4(|g|^2/\gamma) D(t) |b(t)|^2. \tag{20}$$

In a similar fashion we can again apply the idea of slaving by taking into account the first pair in the hierarchy (15). The formal integration of (20) yields

$$D = D_0 - (4|g|^2/\gamma) \int\limits_{-\infty}^{t} \exp[-\gamma_\|(t-\tau)] D(\tau) |b(\tau)|^2 d\tau. \tag{21}$$

This equation can be solved iteratively by a standard procedure for the solution of Volterra integral equations of the second kind. We start with

$$D \approx D^{(1)} + D^{(2)} + \ldots, \tag{22}$$

assuming the contribution of the integral in (21) to be a small quantity. By comparing different orders of magnitude we obtain

$$D^{(1)} = D_0 \tag{23}$$

and

$$D^{(2)} = -(4|g|^2/\gamma) \int\limits_{-\infty}^{t} \exp[-\gamma_\|(t-\tau)] D_0 |b(\tau)|^2 d\tau \tag{24}$$

$$\approx -\frac{4|g|^2}{\gamma\gamma_\|} D_0 |b(t)|^2. \tag{25}$$

In evaluating (24) we have again taken into account consequences of the different roles of slow and fast variables. Inserting the result (25) in the equation for the field mode (19) we finally obtain

$$\dot{b} = -(\kappa - |g|^2 D_0/\gamma)b - 4|g|^4 D_0(\gamma^2\gamma_\|)^{-1} |b|^2 b. \tag{26}$$

Equation (26) represents the basic result which we have obtained from a heuristic application of the slaving principle: The complete macroscopic time dependent action of the complex laser system can be exhaustively understood from an effective equation of motion of one variable, namely the complex field amplitude $b(t)$ of the surviving mode. In synergetics Hermann Haken has coined the notion '*order parameter*' for this most important variable. Its value determines the inversion D and the polarization P through (24) and (18), respectively. They play the role of enslaved variables. The instantaneous value of the order parameter can be considered as a qualitative and quantitative measure of the spontaneous macroscopic ordering of the system. In the case of a single mode laser the newly evolving stable state is observed above the so-called laser threshold which is marked by the condition

$$D_0 = \kappa \gamma |g|^{-2} . \tag{27}$$

2.6   Discussion of the results

The idea of slaving in connection with the occurrence of collective modes has considerably reduced the complicated set of original equations (5–7) to an analytically manageable equation of one complex order parameter $b(t)$. It therefore turns out that the slaving principle not only becomes of fundamental importance from a theoretical point of view but also provides us with a practically applicable tool to handle quite complex systems with comparatively simple nonlinear equations of motion: The spontaneous formation of macroscopic ordering far from thermal equilibrium is systematically mapped by time scale arguments onto the motion of – in general – few order parameters.

There are further general aspects which can be gained from the laser example. Here we especially mention the notion of non-equilibrium potentials – a concept introduced by Graham and Haken (1971a), Graham and Haken (1971b) and extended by Graham and his coworkers (e.g. Graham (1989)). The introduction of non-equilibrium potentials has been achieved through the observation that we can write (26) in the form

$$\dot{b} = -\frac{\partial V}{\partial b^*} , \tag{28}$$

if we choose V (up to an arbitrary constant)

$$V = (\kappa - |g|^2 D_0/\gamma)|b|^2 + 2|g|^4 D_0 (\gamma^2 \gamma_{\|})^{-1} |b|^4 . \tag{29}$$

Equation (28) can be interpreted as representing the overdamped motion of a particle in the potential $V(b, b^*)$. The minima of the potential become the stable stationary points of the system the maxima correspond to unstable ones. V therefore has all the properties of a Lyapunov function of the system. For this reason $V(b, b^*)$ rules the stability (and also the fluctuations) of the system. The shape of the potential V is changed characteristically by the variation of the control parameter D_0 (compare Fig. 10). The new ordered state above the laser threshold is reached through an instability and the minimum finally assumed is selected purely by chance (when, for simplicity, b is considered as a real variable). This reflects the role of the fluctuations. The transition can be classified as a symmetry breaking transition, where symmetry is restored through the action of the fluctuations. Following Haken we can give these observations the rank of principles: Ordered states are created through <u>instabilities</u> far from thermal equilibrium and <u>fluctuations</u> are responsible for which of the possible ordered states of a system will eventually be acquired in a particular realization.

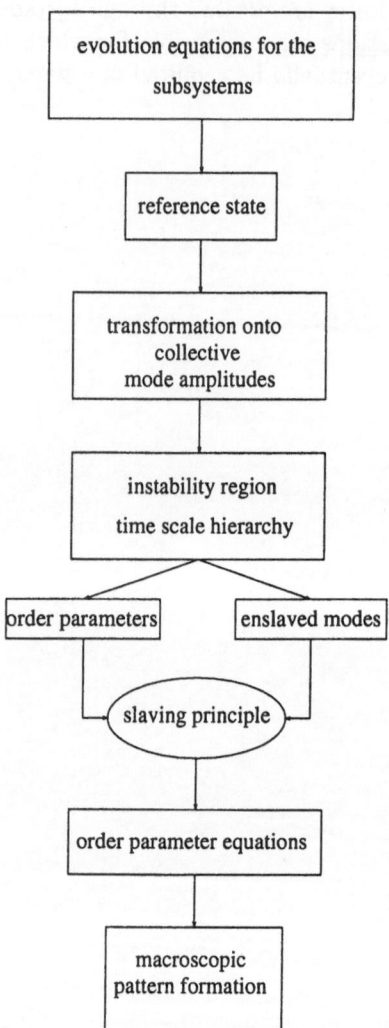

Fig. 10. Potential landscape for different values of control parameter σ.

2.7   Remarks on chaos in laser systems

We have taken the single mode laser instability, which is connected with the transition from a lamp to laser action, as the simplest representative of an instability leading from microscopic chaos (in the sense of Boltzmann) to the spontaneous macroscopic ordering of a system. However, there exist higher order instabilities Haken (1983), Haken (1970), Weiss and Vilaseca (1991). Furthermore there is a possibility to observe chaotic states of the single mode laser. In fact, when we choose the field mode $b(t)$ in (12-14) as a real variable and $P(t)$ as purely imaginary the equations can be mapped onto the Lorenz equations Haken (1975) by appropriately scaling and shifting the variables. When the cavity damping becomes large the conditions for the Lorenz instability can be met and a strange behavior of the laser light is yielded. Taking into account that b, P, and D are collective variables of the system, we may interpret this behavior as macroscopically chaotic, generated by the three collective variables of the system. They can again be considered as order parameters: All other degrees of freedom are enslaved by these few macroscopic variables.

3   Remarks on mathematical methods

3.1   Synergetics - from the mesoscopic level to macroscopic order

In the following we shall present an overview on the general mathematical method of synergetics for a situation where the details of the subsystems are completely known on a mesoscopic scale (compare Fig. 11).

Complete knowledge means that we know all the variables of the subsystems which we denote by $U_i (i = 1, 2 \ldots)$, the evolution of the single subsystems, as well as their mutual interactions. If we put together all the variables into a state vector $\mathbf{U}$ in a state space Γ

$$\mathbf{U} = (U_1, U_2, \ldots), \tag{30}$$

its time evolution is typically governed by an equation of motion of the following general type

$$\dot{\mathbf{U}} = \mathbf{N}(\mathbf{U}, \nabla, \{\sigma\}) + \mathbf{F}. \tag{31}$$

Here $\mathbf{N}$ denotes a nonlinear vector field in $\mathbf{U}$ which may depend on spatial inhomogeneities indicated through the ∇ - symbol as well as a set of external parameters which have been abbreviated by $\{\sigma\}$. $\mathbf{F}$ symbolically summarizes the presence of fluctuations which may be considered as a result of the action of the already supressed microscopic degrees of freedom. We note that this typical form of the evolution equations may be generalized in various ways: We also may include time delay-effects, non-local interactions in space, we can formulate them for time discrete processes, etc.

At this stage the first step of the systematic method is completed. It is this step which has to be performed by the specialized scientific disciplines. The following examples may clarify this statement. In laser physics the state vector is

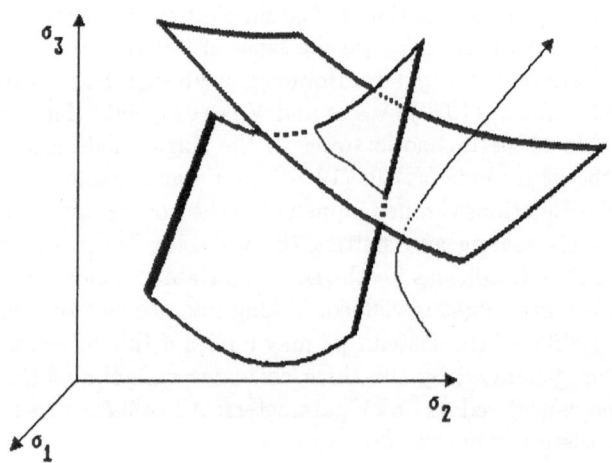

Fig. 11. The general method of synergetics to analyze the behaviour of complex systems.

built from the atomic variables, i.e. the polarization and the inversion of the single atoms and the amplitudes of the light field in a semiclassical approach. The equations of motion for these variables may, as already mentioned, be derived from first principles. In hydrodynamics we can consider the velocity field, temperature field, passive scalars etc. as further examples for variables constituting U and the equations of motion are then governed by conservation laws like the conservation of mass, momentum, energy, particle densities etc. In chemistry we may consider concentrations of the chemicals as appropriate variables and there interactions are provided through the special chemical reaction under consideration. The same still remains true for morphological models in biology where the concentrations and interaction of morphogenes may be considered along simliar lines. It is worthwhile to mention that even in social sciences such state vectors can be identified. In the following we wish to consider all these different systems from the unified view which has been introduced through the development of synergetics.

The second step consists in a further treatment of equations (31). we are not able to analyze the complete global behavior of the solutions of (31) analytically; even if we confine ourselves to the deterministic part which is contained in N. For this reason we shall concentrate ourselves to more simple but relevant subsets of the state space Γ. These are selected as to be of fundamental importance for an understanding of the macroscopic behavior of the underlying system: We consider low dimensional reference states. The most simple ones are stationary states which describe a homogeneous situation in time and space. Such states are usually referred to as the thermodynamic solution branch of the system. However, the method can be considerably generalized, i.e. we may even consider reference states which are not homogeneous in space or in time (periodic and quasi periodic states). For the moment beeing, however, we confine ourselves to a most simple reference state which is homogeneous in time and space. We may construct this state as a solution of

$$\dot{\mathbf{U}} = 0 \qquad \longrightarrow \qquad \mathbf{U} = \mathbf{U}_0. \tag{32}$$

Our third step then consists in the test of the stability of the selected state (32).

Because we may alter external conditions through a change in the external parameters the stationary state $\mathbf{U}_0$ may lose its stability. Mathematically this means that we are looking for instability regions in the parameter space spanned by $\{\sigma\}$. A general systematic method to achieve this goal is provided by a linear stability analysis. That is we analyze the behavior of small deviations $\mathbf{q}$ from the stationary state $\mathbf{U}_0$:

$$\mathbf{U} = \mathbf{U}_0 + \mathbf{q}. \tag{33}$$

We then immediately get the equation of motion for $\mathbf{q}$ from (31):

$$\dot{\mathbf{q}} = L\mathbf{q} + O\left(\|\mathbf{q}\|^2\right). \tag{34}$$

The elements of the linear matrix L are given through

$$L = (L_{ik}) \qquad \text{and} \qquad L_{ik} = \frac{\partial N_i}{\partial U_k}\bigg|_{\mathbf{U}=\mathbf{U}_0}. \tag{35}$$

Obviously the matrix L is still a function of the stationary state $\mathbf{U}_0$, may depend on gradients as well as on the external parameters $\{\sigma\}$. There is a well–known algebraic procedure to solve (34). The hyptesis

$$\mathbf{q} = \mathbf{q}_0 \exp(\lambda t) \tag{36}$$

transforms (34) into a linear algebraic eigenvalue problem from which we may calculate the set of eigenvalues λ_i and the corresponding eigenvectors which we denote by $\mathbf{o}_i$. We assume that they form a complete set. In the space of the external parameters instabilities are then indicated by the condition

$$\text{Re}(\lambda_i) = 0. \tag{37}$$

We may now summarize the information we get from our linear stability analysis. First we find the regions of instability in the space of the external parameters. Secondly we find the eigenvectors which we interpret as the collective modes of the system and finally we obtain locally the directions in Γ along which the stationary state may become unstable.

The fourth step of the systematic treatment consists in a complete nonlinear analysis of (31). We note that we may present the solutions of the full nonlinear equation in the form

$$\mathbf{U} = \mathbf{U}_0 + \mathbf{q} \quad \text{and} \quad \mathbf{q} = \sum_i \xi_i(t)\,\mathbf{o}_i\,. \tag{38}$$

We therefore can transform the original equation (31) into an equation for the amplitudes of the collective modes $\xi_i(t)$:

$$\dot{\xi}_i(t) = \Lambda_{ik}\xi_k(t) + H_i(\{\xi_i\}) + \tilde{F}_i\,. \tag{39}$$

Here Λ is a diagonal matrix or at least of the Jordan canonical form, H_i contains the complete nonlinear terms, and $\tilde{F}_i$ denotes the correspondingly transformed fluctuating forces. We note that the set of equations (39) is still exact.

We now arrive at the central step 5, the application of the slaving principle. As already noted slaving means that in the vicinity of a critical region in the space of external parameters only few modes, the so–called order parameters, dominate the behavior of the complex system on macroscopic scales. Obviously, even the linear stability analysis indicates amplitudes of the collective modes which finally will become the order parameters. Indeed, the linear movement yields a separation in time scales: The modes which become unstable will move on a very slow time scale because of $|\mathrm{Re}(\lambda_u)| = \dfrac{1}{\tau_u}$, where we have identified the index i with u to exhibit unstable behavior. In the vicinity of the critical point τ_u becomes extremely large. On the other hand the modes still remaining stable have a comparably short time scale which we denote by τ_s. Therefore close to the instability region we find a hierarchy in the time scales

$$\tau_u \gg \tau_s\,. \tag{40}$$

Taking into account the nonlinear interaction of stable and unstable modes we expect for the long term behavior of the system that the stable variables are moving in a way which is completely determined through the unstable modes, that is the order parameters. This observation can be put into mathematical terms. We may split the set of $\{\xi_i\}$ into a set of modes which will become unstable and denote it by the vector $\mathbf{u}$ and the remaining set of stable modes which we collect into the vector $\mathbf{s}$. Accordingly we split the set of equations (39) into

$$\dot{\mathbf{u}} = \Lambda_u \mathbf{u} + \mathbf{Q}(\mathbf{u},\mathbf{s}) + \mathbf{F}_u\,, \tag{41}$$
$$\dot{\mathbf{s}} = \Lambda_s \mathbf{s} + \mathbf{P}(\mathbf{u},\mathbf{s}) + \mathbf{F}_s\,. \tag{42}$$

Slaving of the stable modes through the order parameters now means

$$\mathbf{s} = \mathbf{s}_1(\mathbf{u},t) + \mathbf{s}_2\,. \tag{43}$$

(43) expresses the fact that the values of the stable modes are – besides some corrections $\mathbf{s}_2$ which arise from the fluctuations - completely determined by

the instantaneous values of the order parameters $\mathbf{u}$. With the relation (43) the complete set of equations (41) and (42) can be drastically simplified. In fact we may use this result to eliminate the stable modes from the equations of the order parameters to arrive at the equation

$$\dot{\mathbf{u}} = \lambda_u \mathbf{u} + \mathbf{Q}\left(\mathbf{u}, \mathbf{s}_1(\mathbf{u}, t) + \mathbf{s}_2\right) + \mathbf{F}_u. \tag{44}$$

This equation is the final order parameter equation. We observe that the idea of slaving generally leads to a drastic reduction of the degrees of freedom of the system which, however, now yields an exhaustive description of the system on macroscopic scales. The last step then consists in the solution of the order parameter equation (44) and in the identification of the macroscopically evolving ordered states which correspond to definite patterns or functioning of the system. The central problem which still remains to be solved rests in the construction of (43).

3.2 On the movement of the slaved modes

We are concerned with the construction of

$$\mathbf{s} = \mathbf{s}_1(\mathbf{u}, t) + \mathbf{s}_2 \, .$$

Our interest is devoted to the long term behavior of the slaved modes and we can formally integrate (43) in the following way

$$\mathbf{s} = \int_{-\infty}^{t} \exp\left(\Lambda_s(t - \tau)\right) \left[\mathbf{P}(\mathbf{u}, \mathbf{s}) + \mathbf{F}_s\right]_\tau d\tau \, . \tag{45}$$

Through (45) we define the operator $\left(\dfrac{d}{dt} - \Lambda_s\right)^{-1}$:

$$\mathbf{s} = \left(\frac{d}{dt} - \Lambda_s\right)^{-1} \left[\mathbf{P}(\mathbf{u}, \mathbf{s}) + \mathbf{F}_s\right] \, . \tag{46}$$

We consider the purely deterministic case where $\mathbf{F}_s$ vanishes and $\mathbf{s}_2$ correspondingly (generalizations are given in Springer Series in Synergetics (1977–92)). We further introduce the operators

$$\left(\frac{d}{dt}\right) = \mathbf{Q}\frac{\partial}{\partial \mathbf{u}} \tag{47}$$

and in addition

$$\left(\frac{d}{dt} - \Lambda_s\right)^{-1}_{(0)} \mathbf{P} = \int_{-\infty}^{t} \exp\left(\Lambda_s(t - \tau)\right) \mathbf{P}(\mathbf{u}, \mathbf{s})_t \, d\tau \, . \tag{48}$$

A partial integration in (46) is now performed by using the above defined operators in the following way:

$$\left(\frac{d}{dt} - \Lambda_s\right)^{-1} = \left(\frac{d}{dt} - \Lambda_s\right)^{-1}_{(0)} - \left(\frac{d}{dt} - \Lambda_s\right)^{-1} \left(\frac{d}{dt}\right) \left(\frac{d}{dt} - \Lambda_s\right)^{-1}_{(0)}. \quad (49)$$

We use equation (49) to construct a systematic iteration scheme to find the solution (43). To this end we apply the ansatz

$$\mathbf{s} = \sum_{n=2}^{N} \mathbf{C}^{(n)}, \quad (50)$$

where $\mathbf{C}^{(n)}$ is precisely of order n in $\mathbf{u}$. Correspondingly we may define $\mathbf{P}^{(n)}$ and $(d/dt)^{(n)}$. Inserting (49) and (50) into (45) we obtain the solution in the form

$$\mathbf{C}^{(n)} = \sum_{m=0}^{n-2} \left(\frac{d}{dt} - \Lambda_s\right)^{-1}_{(m)} \mathbf{P}^{(n-m)} \quad (51)$$

where we have defined

$$\left(\frac{d}{dt} - \Lambda_s\right)^{-1}_{(m)} = \left(\frac{d}{dt} - \Lambda_s\right)^{-1}_{(0)} \sum \prod_i \left[\left(\frac{-d}{dt}\right)^i \left(\frac{d}{dt} - \Lambda_s\right)^{-1}_{(0)}\right] \quad (52)$$

and the product over i has to be taken in such a way that $i \geq 1$ and $\sum i = m$. The sum means summing up over all different products. (52) provides us with a systematic method to construct (43) and it has meantime been generalized in various ways, especially by taking into account fluctuations, time–lag effects etc.

3.3   The need for phenomenological synergetics

The need for phenomenological synergetics is based on the observation that we meet many complex systems in nature which are built from complicated subsystems which are not known in detail Haken et al. (1983). Many examples of such systems are provided in biology where, for example, we may find single cells as subsystems constituting macroscopic functional systems such as the nervous system, or the like.

The possibility of a phenomenological approach is given by the fact that the macroscopic behavior of a system, as captured in the order parameter equations, can be independent from microscopic details of the subsystems. It has, therefore, definite advantages to make inferences about the order parameter equations from purely macroscopic data. This method can be regarded as a top down approach being complementary to the bottom up approach as presented in Sect. 3. (Fig. 12).

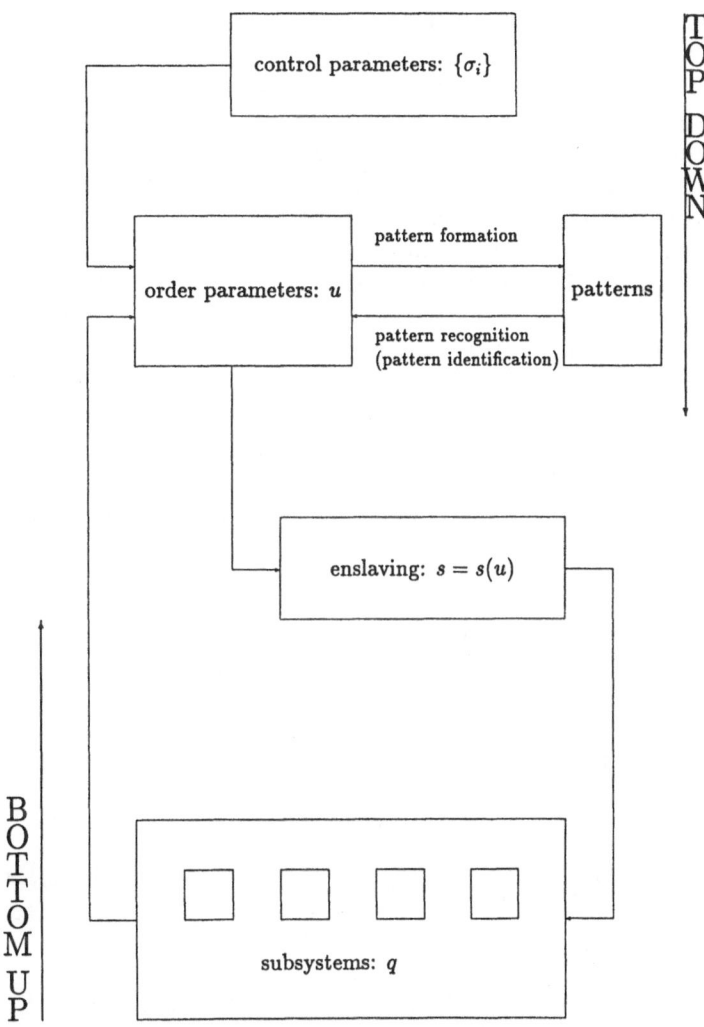

Fig. 12. Top down versus bottom up approach

The phenomenological strategy can be characterized as follows: We try to identify the relevant macroscopic quantities which are the control parameters and the order parameters. One may want to note that the number of control parameters which are altered at the same time, essentially determines the types of instabilities which can be observed on macroscopic scales. We consider a parameter space such as the one drawn in Fig. 13.

The instability regions are provided from the linearized theory of Sect. 3 by the condition $\mathrm{Re}\lambda_i = 0$. In our example (Fig. 13) this equation represents a two dimensional manifold in three dimensional parameter space $\{\sigma\}$. More generally, in an n–dimensional parameter space these are typically manifolds of dimension $n - 1$ or, in other words, of codimension 1. When we now change the control parameters along a 1–dimensional manifold (which is equivalent to the change of one control parameter), we will typically meet a single manifold as indicated in the figure but not a cut of two manifolds. Indeed, the slightest disturbance would remove us from such a cut. We therefore conclude that when we change only one control parameter we typically have the situation that only one eigenvalue in the real case and two conjugate complex eigenvalues in the case of complex eigenvalues can become unstable. Other instabilities will not be observed in reality.

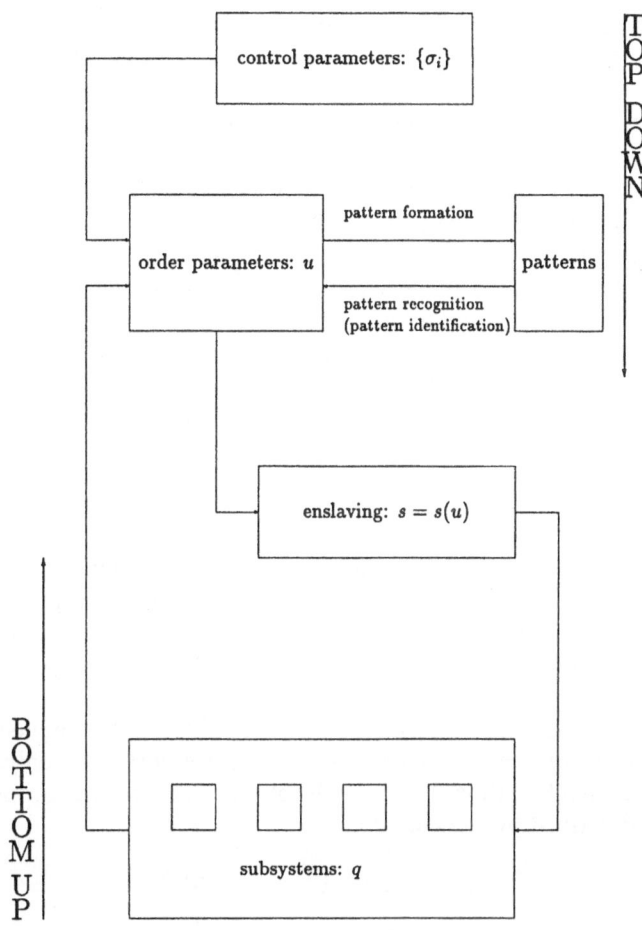

Fig. 13. Critical regions in three dimensional parameter space $\{\sigma\}$.
The two dimensional manifolds have been drawn for two possible order parameters.

It turns out that it is possible to give a precise mathematical form of these possible instabilities in the nonlinear case. This is provided by the mathematical method of normal forms Arnold (1983) which leads to the simplest mathematical structures which qualitatively completely describe the dynamics of our system near the instability point. We should mention, however, that when we change more than one control parameter, the number of different types of possible instabilities increases. This implies that we cannot give the normal form uniquely but have to consider different types of instabilities. In order to test our inferences about the underlying mathematical structure, we have to qualitatively compare predicted macroscopic behavior of the system with actual experimental observations. This method has been applied to, e.g., social problems Wischert and Wunderlin (1993), Wunderlin and Haken (1984), to the coordination of human hand movements Haken et al. (1985), in EEG analysis Friedrich and Uhl (1992), etc. We shall exemplify this approach by applying it to an economical problem, namely the crash and survival of a business enterprise.

A simple model from economics We consider a company and infer an important control parameter to be the amount of capital which constitutes its financial assets. For the company we identify as an important order parameter the making of profit as gained by producing goods or by providing services. In this simplified model, we shall study the nonlinear behavior of the inferred order parameter (making profit) when the control parameter (the amount of capital) is changed.

When the method of normal forms (Arnold (1983)) is applied, the only instability which can occur if we consider one control parameter only, is the so–called saddle–node bifurcation. When we denote the order parameter by u its dynamics is then described by the equation

$$\dot{u}(t) = (\sigma - \sigma_c) - u(t)^2 . \tag{53}$$

Before we discuss the simple mathematical properties of this equation we want to emphasize what we have gained at this step. When we change one control parameter under the above assumptions the whole behavior of the complex system company is completely determined by such a simple equation of motion for the order parameter. It is important to note that eq. (53) is a nonlinear equation which contains much more information than purely linear extrapolation could do. We substantiate this by first looking for the stationary solutions u_0 of this equation, which are characterized by a vanishing time derivative. They are given by the following formula

$$u_0^{\pm} = \pm\sqrt{\sigma - \sigma_c} . \tag{54}$$

We observe that there is no stationary solution if σ is smaller than σ_c. This result which rests in the nonlinearity of the equation, can be interpreted as follows: There is a minimal amount of capital needed so that the company can exist. Beyond σ_c two different solutions emerge, where it turns out that u_0^{+} is a stable

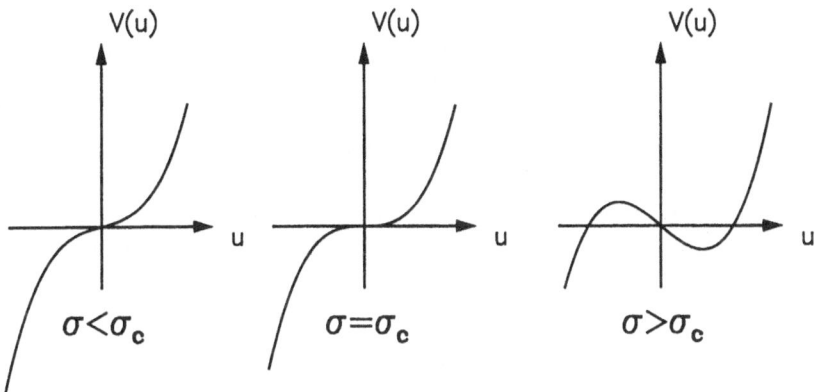

Fig. 14. Potential for the saddle–node bifurcation for different values of σ.

solution (disturbances are damped) whereas u_0^- is an unstable solution (small deviations are enhanced). Again, this has a simple interpretation. A company can only be stationary stable when it makes profits (positive value of the order parameter) and it will become unstable when it produces losses. The nonlinear effect here is concerned with the behavior in the neighbourhood of σ_c. One observes that the stationary state crashes fast with small changes in the control parameter (the slope of our curve goes to infinity).

To get some insight into the dynamical behavior we write (53) as

$$\dot{u}(t) = -\frac{d}{du}V(u)\,. \qquad (55)$$

This allows us, following Haken (1983), Haken (1987), to interpret the dynamics of the order parameter in terms of the overdamped motion of a particle in the potential V which is given by

$$V(u) = -(\sigma - \sigma_c)u + \frac{1}{3}u^3\,, \qquad (56)$$

where we put the arbitrary constant equal to zero. The potential V is drawn in Fig.(14) for three different values of σ.

In the case $\sigma < \sigma_c$ no minimum appears in the potential which means that the company will crash for any initial condition $u(0)$. In the case $\sigma = \sigma_c$ we have a turning point of the potential at the origin with slope zero. This indicates the appearance of a rest point at the origin. The smallest disturbance, however, still yields a crash of the company, i.e., our particle escapes into $-\infty$. The situation changes dramatically when we consider the third case $\sigma > \sigma_c$. The stationary points of the bifurcation diagram are now represented by the minimum (u_0^+) and the maximum (u_0^-) of the potential curve. It remains of relevance to note that not each initial condition of the order parameter leads to the stable minimum and a crash still remains possible. Again, this has a simple and nice interpretation. Consider a situation where σ is smaller then σ_c and the particle is starting to escape to $-\infty$. If we were to avoid a crash we have to very quickly feed capital into the company in order to get a situation which is represented by our third case. The conclusion, then, is that it is not always possible to avoid the crash. This is a property of the nonlinear behavior of the order parameter. We observe that our nonlinear eq. (55) does indeed qualitatively reflect phenomena which can be observed in reality.

4   Delay–induced instabilities

Here we shall present an important generalization of the synergetic method developed in section 3 to a class of functional differential equations. As a special example we consider so–called nonlinear delay equations, where we can observe delay–induced instabilities Wischert (1993), Wischert et al. (1994).

Delay–induced instabilties provide a powerful tool in the investigation of regular and irregular behavior observed in quite different disciplines. Delay effects especially appear in radio engineering sciences, in the domain of optical bistable devices and in engineering sciences. They play an important role in physiological control systems and have also been applied to economic systems and to cognitive sciences.

4.1   Some properties of delay differential equations

The dynamical behavior of a nonlinear system is as usual described by a state vector $\mathbf{U}(t)$ in an n-dimensional state space Γ. When the time evolution of the state vector is influenced by a delay the autonomous delay differential equation assumes the form:

$$\dot{\mathbf{U}}(t) = \mathbf{N}\left(\mathbf{U}(t), \mathbf{U}(t-\tau), \{\sigma_i\}\right). \tag{57}$$

Again $\mathbf{N}$ denotes a nonlinear vector field which depends on the state vector $\mathbf{U}$ at the times t and $t-\tau$, where τ stands for the time delay, and a parameter set $\{\sigma_i\}$ which serves for measuring external influences on the system. Furthermore we assume the control parameters $\{\sigma_i\}$ being kept fixed during the observation time so we can omit them in our notation.

To gain solutions of eq. (57) at times $t \geq 0$ it is necessary to define the state vector $\mathbf{U}(t)$ in the entire interval $[-\tau, 0]$. Therefore we have to consider eq. (57) together with the initial condition

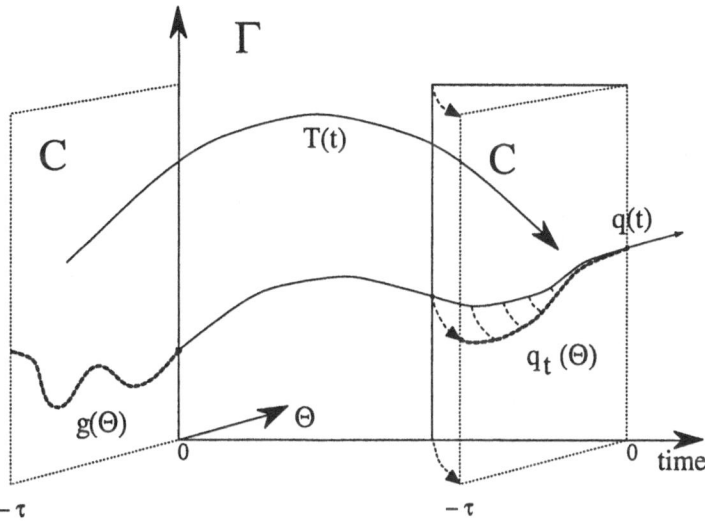

Fig. 15. Folding mechanism relating a history of $\mathbf{q}(t) \in \Gamma$ to a single point $\mathbf{q}_t \in \mathcal{C}$.

$$\mathbf{U}(\Theta) = \mathbf{g}(\Theta), \qquad -\tau \leq \Theta \leq 0, \tag{58}$$

where $\mathbf{g}$ is a given continuous initial vector valued function in a suitable function space $\mathcal{C}$.

Here we have to remark, that the initial value problem given by eqs. (57) and (58) represents an unsatisfactionary situation in the sense, that the vector valued function $\mathbf{g}$ is mapped onto a trajectory in the n-dimensional state space Γ. Such a mapping from an infinite dimensional function space onto a finite dimensional vector space is accompanied by a considerable loss of information. Different initial vector valued functions can lead to crossing of corresponding trajectories in Γ. This means that the uniqueness of solutions cannot be assured when we restrict our attention to the state space Γ.

To get rid of this problem we have to reformulate our description of the delay system. This can be performed by extending the finite dimensional state space Γ to an infinite function space $\mathcal{C}$ where the initial vector valued function $\mathbf{g}$ is defined. This point of view enables us to describe the state of the delay system at time t by an extended state vector $\mathbf{U}_t \in \mathcal{C}$ and to refer to $\mathcal{C}$ as extended state space. We therefore construct $\mathbf{U}_t$ by adjusting the trajectory $\mathbf{U}(t) \in \Gamma$ in the interval $[t - \tau, t]$ according to the prescription (compare Fig. 15)

$$\mathbf{U}_t(\Theta) = \mathbf{U}(t + \Theta), \qquad -\tau \leq \Theta \leq 0. \tag{59}$$

We observe that we can now retain uniqueness by considering the motion of $\mathbf{U}_t(\Theta)$. In order to formulate the evolution equation in the extended state space we define a time evolution operator $\mathcal{T}(t)$ in the following way

$$\mathbf{U}_t(\Theta) = (\mathcal{T}(t)\mathbf{g})(\Theta), \qquad -\tau \le \Theta \le 0. \tag{60}$$

The uniqueness of this mapping implies that $\mathcal{T}(t)$ has the properties of a semigroup.

The original initial value problem can now be formulated in the form

$$\frac{d}{dt}\mathbf{U}_t(\Theta) = (\mathcal{A}\mathbf{U}_t)(\Theta), \qquad -\tau \le \Theta \le 0. \tag{61}$$

The $\mathcal{A}$ represents the infinitesimal generator defined by

$$(\mathcal{A}\mathbf{U}_t(\Theta)) = \lim_{\varepsilon \to 0} \frac{1}{\varepsilon}\left[(\mathcal{T}(\varepsilon)\mathbf{U}_t(\Theta) - \mathbf{U}_t(\Theta)\right]. \tag{62}$$

A lengthy but straightforward calculation which has been performed in Wischert (1993), Wischert et al. (1994) and is based on identities like

$$\mathbf{U}(t-\tau) = \int\limits_{-\tau}^{0} d\Theta \delta(\Theta + \tau)\,\mathbf{U}_t(\Theta),$$

leads to the following result for the infinitesimal generator

$$(\mathcal{A}\mathbf{U}_t)(\Theta) = \begin{cases} \dfrac{d}{d\Theta}\mathbf{U}_t(\Theta), & -\tau \le \Theta \le 0, \\ \mathcal{N}[\mathbf{U}_t(\cdot)], & \Theta = 0. \end{cases} \tag{63}$$

The vector valued functional $\mathcal{N}$ can be expanded in powers of $\mathbf{U}_t$ according to

$$\mathcal{N}[\mathbf{U}_t(\cdot)] = \sum_{k=1} \mathcal{N}^{(k)}[\mathbf{U}_t(\cdot)], \tag{64}$$

with

$$\mathcal{N}_i^{(k)}[\mathbf{U}_t(\cdot)] = \int\limits_{-\tau}^{0} d\Theta_1 \dots \int\limits_{-\tau}^{0} d\Theta_k\, w_{i,j_1,\dots,j_k}^{(k)}(\Theta_1,\dots,\Theta_k)\,\mathbf{U}_{t,j_1}(\Theta_1)\dots\mathbf{U}_{t,j_k}(\Theta_k). \tag{65}$$

The matrix valued $w^{(k)}$ are combinations and products of δ-functions which result from the identities mentioned above. We note, however, that the form of this functional equation could also describe more general delay systems.

The result (61) allows a systematic application of synergetic concepts. We note that this was not possible in the original formulation of the equations given in (57).

4.2   Stability analyses of the reference state

We assume that there exisits a stationary state $\mathbf{U}_0$ which is homogeneous in time. The state must be a solution of the equation

$$\mathcal{N}\left[\mathbf{U}_0\right] = 0. \tag{66}$$

In the next step we analyse the behavior of the system in a certain neighborhood of the chosen reference state by considering small deviations $\mathbf{q}_t$ from $\mathbf{U}_0$. Inserting

$$\mathbf{U}_t(\Theta) = \mathbf{U}_0 + \mathbf{q}_t \tag{67}$$

into eq. (61). We obtain in the linear approximation

$$\frac{d}{dt}\mathbf{q}_t(\Theta) = \left(\mathcal{A}_L \mathbf{q}_t\right)(\Theta), \tag{68}$$

$$\left(\mathcal{A}_L \mathbf{q}_t\right)(\Theta) = \begin{cases} \dfrac{d}{d\Theta}\mathbf{q}_t(\Theta), & -\tau \le \Theta \le 0, \\ \mathcal{L}[\mathbf{q}_t(\cdot)], & \Theta = 0, \end{cases} \tag{69}$$

$$\mathcal{L}[\mathbf{q}_t(\cdot)] = \int\limits_{-\tau}^{0} d\Theta\, w(\Theta)\, \mathbf{q}_t(\Theta). \tag{70}$$

$\mathcal{A}_L$ denotes the infinitesimal generator restricted to the linear case and the vector valued functional $\mathcal{L}$ is the linear approximation to $\mathcal{N}$ in the neighborhood of $\mathbf{U}_0$. Its matrix valued density w is given as functional derivative of $\mathcal{N}$ evaluated at the reference state

$$w(\Theta) = \frac{\delta \mathcal{N}[\mathbf{q}_t(\cdot)]}{\delta \mathbf{q}_t(\Theta)}\bigg|_{\mathbf{q}_t=0}. \tag{71}$$

In order to get the linear eigenvalues we put the ansatz

$$\mathbf{q}_t(\Theta) = \boldsymbol{\phi}^\lambda(\Theta)e^{\lambda t} \tag{72}$$

into the linear evolution equation (68) and obtain the following eigenvalue problem for the infinitesimal generator $\mathcal{A}_L$

$$\left(\mathcal{A}_L \boldsymbol{\phi}^\lambda\right)(\Theta) = \lambda \boldsymbol{\phi}^\lambda(\Theta). \tag{73}$$

Evaluating this eigenvalue problem in the interval $[-\tau, 0)$ and taking into account the definition of the infinitesimal generator in eq. (69) we obtain for the right-hand eigenfunctions $\boldsymbol{\phi}^\lambda$ the solution

$$\boldsymbol{\phi}^\lambda(\Theta) = \boldsymbol{\phi}^\lambda(0)e^{\lambda\Theta}, \tag{74}$$

for arbitrary values of λ. The eigenvalues are determined when we wish to guarantee that eq. (74) satisfies the eigenvalue problem eq. (73) for the single value $\Theta = 0$. Here the eigenvalue equation assumes the form

$$W(\lambda)\boldsymbol{\phi}^\lambda(0) = \lambda \boldsymbol{\phi}^\lambda(0), \tag{75}$$

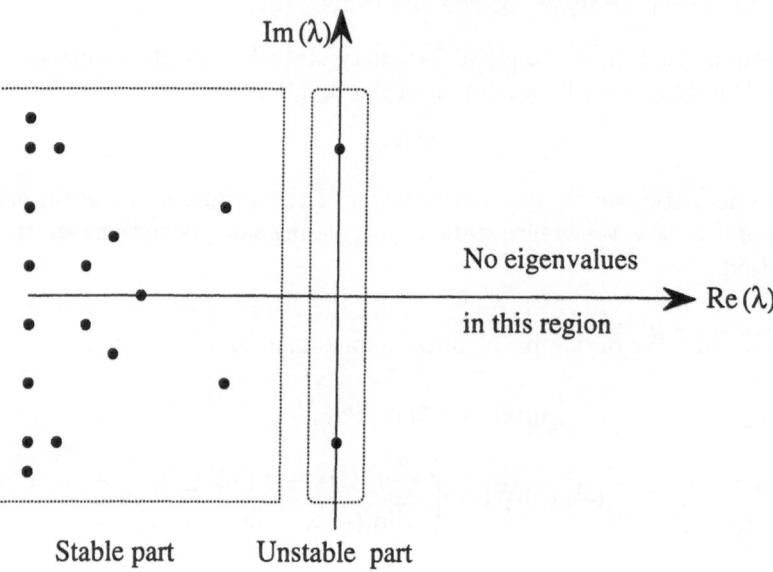

Fig. 16. Schematic representation of the spectrum of a delay differential equation at an instability.

where $W(\lambda)$ is given by

$$W(\lambda) = \int\limits_{-\tau}^{0} d\Theta w(\Theta) e^{\lambda\Theta}. \qquad (76)$$

The corresponding characteristic equation for the eigenvalues λ is given by

$$\det(W(\lambda) - \lambda I) = 0, \qquad (77)$$

where I denotes the unity matrix.

Without proof we summarize some properties of the solutions to the characteristic equation. They correspond to a pure point spectrum (Fig. 16) with an infinite number of eigenvalues. In each finite strip, however, which is parallel to the imaginary axis only a finite set of eigenvalues is located. The eigenvalues accumulate for $\mathrm{Re}(\lambda) \to -\infty$. Finally we note that the real part of the eigenvalues is bounded from above.

In order to formulate the adjoint eigenvalue problem we have to define a suitable scalar product. It turns out that the situation in our case is more sophisticated: We are only able to define a bilinear form which only partly fulfills the properties of a scalar product. However, it can be shown that this bilinear form allows for the definition of projectors which guarantee for the projection onto invariant subspaces of the linear problem. The bilinear form has the explicit form

$$(\psi^t, \phi) = \psi^t(0)\phi(0) - \int\limits_{-\tau}^{0} \int\limits_{0}^{\Theta} ds\psi^t(s - \Theta)w(\Theta)\phi(\Theta) \tag{78}$$

for all $\psi^t \in C$ and $\phi \in C$.

The adjoint eigenvalue problem can now be formulated along similar lines as the original one. We find that the righthand and lefthand eigenvectors corresponding to the original and the adjoint problem, respectively form a biorthogonal set which fulfills the relations

$$\left(\psi^{t\lambda}, \phi^{\mu}\right) = \delta_{\lambda\mu}, \tag{79}$$

when they have been properly normalized.

4.3   Nonlinear treatment near instabilities

In general it appears to be impossible to present a general solution method for nonlinear functional differential equations. However, in the case of ordinary differential equations the concept of order parameters and enslaved modes (see section 3) provides for a powerful tool to formulate an equivalent simpler nonlinear problem which allows for the discussion of qualitative properties close to instabilities. This motivates us to generalize this concept to functional differential equations.

We shall start from the complete nonlinear problem for the deviations from the reference state $q_t(\Theta)$. Splitting the vector valued functional into its linear and strictly nonlinear part we obtain the functional differential equation

$$\frac{d}{dt}q_t(\Theta) = (A_L q_t)(\Theta) + X_0(\Theta)\mathcal{N}[q_t(\cdot)]. \tag{80}$$

Here X_0 denotes a matrix valued function with the properties

$$X_0(\Theta) = \begin{cases} 0, & -\tau \leq \Theta < 0 \\ I, & \Theta = 0. \end{cases} \tag{81}$$

In the following we shall discuss this equation close to an instability. We consider a situation where the control parameters are chosen in a way that the spectrum of the infintesimal generator is bounded from above by the imaginary axis. When we now appropriately change the control parameters two conjugate complex eigenvalues may cross the imaginary axis. Indeed we know from the spectral properties of the linear infinitesimal generator that only a finite number of modes can become unstable.

We can now perform a decomposition of the state space C into a finite dimensional subspace $\mathcal{U}$ spanned by the unstable modes and an infinite dimensional subspace S corresponding to the remaining stable modes. To this end we define the projector

$$\mathcal{P}_{u\cdot} = \Phi_u(\Theta)\left(\Phi_{u\cdot}^t, \cdot\right). \tag{82}$$

Here the matrix $\Phi_u(\Theta)$ is defined by

$$\Phi_u(\Theta) = \left(\phi^{\lambda_1}(\Theta)\cdots\phi^{\lambda_m}(\Theta)\right). \tag{83}$$

The adjoint matrix Ψ_u^t is defined correspondingly. In the subspace of the unstable modes they clearly fulfill the relation

$$\left(\Phi_u, \Psi_u^t\right) = I. \tag{84}$$

We note that the projection onto the stable directions can be performed by a projector $\mathcal{Q}_s$ which is given by

$$\mathcal{Q}_s = (I - \mathcal{P}_u). \tag{85}$$

In correspondence to the projection we can split the state vector into two parts

$$\mathbf{q}_t(\Theta) = \mathbf{U}_t(\Theta) + \mathbf{s}_t(\Theta). \tag{86}$$

By the solution of the linear problem the following ansatz is suggested for the unstable part, (compare the general discussion presented in Wischert et al. (1994)).

$$\mathbf{U}_t(\Theta) = \Phi_u(\Theta)\mathbf{u}(t). \tag{87}$$

By using the properties of the projectors we end up with the following set of equations

$$\frac{d}{dt}\mathbf{u}(t) = \Lambda_u \mathbf{u}(t) + \Psi_u^t(0)\mathcal{N}\left[\Phi_u(\cdot)\mathbf{u}(t) + \mathbf{s}(t)\right], \tag{88}$$

$$\frac{d}{dt}\mathbf{s}_t(\Theta) = (\mathcal{A}_L\mathbf{s})(\Theta) + \left[X_0(\Theta) - \Phi_u(\Theta)\Psi_u^t(0)\right]\mathcal{N}\left[\Phi_u(\cdot)\mathbf{u}(t) + \mathbf{s}(\cdot)\right]. \tag{89}$$

We now introduce the fundamental concept of slaving by using the ansatz

$$\mathbf{s}_t(\Theta) = \mathbf{h}(\Theta, \mathbf{u}(t)). \tag{90}$$

Here we assume as usual that $\mathbf{h}$ and its first derivative with respect to $\mathbf{u}$ vanish at $\mathbf{u} = 0$. One now can give an approximate solution for $\mathbf{h}$ by using the method of adiabatic elimination (Wischert et al. (1994)). The final result for the order parameter equation then reads

$$\frac{d}{dt}\mathbf{u}(t) = \Lambda_u \mathbf{u}(t) + \Psi_u^t(0)\mathcal{N}\left[\Phi_u(\cdot)\mathbf{u}(t) + \mathbf{h}(\cdot, \mathbf{u}(t))\right]. \tag{91}$$

Our result consists in an approximate order parameter equation which no longer contains any delay terms. This means that we could drastically simplify our set of delay equations when we are close to an instability. The original functional differential equation has been reduced to an ordinary differential equation. We note that this became only possible by using the concept of extended state space.

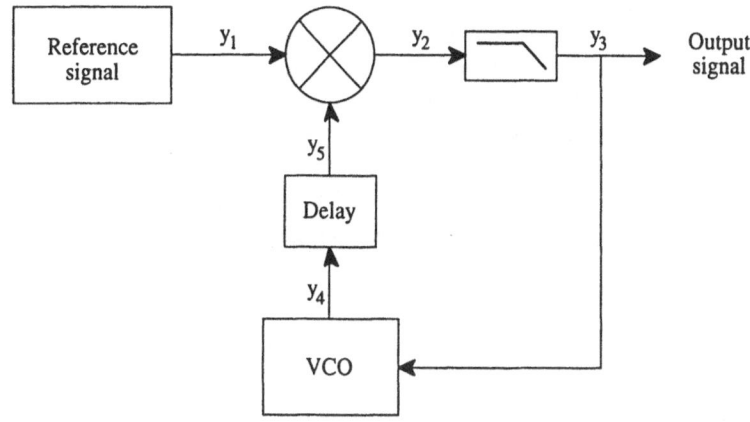

Fig. 17. A PLL–circuit as a realiziation of a delay differential equation.

4.4 An example

Here we mention an example which can be physically realized by a first order phase–locked loop (compare Fig. 17).

It can be shown that the phase variable ϕ which measures the difference of the phase of the two oscillators indicated in the figure obeys an equation of the form

$$\dot{\phi}(t) = -K \sin [\phi(t - \tau)] \tag{92}$$

when both oscillators are in resonance. K denotes a constant which represents the properties of the phase–locked loop. To analyze this equation we choose as a reference state

$$\phi_0 = l\pi \tag{93}$$

where l is an arbitrary integer. The stable stationary states $\phi_0 = 0 \ (\text{mod} 2\pi)$ loose their stability when the delay time τ exeeds the value $\tau = \pi/2K$. Applying the general methods and using a rotating wave approximation we find the following order parameter equation for the complex order parameter $u(t)$

$$\frac{d}{dt}u(t) = \lambda^+ u(t) - b |u(t)|^2 u(t) \tag{94}$$

which can be identified as the normal form of a Hopf bifurcation. A detailed calculation is given in Wischert (1993), Wischert et al. (1994).

5 Conclusions

We have presented the general methods of synergetics in connection with the central problem of self–organization. The example of laser action, where the evolution equations for the microscopy are known, was worked out in some detail and showed that the ideas of synergetics can be tested. An example from economy was given to introduce the idea of a 'phenomenological' synergetics, i.e., the synergetic understanding of systems with a (largely) unknown microscopy. These examples, and the plethora of other examples in the literature, lead us to the conclusion that the methods of synergetics offer a powerful starting point in trying to understand the behavior of complex systems from a universal point of view. It is our impression that this idea has gained rather general acceptance over the last decades.

References

Arnold, V.I. (1983): *Geometrical Methods in the Theory of Ordinary Differential Equations* (Springer, New York)

Chanderasekhar, S. (1981): *Hydrodynamic and Hydromagnetic Stability* (Dover Publications Inc., New York)

Friedrich, R., Bestehorn, M., Haken, H. (1990): Int. J. Mod. Phys. **B 4**, 365

Friedrich, R, Uhl, C. (1992): Synergetic analysis of human electroencephalograms: Petit–mal epilepsy, in *Evolution of Dynamical Structures in Complex Systems*, eds. R. Friedrich, A. Wunderlin (Springer, Berlin)

Graham, R. Haken, H. (1971): Z. Physik **243**, 289

Graham, R., Haken, H. (1971): Z. Physik **245**, 141

Graham, R. (1989): *Noise in nonlinear dynamical Systems* eds. F. Moss, P.V.E. Clintock (Cambridge University Press, New York)

Haken , H. (1983): *Synergetics: An Introduction* (Springer, Berlin)

Haken, H. (1987): *Advanced Synergetics* (Springer, Berlin)

Haken, H. (1985): *Light, Vol. 2, Laser Light Dynamics*, (North-Holland Physics Publishing)

Haken, H. (1970): *Laser Theory*, in Encyclopedia of Physics, Vol. XXV/2C, ed. S. Flügge (Springer, Berlin)

Haken, H., Kelso, J.A.S., Bunz, H. (1985): Biol. Cybern. **51**, 347

Haken, H., Wischert, W., Wunderlin, A., Meijer, O. G. (1993): Introduction to Synergetics. in *Some Physico–chemical and Mathematical Tools for Living Systems Understanding*, eds. H. Greppin, M. Bonzon, and R. Degli Agosti (University of Geneva)

Haken, H. (1975): Z. Physik **B 21**, 105

Landau, L. D., Lifshitz, E. M. (1952): *Course of Theoretical Physics*, Vol. V (Pergamon Press, London)

Springer Series in Synergetics, (Vol. 1–57), ed. H. Haken, Springer, Berlin (1977–92)

Weiss, C. O., Vilaseca, R. (1991): *Dynamics of Lasers* (Weinheim)

Wischert, W. (1993): *Anwendung synergetischer Konzepte auf Selbstorganisationsprozesse in zeitlich verzögerten Systemen*, Reihe Physik, Verlag Shaker, Aachen

Wischert, W., Wunderlin, A. (1993): On the application of synergetics to social systems. In *Interdisciplinary Approaches to Nonlinear Complex Systems*, ed. H. Haken (Springer, Berlin)

Wischert, W., Wunderlin, A., Pelster, A., Olivier, M., Groslambert, J. (1994): Delay–Induced Instabilities in Nonlinear Feedback Systems. Phys. Rev. E. **49**, 203

Wunderlin, A. (1987): On the Slaving Principle. In *Lasers and Synergetics*, eds. R. Graham and A. Wunderlin (Springer, Berlin)

Wunderlin, A., Haken, H. (1984): Some applications and models of synergetics to sociology. In Springer Series in Synergetics: *From Microscopic to Macroscopic Order*, ed. E. Frehland (Springer, Berlin)

Discussion of Arne Wunderlin's Paper

Batterman, Chirikov, Miller, Schurz, Thaler*, Weingartner, Wunderlin

Chirikov: What is the - how did you call it - "synergetic computer", you mentioned at the beginning?

Wunderlin: The idea of a synergetic computer is based on the methodes of synergetics, namely the concept of order parameters and slaving. As an example we consider a device which is trained to recognize and discriminate, for example, faces of several people. These faces are implemented into the device by putting a grid on the faces of say 60 to 60 pixels and attribute a certain grey value to each of these pixels. These pixels are considered as representing a high dimensional vector space. Within this vector space the pictures which have to be recognized form a low dimensional subspace. This low dimensional subspace is identified with the space of the order parameters. When a picture is offered to that device a gradient dynamic is implemented to the device which drives the corresponding state vector of this picture quickly into the subspace of the order parameters and in that subspace to a minimum which just represents the offered pattern. It becomes obvious that the device can act as an associative memory. Differences between the offered pattern and the pattern which can be recognized are ruled out by the gradient dynamics. The latter can be considered to note that one is able to implement invariances to the device which are concerned with a shift and a rotation of the patterns which have to be recognized.

Thaler: What do you mean by chaotic behavior and how do you measure it?

Wunderlin: We consider here low-dimensional chaos. This can be tested and characterized by the dimension of the attractor, the positive Lyapunov-exponents and by the universal scenario which yields chaotic behaviour (if it exists).

Miller: Can you say something more about the higher level laws, which emerge - to use that word - and are not deducible from the microscopic theories? Are there local laws pertaining to particular systems, or are they universal and applying to all systems of the same type? I am interested in whether there could emerge in one system something general, but in another system something quite different. Is that what you have in mind?

Wunderlin: What happens here is that when one starts from the microscopic level by this systematic methods one cannot derive these macroscopic classification laws. This classification is based on the notion of topological equivalence. This idea of topological equivalence is really a new form of law - I think - which you cannot deduce from the microscopic properties but which are consistent with the macroscopic properties. The corresponding forms are a list of all these possible universal law forms. And that it is possible that these normal forms qualitatively describe the critical behaviour of a large class of systems. Not only that class but also the representation by classes itself.

---

* Maximilian Thaler, Institut für Mathematik, Universität Salzburg, Austria.

Miller: Statistical laws are normally thought to be laws that cannot be derived from the microscopic basis.

Wunderlin: There are models how to do it.

Thaler: What do you mean by intermittency?

Wunderlin: Intermittency describes a pathway to chaos. Roughly one may say that one has a regularly behaving system and starts to change an external parameter. Then the regular behaviour may become unstable and burst of irregular behaviour appear. After a burst the system behaves again regularly up to the moment when a new burst occurs. Changing the parameter further the time intervalls of regular behaviour become shorter and shorter and the final behaviour is completely chaotic.

Chirikov: But this is a particular type of intermittency. Could you say in general that the intermittency is random patterns which is a typical situation in a complex system. This is essential for what you call intermittency, you agree?

Wunderlin: Yes, I would expect that.

Schurz: I have a conceptual question: Assume a system of the sort you described consisting of subsystems being open you have some control parameters which you can change in degree. Do you think that every system of that kind exhibit the behavior you described, namely that there is first microscopic chaos and then there are macroscopic ordering structures and then there is a low dimension chaos? Do you think that it holds generally for all kinds of systems in nature or do you think that for some systems no macroscopic ordering will develop?

Wunderlin: We can only give our necessary conditions for the chaos but we do not have sufficient conditions.

Schurz: What are these necessary conditions?

Wunderlin: One example we saw when there are parameter regions where you miss the ordered states. It directly goes into chaotic behaviour. But I think nobody is able to give sufficient conditions.

Schurz: It would be philosophically interesting because it is connected with the questions for instance: is it an accident or not an accident that there exists intelligence and evolution in the universe?

Miller: There could be an art in choosing control parameters, for instance. Is it more than that? A casual description says that systems that start off very similar may evolve very differently. The question is: what is it for them to start off similarly? One might very well say that because the systems evolve so differently they clearly were different in the beginning. In what way does one get the intuitive feeling, that the chosen control parameters are the right ones?

Wunderlin: The reason why I called this art is the observation that the medical doctor with great experience can identify very well an EEG. And we have a lot of difficulties to reproduce this. The idea was to construct an apparatus which makes it much more probably that the doctor or at least we are independent of the quality of the doctor and therefore you must find the leading modes which behave usually chaotic. But which modes are mainly important for these dynamics? What we take here is a method developed by Karhunen and Loeve which allows from a linear approximation to find the most prominent modes of the system.

Miller: Are there then examples where it is not known how to do this?

Wunderlin: Yes, more than examples where it is known.

Miller: It sounds self-reinforcing in a way, appealing to the work of experienced people and so on.

Wunderlin: There are also people who try to apply this to social processes, for example the formation of opinion might also have such critical situations.

Miller: I am interested to see whether the different guesses could be tested. Are there crucial experiments?

Wunderlin: I do not know about them in social systems. But what I want to emphasize is that you can make predictions in the form of well-defined possibilities which appear to me more valuable than a linear extrapolation.

Weingartner: You said that these new laws which emerge from the synergetics on the macroscopic level - they are consistent, you say - with the basic laws of quantum mechanics or on the microscopic level at least. I am just wondering what you mean by "consistent". Your claim is stronger of course if there is already a certain dependency here. But if they are quite independent then the claim that they are consistent with each other is not very strong because it would just mean so far they don't touch each other or so far we cannot connect them or something like that. Or is it just so that so far no difficulty has been discovered, so far no consequence of these new laws was coming into conflict with what we know as well corroborated microscopic laws. So I was just wondering whether you could be a little bit more detailed on this point.

Wunderlin: For example if you consider statistical mechanics and compare this to these macroscopic motions then you will see that these motions are possible on the level of statistical mechanics but are very very rare. So it means that you need the new laws on the microscopic scale but the macroscopic introduces really new things. For example, this law of equi-partition says in thermodynamics roughly all degrees are freedom - are exited in the same way. And here you have another law where you are far from equilibrum where a few degrees are exited very strongly and others not. So these are notions which are consistent with the microscopic dynamics but are very rare if I use the laws of statistical mechanics.

Weingartner: Is it just similar to the effect that you discover sometimes hidden symmetries - I mean symmetries that are hidden in such a way that they are not used but then you find out they can be used, they are allowed. So in some rare cases they are in fact used.

Wunderlin: I never tried to look in this direction.

Weingartner: A very simple example from biology would be that snails have a certain spiral, but very rarely the other way round. So there it is a so-called hidden symmetry which permits both directions but which is usually not used or only very rarely (perhaps because of special asymmetric initial conditions within the evolution). And so in your case by analogy statistical mechanics would provide the (hidden) symmetric laws behind the apparant symmetry-breaking of motions on the macroscopic level which occur very rarely, maybe because of some asymmetric initial conditions.

Wunderlin: I'll look for that ...

Schurz: Maybe you mean with "consistent" that there exist trajectories on possible initial conditions which describe this macroscopic structures. But you don't know the initial conditions here.

Miller: But that does not sound like the emergence of new laws.

Schurz: This is right, it would imply that are not really new laws, new fundamental laws.

Miller: Doubtless the microscopic laws do not say everything that there is to be said about the system. That does not imply that we need new laws in addition to the microscopic laws.

Wunderlin: ... New laws in the sense that they cannot be derived from the microscopic level.

Miller: But they could just be the microscopic initial conditions.

Chirikov: You do derive this instability, and the result of instability from an equation.

Wunderlin: But for this we need a experimental verification. From the microscopic calculations we cannot deduce the existence of these laws. You always will arrive at a normal form but cannot derive the whole consequences of topological equivalence.

Chirikov: Do you think it is some principal difficulty or you just had not enough time to do it?

Wunderlin: No, I really believe this concept originally introduced by Poincaré is a mathematical concept and what we try to do is to give it a basis in natural sciences.

Batterman: Do the normal forms describe a whole class of systems which are structurally stable?

Wunderlin: It depends on the degeneracy given by symmetries.

Batterman: Are these normal forms similar to normal forms that you get describing catastrophies. But in catastrophe theory they've got a pretty high classification.

Wunderlin: Yes, this is a very special case where you have a potential dynamics. We looked also for normal forms where no potential conditions are valid.

Chirikov: At the beginning you defined this synergetic system of models as some complicated open systems. Does this mean that the fundamental equations of the synergetic theory are always, by definition, stochastic equations including some noise? Do you start with the stochastic equation and then try to calculate consequences?

Wunderlin: Yes.

Chirikov: You did not begin with dynamical equations. You don't use that originally.

Wunderlin: Originally not, but we try to approximate it macroscopically by dynamical equations. And often we can justify that the fluctuations can be neglected.

Chirikov: Now it is clear.

Subject and Author Index

Lecture Notes in Physics

For information about Vols. 1–444
please contact your bookseller or Springer-Verlag

New Series m: Monographs